Indoor Air Pollution Control

Thad Godish

LEWIS PUBLISHERS

Library of Congress Cataloging-in-Publication Data

Godish, Thad.
  Indoor air pollution control / by Thad Godish.

    p.   cm.
  Bibliography: p.
  Includes index.
  ISBN 0-87371-098-3
  1. Indoor air pollution.   I. Title.
  TD883.1.G63 1989   89-33791
  697--dc20   CIP

LEWIS PUBLISHERS, INC.
121 South Main Street, Chelsea, Michigan 48118

PRINTED IN THE UNITED STATES OF AMERICA

To the loving memory of immigrant parents

Unschooled, they inspired me to sip
from the font of knowledge.

Preface

Indoor air pollution is a relatively new public health concern, only a decade or so old. Research activities that characterize indoor air contaminants and contamination problems, define health effects and causal relationships, and determine the efficacy of control measures have expanded concomitant with increased public and governmental awareness of indoor air quality problems. Asbestos, formaldehyde, radon, and "sick buildings" have elevated indoor air quality to the status of a major public health and environmental problem.

As research has expanded and the results been published, a need for reference works that provide an overview and a distillation of important concepts and findings has emerged. Early publications on indoor air quality include those of Meyer and Wadden and Scheff, as well as several editions that review the literature on a contaminant-by-contaminant basis.

Significant treatment of control principles and practices is notably lacking in previous works on this subject. This has been due for the most part to the relative immaturity of the indoor air quality field and the paucity of studies in the area of control measures and their efficacy.

This book has been written primarily as a reference work. It contains an overview or definition of the problem and expanded discussions of source control measures for specific contaminants such as asbestos, combustion-generated pollutants, radon, formaldehyde, volatile organic compounds, pesticides, and biogenic particles. It also reviews public policy and regulatory issues associated with the problem.

Beyond these reference aspects of the book, the author expands the utility of his work by the inclusion of chapters dedicated to problem solving. Chapter 8 focuses on air quality diagnostics. It is designed to assist readers in identifying specific indoor air contamination problems in residences and public buildings. Chapter 9 describes suggested mitigation practices on a case history basis. Chapters 8 and 9 are practical guides to identifying and solving real-world problems. As such, they complement the theory and principles, the primary focus of the reference part of the book.

The book is intended for a variety of audiences, including public health and environmental professionals, industrial hygienists, consultants, architects, and academics. It may prove particularly valuable to physicians who wish to become familiar with indoor air quality problems experienced by patients and to include indoor air quality principles and applications in clinical practice.

As previously described, this book is intended both as a reference text and as a practical guide in problem solving. It is the author's fervent wish, however, that it will go further. There is no longer any doubt that indoor air quality is a significant environmental and public health problem. With its evolving maturity, indoor air quality will soon become an academic subject serving the training needs of environmental health/public health sanitarians, industrial hygienists, air quality scientists, and possibly architects as well. Such a course, entitled "Indoor Air Quality Management," already exists at Ball State University; the author hopes it will be a forerunner of new courses at other institutions. *Indoor Air Pollution Control,* with its well-developed overview of the problem, expanded treatments of ventilation, air cleaning, and source control principles, discussion of public policy and regulatory issues, and emphasis on practical aspects of problem solving, should be the textbook of choice.

Acknowledgments

Book writing is a long and sometimes arduous task. The final product reflects the energies, talent, and commitment of a variety of individuals. I am particularly grateful to Hank McDermott, who reviewed Chapters 5 and 6, to Howard Heskett, who reviewed the entire book, and to my wife, Diana, who reviewed, copyedited, proofed, and encouraged. I acknowledge, too, the word processing assistance and skills of my daughters, Larissa and Leta, who translated my cramped longhand into readable type.

THAD GODISH is the Director of the Indoor Air Quality Research Laboratory and Professor of Natural Resources at Ball State University in Muncie, Indiana. He received his doctorate at Pennsylvania State University, where he was a predoctoral PHS Air Pollution Special Fellow and a postdoctoral scholar with the Center for Air Environment Studies.

Dr. Godish has conducted research and public service activities in a variety of air pollution–related areas, most notably indoor air pollution and plant toxicology. The former serves as the focal point of his research and public service activities at Ball State. He is nationally known for his research on formaldehyde.

He has had extensive consulting experience through his firm, Indoor Air Quality Services, Inc. This consulting practice focuses on conducting indoor air quality investigations and providing air testing services to homeowners and managers of public-access buildings. Dr. Godish also has considerable experience as an expert witness in product liability claims associated with indoor air pollution. In addition, he is the author of *Air Quality*, published in 1985.

Dr. Godish is a member of Sigma Xi, the national honorary research society, and is a Fellow of the Indiana Academy of Science. He is a member of the Air Pollution Control Association, the American Industrial Hygiene Association, the National Environmental Health Association, and the American Association for the Advancement of Science.

Table of Contents

x

1 PROBLEM DEFINITION

The contamination of indoor air by a variety of toxic and/or hazardous pollutants has in the last decade become increasingly recognized as a serious (or potentially serious) public health problem. Several factors combined to produce this awareness:

1. Questions were raised in the late 1970s about the presence of friable asbestos in public schools.

2. Numerous complaints were received from owners of urea-formaldehyde foam–insulated houses of odor and acute irritating symptoms.

3. Hundreds of outbreaks of illness among occupants of new or recently remodeled offices, schools, and other public access buildings were reported.

4. The potential for energy conservation measures to increase levels of indoor contaminants was recognized by governmental authorities who advocated them.

5. Homeowners began seeking alternatives (e.g., kerosene heaters and wood-burning stoves) to the use of what they perceived as high-cost central heating systems.

6. The environmental nature of allergies and asthma became better understood.

7. It became apparent that radon contamination of residences was not just a problem of uranium mill waste tailings and abandoned phosphate strip mines but a problem affecting millions of homes in the United States, Canada, and Northern Europe.

Indoor air pollution is an age-old problem, dating back to prehistoric times when humans came to live in enclosed shelters. It began, no doubt, with the stench of humans themselves. It began, too, when the utility of fire was discovered and brought indoors for cooking and heating. What is new about the problem of indoor air pollution is our emerging awareness of it and the increasing attention given to its various dimensions by research scientists and regulatory authorities. Within this context, indoor air pollution, or indoor air quality, is a "new kid on the block" and, as we have come to learn, a very big kid indeed. In terms of documentable health effects it appears to be enormously more significant than ambient (outdoor) air pollution, a problem for which the implementation of control measures has cost the United States over $200 billion in the past decade.

INDOOR/OUTDOOR RELATIONSHIPS

During alert, warning, and emergency stages of episodes (periods of high ambient air pollution associated with inversions), episode control plans call for air pollution and public health authorities to advise citizens in the affected area to remain indoors, particularly those at special risk—those, for example, who have existing respiratory or cardiovascular disease. This advice is based on the premise that pollutant concentrations will be lower indoors.

For two of the pollutants of major concern under episode conditions, such advice is in fact appropriate. For sulfur dioxide (SO_2) and ozone (O_3), both reactive gases, indoor concentrations are only a fraction of ambient (outdoor) levels. A typical relationship between SO_2 concentrations indoors and in the ambient environment can be seen in Figure 1.1.[1] Indoor/outdoor ratios of 0.3-0.5 are typical for areas where ambient levels of SO_2 are moderate to high. Indoor/outdoor ratios (0.7-0.9) are larger and approach unity when ambient SO_2 levels are low.[2] With few exceptions the predominant source of SO_2 indoors is infiltration through the building envelope from the ambient environment. An exception to this is households using kerosene heaters with high-sulfur (around 0.30%) fuel. Indoor/outdoor ratios of O_3 in houses (0.1-0.2) are substantially lower than unity, indicating that O_3 reacts with components of the building envelope or is scavenged by other indoor contaminants.[3] In mechanically ventilated buildings indoor/outdoor ratios are on the order of 0.6-0.7.[4] Of particular note in the former regard are studies conducted by the author that demonstrate that formaldehyde at levels common to residences rapidly scavenges O_3 indoors.[5] The ambient environment is in most instances the primary source of O_3 indoors. Ozone, however, may be elevated above background levels by emissions from electronic air cleaners, ion generators, and photocopy machines.

Indoor concentrations of combustion-generated contaminants such as nitrogen oxides (NO_x), carbon monoxide (CO), and respirable particles (RSP) have widely varying indoor/outdoor ratios. This variation is due in good measure to the fact that both indoor and outdoor environments can contribute (under varied circumstances) to indoor levels. For nitrogen dioxide (NO_2), a relatively reactive gas, indoor/outdoor ratios of less than or near unity are

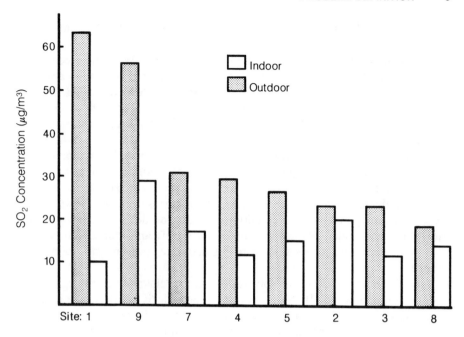

Figure 1.1. SO$_2$ concentrations in both indoor and outdoor environments at eight monitoring locations in St. Louis, Missouri.[1]

typical of residences that have no indoor sources.[6] In residences with gas cooking stoves and/or unvented gas or kerosene heaters, indoor NO$_2$ levels often exceed those in the ambient environment. Because of its low reactivity, indoor/outdoor ratios of CO in the absence of indoor sources typically approach unity.[7] Under episode conditions, CO levels in the indoor environment may not differ from ambient levels. Ratios above unity are associated with the use of gas cooking stoves and ovens, unvented gas and kerosene space heaters, and flue gas spillage.[6] A broad range of indoor/outdoor ratios of respirable particles has been reported.[8,9] In general, however, RSP levels are considerably higher indoors than they are in the ambient environment. The primary reason for this is tobacco smoking, which is the single most significant contributor to indoor RSP levels.

Organic compounds, here described as formaldehyde, volatile organic compounds (VOCs), and pesticides, often occur indoors at concentrations that exceed ambient levels by several fold or, in the case of formaldehyde, one to two orders of magnitude. Because of the widespread use of organic solvents in building materials, paints, adhesives, and furnishings, indoor nonmethane hydrocarbon levels have been seen to exceed ambient levels by 1.5–1.9 times.[9] Concentrations of specific VOCs may exceed ambient levels by a factor of two or more and by as much as a factor of 10.[10,11] Because of pesticide use indoors or under the building substructure, levels of such pesticides as chlordane, heptachlor, and chlorpyrifos are often several orders of magnitude higher indoors.[12]

Radon is ubiquitous in the ambient environment. With the exception of high-ventilation conditions, when indoor/outdoor ratios should approach unity, indoor/outdoor ratios of radon tend to be high to very high. Indoor/outdoor ratios may vary from as low as 2–3 to a high of 1000 or more.[13,14] What is unique about radon is that the predominant source is exterior to the building envelope.

Indoor/outdoor ratios of biogenic particles such as mold and bacteria can vary significantly depending on the prevalence of indoor and outdoor sources and the season of the year. In the midwest, mold levels are significantly higher outdoors during the summer months, with correspondingly low indoor/outdoor ratios. Under winter conditions, indoor/outdoor ratios may be considerably above unity, depending on whether significant internal sources exist and whether snow cover is present.[15] Dust mite antigen has no outdoor source.

Differences in indoor/outdoor ratios have special relevance to controlling indoor contamination. If the indoor/outdoor ratio is below 1, the effect of applying general ventilation for contaminant control is increased indoor concentrations of those particular contaminants, notably SO_2 and O_3 and, in summer months, mold spores. On the other hand, when indoor/outdoor ratios are high, considerably exceeding 1, the use of general ventilation for contaminant control would be particularly appropriate. This would be applicable to combustion-generated contaminants, formaldehyde, and radon.

PERSONAL POLLUTION EXPOSURES

As seen in the previous section, indoor concentrations of specific contaminants may be significantly higher indoors than in the ambient environment. This fact becomes more important when coupled with a consideration of exposure or potential exposure duration. Studies of human activity indicate that, on the average, we spend approximately 22 hr/day indoors, about 16 in our homes.[16] We receive most of our personal pollution exposure indoors. Such exposures, for the general population, are probably of more public health importance than exposures to ambient air.

Studies comparing ambient, indoor, and total personal exposures to RSPs have been reported for residents of Topeka, Kansas.[17] As can be seen from Figure 1.2, indoor exposures were higher than ambient exposures, with highest values reported for 24-hr total personal exposures.

One of the most significant efforts to characterize personal air pollution exposures has been carried on by Wallace et al. under the EPA Total Exposure Assessment Methodology (TEAM) study between 1979 and 1986.[10,11] Twenty VOCs were measured in personal air, outdoor air, drinking water, and breath of approximately 400 residents of the cities of Bayonne and Elizabeth, New

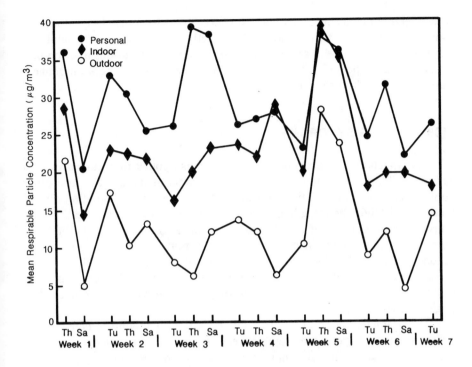

Figure 1.2. Indoor, outdoor, and total personal exposure to respirable particles in Topeka, Kansas residents.[17]

Jersey, Greensboro, North Carolina, and Devil's Lake, North Dakota. The Bayonne/Elizabeth area is notable for its concentration of chemical plants and refineries. Median concentrations of 11 VOCs are summarized for personal and outdoor air samples in Table 1.1 for Bayonne/Elizabeth residents. In all

Table 1.1   Weighted Median VOC Concentrations ($\mu g/m^3$) in Personal and Outdoor Air Samples[10]

Compound	Fall 1981		Summer 1982		Winter 1983	
	Personal	Outdoor	Personal	Outdoor	Personal	Outdoor
1,1,1-Trichloroethane	17.0	4.6	9.3	5.1	22.0	2.3
Benzene	16.0	7.2				
m-, p-Xylene	16.0	9.0	13.0	10.0	22.0	3.0
Carbon tetrachloride	1.5	0.9	0.9	0.7		
Trichloroethylene	2.4	1.4	2.8	0.8	1.6	0.1
Tetrachloroethylene	7.4	3.1	5.7	2.6	8.2	4.5
Styrene	1.9	0.7	1.3	0.4	1.5	0.2
p-Dichlorobenzene	3.6	1.0	2.8	1.1	5.0	1.2
Ethylbenzene	7.1	3.0	4.6	2.2	6.8	1.3
o-Xylene	5.4	3.0	5.2	3.6	8.0	1.0
Chloroform	3.2	0.6	0.8	0.1	2.2	0.1

cases, personal concentrations and exposures exceeded those in outdoor air, usually by a factor of two or more, and sometimes by an order of magnitude. High correlations were observed for breath VOC concentrations and the previous 12-hr air exposure. Correlations between breath and outdoor air concentrations were seldom significant.

SICK BUILDING SYNDROME

The appellation "sick building syndrome" is applied to occurrences of a variety of illness symptoms reported by occupants of large office and other public access buildings. This problem has also been called "building-related illness" and "tight building syndrome." In problem buildings, there are an unusually high number of individuals (in some cases > 30% of the building population) who complain of symptoms of a nonspecific nature, including headaches, unusual fatigue, eye, nose, and throat irritation, and shortness of breath. A more specific symptom constellation (fever, headache, myalgia) is associated with building-related cases of hypersensitivity pneumonitis and Legionnaires' disease.

The characterization of building-related health complaints as being due to a sick building suggests that some aspect of the building structure (i.e., the heating, ventilation, and air conditioning [HVAC] system, furnishings, equipment, or a tight thermal envelope) is responsible. The term "tight building syndrome," once commonly used, implied that the indoor air quality and occupant illness problems were due to energy-conserving design. It was at best a very simplistic characterization of the problem.

Studies of sick buildings have been reported by a number of investigators. Reports have often been anecdotal in nature, relating the experience of consultants and various investigative teams. In most cases, the focus of sick building investigations has been to identify the pollutant or pollutants responsible by air quality monitoring. Air quality monitoring has often proven to be a fruitless exercise, as concentrations of contaminants measured are usually considered to be too low to have caused the illness complaints.

One of the most notable attempts to characterize the nature of sick building problems was that of Wallingford and colleagues[18,19] at the National Institute of Occupational Safety and Health (NIOSH). From 1978 through 1985, NIOSH conducted 356 health hazard evaluations on public access buildings for which illness complaints had been reported. The frequency of health complaints as a percentage of buildings investigated is summarized in Table 1.2. Symptoms with reported frequencies greater than 50% were (in increasing frequency) sinus congestion, fatigue, headache, dry throat, and eye irritation. Frequencies of ascertained causes are summarized in Table 1.3. The major building problems were placed in three categories (in increasing frequency): microbial contamination, chemical contamination, and inadequate ventilation.

Table 1.2 Prevalence of Health Complaints in Buildings Investigated by NIOSH Health Hazard Investigation Teams[18]

Health Complaint	% of Buildings
Eye irritation	81
Dry throat	71
Headache	67
Fatigue	53
Sinus congestion	51
Skin irritation	38
Shortness of breath	33
Cough	24
Dizziness	22
Nausea	15

Inadequate ventilation was described as the primary cause of illness complaints and poor air quality in approximately 50% of the buildings investigated. The determination of inadequate ventilation was based on measurements of CO_2 levels which exceeded 1000 ppm. Ventilation problems commonly included (1) insufficient outdoor air supplied to building spaces, (2) poor air distribution and mixing, (3) temperature and humidity extremes or fluctuations, and (4) air filtration problems caused by improper or inadequate maintenance of HVAC systems

Ventilation problems were observed in many instances to have been exacerbated by certain energy-conserving measures, including (1) reducing or eliminating the introduction of outdoor air, (2) reducing infiltration and exfiltration, (3) lowering thermostats in winter and raising them in summer, (4) eliminating humidification or dehumidification systems and (5) early shutdown and late start-up of ventilation systems.

Chemical contamination associated with the building fabric, sources within the building, and contaminants drawn in from the outside accounted for 34% of the air quality problems investigated.[19] Sources of contaminants associated with the building fabric included formaldehyde from pressed wood products, fibrous glass from lined ventilation ducts, various organic solvents from glues and adhesives, and acetic acid used as a curing agent in silicone caulking.

Table 1.3 Frequency of Indoor Air Quality Problems Ascertained from NIOSH Health Hazard Evaluations[18]

Problem Type	Number	%
Building fabric contamination	14	4
Microbial contamination	19	5
Outdoor contamination	38	11
Indoor contamination	67	19
Inadequate ventilation	179	50
Unknown	39	11
	356	100

Sources within the building space were observed to be the cause of indoor air quality problems in one out of every five buildings investigated. These included ozone from copying machines, methyl alcohol from spirit duplicators, butyl methacrylate from signature machines, trinitrofluorenone from laser printers, ammonia and acetic acid from blueprint machines, and formaldehyde from carbonless copy paper. Other chemical contamination problems included improperly applied pesticides, boiler additives such as diethyl ethanolamine, improperly diluted cleaning agents such as rug shampoo, tobacco smoke, combustion gases from sources common to cafeterias and laboratories, and cross-contamination from poorly ventilated sources that leaked into other air-handling units.

Sources of chemical contamination external to building spaces were identified as major problems in 11% of the NIOSH investigations. Basically, these were problems of entrainment or reentrainment of contaminants in the building air supply. They included motor vehicle exhaust, boiler flue gases, and previously exhausted contaminated air. Carbon monoxide from basement parking garages was observed to be a common problem. Other contamination problems associated with outdoor sources included construction contaminants such as asphalt, solvents, dusts, gasoline, and sewage vapors infiltrating into basements.

Microbial contamination was observed to be a relatively infrequent cause of office building problems. When it did occur, it was associated with water damage to carpets or furnishings and standing water in HVAC system components.

A variety of other causal or potentially causal factors have been reported or suggested as the cause of illness syndromes in office buildings. These have included vibration,[20] emissions of benzyl chloride and benzal chloride from vinyl floor tiles plasticized with butyl benzyl phthalate,[21] VOCs associated with the installation of new carpeting,[22] total VOC levels resulting from emissions from building materials,[23] cyclopentadienes associated with carpeting adhesives,[24] organic dusts associated with badly cleaned wall-to-wall carpeting,[25] dimethyl acetamide (a potent lachrymator) emissions from plastic panels used as office dividers,[26] air ions and electrostatic fields,[27-29] deposition of air ions by video display terminals,[30] and the so-called fleece/shelf factors (i.e., materials capable of depositing/adsorbing/accumulating pollutants).[31]

The sick building syndrome is also suspected to be a result of psychosocial factors in whole or in part.[32,33] Psychosocial factors may include management/employee relationships, job stress, the susceptibility of employees to suggestion from fellow workers, etc. Morriss and Hawkins[34] have conducted studies in which a significant relationship was observed between complaints of stress and sick building symptoms (SBS) while no association was observed between the incidence of SBS and complaints about air quality. They were not able, however, to conclude whether stress caused the SBS symptoms or sick buildings caused stress. Significant correlations between SBS complaints and psychosocial factors have been reported in the Danish town hall study.[35] Significant factors included satisfaction or dissatisfaction with colleagues, the quantity of work expected, high work speed, and low status.

Though emotional factors are likely to play some role in most SBS investigations, the accusation of hysteria and a conclusion that the problem is "all in their heads" are most often born of emotion itself and not of good science. Management which refuses to face up to the fact that building-related health problems exist, and consultants and investigators who have failed to identify and correct the problem, often seek refuge in attributing the problem to "psychological factors." Most reports of SBS are based on investigations of complaints. They therefore have an inherent bias which makes it difficult to determine the prevalence of such problems as the potential for psychosocial bias in symptom reporting. Several studies have been reported recently which have attempted to survey workers in office buildings on a random basis. Woods et al.[36] surveyed by telephone worker perceptions of indoor air quality and its effects on comfort and well-being in a stratified random sample of 600 U.S. office workers. Notably, nearly one-fourth of office workers surveyed expressed dissatisfaction with air quality in their office environment and one-fifth perceived that office air quality affected their performance.

Office employees working in open areas were 1.5 times as likely to report poor air quality and diminished performance. Despite the fact that SBS is often perceived to be a "new building" problem, office employees in buildings more than 20 years old were 1.5 times more likely to perceive poor indoor air quality and claim reduced work performance than employees in buildings less than 10 years of age. Perception of air quality and reduced work performance were also affected by the type of HVAC equipment and air distribution systems. Dissatisfaction with air quality was approximately 1.5 times higher in buildings in which spaces were heated locally as compared to central heat. Fifteen percent of the 600 office workers were dissatisfied with air quality in buildings with constant-volume HVAC systems, 29% in buildings with variable-air-volume systems, and 40% where HVAC systems were cycled on and off.

Symptoms described by affected respondents were characteristic of SBS and included mucous membrane (eye, nose, and throat) irritation, dry skin, headache, nausea, fatigue, and lethargy. Air quality factors mentioned frequently by study participants included poor ventilation, temperature, cigarette smoke, and humidity.

One of the most extensive studies of the SBS phenomenon has been that of Danish town halls.[35] The study, which included 4369 employees in 14 town halls and 14 affiliated buildings, was designed to determine the influence of various environmental factors on the prevalence of sick building–type symptoms. Results of analyses between the prevalence of mucous membrane irritation and general symptoms (i.e., headache, abnormal fatigue, and malaise) are summarized in Table 1.4. Significant risk factors for mucous membrane irritation appeared to be sex (it is more frequently reported by females), hay fever, job category, hours at the office, handling carbonless copy paper, photocopy-

Table 1.4   Analyses of Associations Between Work-Related Symptoms in Office Workers and Personal Characteristics, Lifestyle, and Residential, Job-Related, and Psychosocial Parameters in Danish Town Halls[35]

Parameter		Symptoms (odds ratio + 95% CL)	
		Mucous Membrane Irritation	General Symptoms
Sex	Males	1.0	1.0
	Females	1.6 (1.3–2.1)	1.8 (1.5–2.3)
Hay fever	No	1.0	NS[a]
	Yes	1.6 (1.2–2.1)	
Migraine	No	NS	1.0
	Yes		1.8 (1.4–2.2)
Smoking	<10 g/day	NS	1.0
	≥10 g/day		1.3 (1.1–1.6)
Coffee	0 cups/day	1.0	
	1–6 cups/day	0.6 (0.5–0.8)	NS
	>6 cups/day	0.8 (0.6–1.1)	
Indoor climate problems in residence	No	NS	1.0
	Yes		1.6 (1.3–2.2)
Job category	Mayor or director	1.0	1.0
	Principal	2.5 (1.0–6.1)	1.1 (0.5–2.1)
	Head clerk	2.5 (1.1–5.8)	1.2 (0.6–2.3)
	Clerk	3.1 (1.4–7.3)	1.6 (0.8–3.0)
	Probationary clerk	1.7 (0.7–4.2)	1.9 (0.9–3.8)
	Social worker	1.8 (0.7–4.5)	2.1 (1.0–4.3)
	Technical assistant	1.1 (0.3–3.7)	0.8 (0.3–2.2)
	Engineer or architect	1.7 (0.7–4.4)	1.0 (0.5–2.1)
Working hours	<20 hrs/week		1.0
	20–39 hrs/week	NS	
	≥40 hrs/week		1.6 (0.8–3.5)
Hours at office	<6 hrs/day	1.0	
	7 hrs/day	1.5 (1.1–1.9)	NS
	≥8 hrs/day	1.1 (0.9–1.4)	
Handling carbonless paper	Monthly or less	1.0	1.0
	Weekly or daily ≤25 forms	1.3 (1.1–1.6)	1.4 (1.1–1.7)
	Weekly or daily >25 forms	1.3 (1.1–1.6)	1.6 (1.1–1.7)
Photocopying	Monthly or less	1.0	
	Weekly or daily ≤1 hour	1.0 (1.1–1.7)	NS
	Weekly or daily >1 hour	1.5 (1.1–2.0)	
VDT work	Monthly or less	1.0	
	Weekly or daily ≤1 hour		NS
	Weekly or daily >1 hour	1.5 (1.2–2.1)	
Varied work	Yes	NS	1.0
	No		1:3 (1.1–1.6)
Satisfaction with one's supervisor	Yes	1.0	NS
	No	1.7 (1.1–1.7)	
Satisfaction with colleagues	Yes		1.0
	No		2.0 (1.6–2.6)
Quantity of work limits job satisfaction	No	1.0	1.0
	Yes	1.4	1.7 (1.4–2.1)
Lack of influence and high work speed	No		1.0
	Yes		1.4 (1.1–1.7)

[a]NS = not significant; p > 0.01.

Table 1.5   Physical Factors Which Separately Were Significantly Correlated to SBS Symptoms in the Danish Town Hall Study[37]

Level 1	Level 2	Level 3
Air temperature	Area of office	Age of building
Air velocity	Volume of office	Type of ventilation system
Static electricity	Number of work places	Floor covering material
Floor dust (weight)	Number of persons	
Allergenic potential of floor dust	Fleece factor	
Airborne bacteria	Shelf factor	
Sound pressure		

ing, working at a video display terminal, and two psychosocial factors (dissatisfaction with supervisor and quantity of work expected). Risk factors for general symptoms included sex (it is more frequently reported by females), migraine headache, smoking, indoor climate problems, job category, working hours, handling carbonless copy paper, and psychosocial factors (dissatisfaction with colleagues, quantity of work expected, work speed, and low status).

Investigators in the Danish town hall study evaluated a variety of physical factors which separately or in combination could describe differences in the prevalence of SBS symptoms. Factors which separately were significantly correlated with symptom prevalence are summarized in Table 1.5.[37] When analyzed further, the combinations of parameters in Table 1.6 were observed to best describe the prevalence of symptoms. The lower the level value in Tables 1.5 and 1.6, the more direct the effect the parameter has on the prevalence of symptoms. Most notable are the allergenic potential of floor dust, the air temperature, fleece factor, and shelf factor. The fleece factor is the amount of fleecy surfaces (e.g., textile floor coverings, curtains, and textile chair covers) divided by the volume of the room; such surfaces are thought to collect pollutants. The fleece factor is not by itself directly responsible for indoor air quality problems; rather, it is likely that it is simply an indicator. Shelf factor is the number of open, filled-up shelves divided by room volume; it is suggested

Table   1.6   Physical   Factors   Which   in   Combination   with   Each   Other   and Psychosocial/Work-Related Parameters Were Significantly Correlated to SBS Symptoms in the Danish Town Hall Study[37]

SBS Symptoms	Level 1	Level 1 + 2	Level 1 + 2 + 3
Mucous membrane irritation	Floor dust (weight), allergenic potential of floor dust	Allergenic potential of floor dust, fleece factor and shelf factor or Level 1	Allergenic potential of floor dust (alt, floor covering material), shelf factor, fleece factor or Level 1
General symptoms (headache, fatigue, malaise)	Allergenic potential of floor dust and air temperature	Fleece factor and number of work places or Level 1	Fleece factor and number of work places or Level 1

to be an expression of the buffer capacity in relation to room humidity. The strong correlation between the prevalence of SBS symptoms and the fleece/shelf factors is suggested to support the premise that it is the sum of pollutants released from these surfaces which is of greatest importance to SBS.[31]

Alternative interpretations of the strong correlations observed between SBS symptoms and the fleece/shelf factors in the Danish town hall study would include the following:

1. A high fleece factor is associated with textile floor coverings which, because of large amounts of surface area, emit large quantities of VOCs, particularly when they are new.
2. Textile materials such as carpeting serve as a medium or produce a microclimate for the production of biogenic particles.
3. A high shelf factor is associated with a large surface area of printed material and therefore a high potential for VOC emissions from ink solvents.

Many writers tend to imply that building-related health problems are somehow all interrelated. It is not uncommon, however, to identify more than one building-related health complaint and cause during an onsite investigation. The resolution of building-related health problems requires recognition that an SBS-type complaint often does not represent a single generic problem. Failure to recognize this can result in the application of control measures that only partly resolve a problem.

ASBESTOS

Asbestos as a potentially serious indoor contamination problem came to public attention in the United States in 1978 when health authorities recognized that friable asbestos-containing material (ACM) used as a fire/heat retardant and acoustical plaster in schools and other public buildings had the potential for releasing a significant quantity of asbestos fibers into indoor air. The use of friable asbestos as a fire/heat retarding insulation in school buildings and other large buildings was banned by the U.S Environmental Protection Agency (EPA) in 1973 under the National Emission Standards for Hazardous Pollutants (NESHAP) provisions of Section 12 of the 1970 Clean Air Act Amendments.[38] The ban was intended to reduce community exposures to asbestos fibers from building demolition. At the time of the NESHAP ban, risks to building occupants were not addressed. This aspect of the problem lay nascent for almost five years, and may have gone on unaddressed except for the fact that friable asbestos materials were widely used during a boom in the construction of school buildings from 1950 to 1973.

Characteristics and Applications

Asbestos is a collective term for a variety of asbestiform minerals that satisfy particular industrial-commercial needs. The term is used in reference to well-developed and thin, long-fibered varieties of certain minerals. These include chrysotile, anthophyllite, riebeckite, cummingtonite-grunerite, and actinolite-tremolite. Commercial names for the first four of these minerals, respectively, are chrysotile, crocidolite, anthophyllite, and amosite. These minerals differ in their chemical composition, but all are silicates.

Asbestos comprises two mineral groups: serpentine and amphibole. Chrysotile, the most common asbestos mineral, is a member of the serpentine group. Chrysotile has a layered silicate structure with the layers curled up to form scrolls or tubes. Amphibole asbestos minerals have a crystalline structure characterized by parallel chains of crystal tetrahedra.[39]

Asbestos fibers are characterized by their small diameter, high length-to-width ratio, and smooth and parallel longitudinal faces. They have great strength and flexibility, with the tensile strength of commercial quality fibers 20–50 times greater than nonasbestiform crystals of the same mineral. The strength of asbestos fibers per unit of cross-sectional area increases as the diameter decreases; the smaller the fiber diameter, the greater the strength.

Commercial use of silicate mineral fibers described as asbestos has been extensive, with over 3000 applications. Major applications include fireproofing, thermal and acoustical insulation, acoustical plaster, friction products such as motor vehicle brake shoes, reinforcing material in cement water pipe, and roofing and floor products. Chrysotile accounts for approximately 95% of the asbestos sold in the United States.

Asbestos may be present in products in the bound or partially bound form. In most applications, including floor tiles, asbestos cements, roofing felts, and shingles, asbestos fibers are firmly bonded and are unlikely to become airborne unless the product is cut, drilled, sanded, broken, or abraded in some way.[40]

Asbestos products that present the greatest concern are (1) those of a friable nature sprayed or troweled on walls, ceilings, and structural components of nonresidential buildings and (2) insulating materials applied to pipes, boilers, tanks, and other equipment. Friable asbestos applications include fireproofing, thermal insulation, noise control, and decorative surfaces. The asbestos content of surfacing materials varies from 1% to 95%. Such materials are typically bound more or less loosely with sodium silicate, portland cement, and a variety of organic binders. An asbestos-containing product is deemed friable when it can be crushed by the application of hand pressure.[40]

Figure 1.3. Sprayed-on friable asbestos on building beam.

Indoor Sources

Asbestos in both friable and nonfriable forms can be found in most large public access buildings, including schools, office buildings, commercial buildings, etc. It can be found in residences as well.

In schools and other public access buildings constructed prior to 1973, asbestos surfacing materials were sprayed or troweled on a variety of building surfaces, most notably ceilings and steel beams. In the latter case, friable asbestos materials were used to prevent buildings from collapsing during a fire. Such an application can be seen in Figure 1.3. In other instances, friable asbestos materials were applied to ceilings as acoustical plaster. Fiber release from surfacing materials occurs as they age and the binding material weakens. This may result in continuous fiber erosion or a delamination from the substrate. Delamination may cause large episodic releases of asbestos fibers to occur. Similar episodic releases may result when friable asbestos surfacing material is disturbed and becomes damaged. Damage may result from maintenance or renovation activities; vibration from mechanical equipment, sound, or athletic events; deprecatory activity by occupants (students beating on ceilings); water incursions; etc. Significant fiber release may also occur when drop ceiling panels are disturbed after asbestos residue has accumulated on them.[40]

Asbestos materials applied as thermal insulation around steam lines and boilers present less risk of fiber release into occupied spaces and exposure to

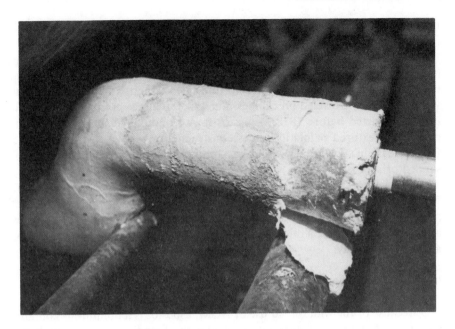

Figure 1.4. Asbestos thermal insulation.

humans. Thermal insulation (Figure 1.4) is less risky because it is usually covered with a rigid cloth or paper material; it is often inaccessible to most building occupants. For purposes of asbestos inspection of school buildings, EPA considers it to be non-friable until it becomes damaged.[41]

Vinyl floor covering may also contain asbestos fibers. Because of its bound nature it has not historically been considered to pose a significant problem of asbestos fiber release into the air of schools and other public buildings. However, in studies conducted in a building in Paris, Sebastien et al.[42] observed elevated concentrations of chrysotile fibers in air samples, despite the fact that asbestos sprayed on ceilings contained only crocidolite. The only material containing chrysotile fibers was the vinyl floor tiles. Airborne chrysotile asbestos was suggested to have been caused by the surface erosion of vinyl floor tile by occupant traffic.

Asbestos-containing materials with the potential for fiber release can also be found in residences. These include pipe insulation, stove/furnace insulation, wall/ceiling insulation, patching compounds, textured paint, vinyl floor covering, walls, ceilings, roofing, shingles, and siding.[43]

Asbestos has been used as thermal insulation on steam lines, hot water pipes, and furnace ducts in many houses built in the last 30–40 years. Such materials are usually covered with a rigid fabric or paper and generally pose little threat of fiber release unless they are physically damaged. Asbestos has also been used to insulate wood-burning stoves and oil, coal, and wood furnaces. It is usually contained in cement sheets, stiff paperboard, or paper in these applications. It may also be in gaskets.

Asbestos in residences may also be found in walls and ceilings. In some homes built between 1930 and 1950, asbestos insulation was sandwiched between exterior and interior walls. In rare cases, asbestos surfacing materials were sprayed or troweled on ceilings much like in larger nonresidential buildings constructed before 1978. Asbestos may be contained in patching compounds or textured paints in homes constructed or renovated prior to 1975. Fiber release in such homes would be expected to occur if such materials were sanded or scraped.

Vinyl floor tiles may also contain asbestos. The greatest potential for fiber release would occur if old flooring in a home were destructively removed and the surface beneath it were sanded. Asbestos-containing roofing, shingles, and siding should pose a minimal risk of occupant exposure, as these materials are used externally.

Indoor Exposures

Because of the widespread use of ACM, asbestos is a ubiquitous contaminant of both indoor and ambient air. A variety of studies has been conducted to assess levels of asbestos particles in indoor and ambient air. Levels expressed on a mass basis and a calculated equivalency in fibers per cubic centimeter (f/cm^3) are summarized in Table 1.7.

Asbestos fibers in the air of schools with asbestos-containing surfaces are one to two orders of magnitude higher than ambient levels in U.S. cities. Highest median concentrations have been reported for schools with damaged surfacing materials.

Data in Table 1.7[44-48] expressed in fibers per cubic centimeter were calculated on the assumption that there would typically be 2000 f/cm^3 in workplace air and fiber size would be in excess of 5 μm in length. Asbestos samples collected in remote ambient air, however, indicate fiber concentrations 35 times greater when all fibers, irrespective of size, are considered. The choice of the lower fiber concentration for conversion purposes reflects a judgement that the larger fiber sizes are of greater health risk.[39]

Studies of asbestos particle levels in public schools that report particle concentrations irrespective of particle size have been conducted in six Colorado public schools (containing asbestos surfacing materials).[49] These data are presented in Table 1.8. Fiber levels compared to those in Table 1.7 are seen to be from one to more than two orders of magnitude higher. If all fibers, irrespective of size, had the same cancer-inducing risk, then the asbestos building contamination problem would be much more serious.

Quantitative estimates of exposure to asbestos in nonoccupational environments are difficult to make. One approach is to express exposure in the same conceptual context as for the workplace, that is, the cumulative dose. The cumulative dose is obtained by multiplying the average concentration of

Table 1.7   Levels of Asbestos Fibers in Buildings and Ambient Air in the United States

| | Concentrations | | | | |
| | ng/m^3 | | f/cm^{3a} | | |
Sample Type	Median	90th Percentile	Median	90th Percentile	Reference No.
Ambient	0.9	9.8	3×10^{-5}	3.3×10^{-4}	44
Ambient: 48 U.S. cities	1.6	6.8	5×10^{-5}	2.3×10^{-4}	45
Ambient: 5 U.S. cities	6.7	31.9	2.2×10^{-4}	1.1×10^{-3}	46,47
U.S. schools: cementitious asbestos	7.9	19.1	2.6×10^{-4}	6.4×10^{-4}	46,47
U.S. buildings: friable asbestos	19.2	96.2	6.4×10^{-4}	3.2×10^{-3}	46,47
U.S. schools: asbestos surfaces	62.5	550	2.1×10^{-3}	1.8×10^{-3}	44
U.S. schools: damaged asbestos surfacing materials	121.5	465	4.0×10^{-3}	1.5×10^{-2}	48

[a]Based on conversion factor of 30 μg/m^3 = 1 f/cm^3.

Table 1.8   Concentrations of Asbestos Fibers in Colorado Public Schools Determined by Transmission Electron Microscope Coupled with Selected Area Electron Diffraction and Energy Dispersive X-Ray Analyses[49]

Building #	Sample Location	Concentration (f/cm3)
1	gym	0.143
	corridor	0.048
	ambient	0.012
3	gym	0.021
	gym	0.298
	ambient	0.050
4	gym	0.061
	classroom	0.782
	ambient	0.043
5	Level 2	0.004
	Level 3	0.006
	ambient	0.004
6	art	0.007
	ambient	0.001

fibers in workplace air by the number of years employed in that environment. The workplace cumulative dose only includes fibers $> 5 \mu m$ in length, counted by phase contrast microscopy. It does not take into account rate per unit time, duration of exposure, and age at exposure. Each of these may be significant in determining health effects, most notably the age at exposure.[39]

A second approach to estimating exposure is the use of the concept of "lifetime fibers." Such exposures are determined by integrating fiber concentrations ($> 5 \mu m/cm^3$) and the intake rate from a source or sources over time. Based on this exposure concept, the Committee on Nonoccupational Health Risks of the National Research Council (NRC) has estimated that 2 million to 6 million schoolchildren exposed for 1000 hr/year for 12 years would receive a lifetime fiber dose of 3×10^6 to 3×10^8. Because of more years of exposure, teachers would receive higher lifetime fiber doses from the same exposure concentration. Estimates of individual asbestos exposures for schoolchildren, teachers, a variety of community exposures, and those occupationally exposed are summarized in Figure 1.5. Note that the typical level of uncertainty of these estimates is in the range of two orders of magnitude (that is, the upper uncertainty limit is 100 times greater than the lower).

Health Hazards

The human health hazards of asbestos are well known, particularly in the context of occupational exposures. Four diseases have been clearly associated with occupational exposures to asbestos: lung cancer, mesothelioma, asbestosis, and nonmalignant pleural disease. Mesothelioma is a cancer of the lining of the chest or abdominal cavity. It is apparently 100% fatal. Asbestosis is a progressive, nonmalignant fibrosis of the lung which results in severe disability and death. Pleural disease is characterized by diffuse pleural thickening and effusions and the formation of fibrous and calcified plaques.[39]

Evidence to establish a causal relationship between asbestos exposure and asbestos-related disease comes from clinical observations, epidemiological studies, and laboratory animal exposures. Clinical and pathological observations of a pneumoconiosis-like syndrome (asbestosis) among asbestos workers was first observed early in this century. This relationship was later confirmed in epidemiological studies.[50] Epidemiological studies of cohorts have demonstrated a high incidence of both lung cancer and mesothelioma associated with asbestos exposures as compared to controls.[51,52] Exposures of laboratory animals to asbestos particles have demonstrated that asbestos can cause lung cancer and mesothelioma.[53]

Individuals occupationally exposed to asbestos particles have a lung cancer risk five times greater than nonsmoking individuals who are not occupationally exposed. The effects of cigarette smoking and occupational exposure to asbestos are multiplicative; the combined risk is 50 times greater than risk to the nonexposed nonsmoker.[54] An increased risk of lung cancer is associated with all major types of commercial asbestos.

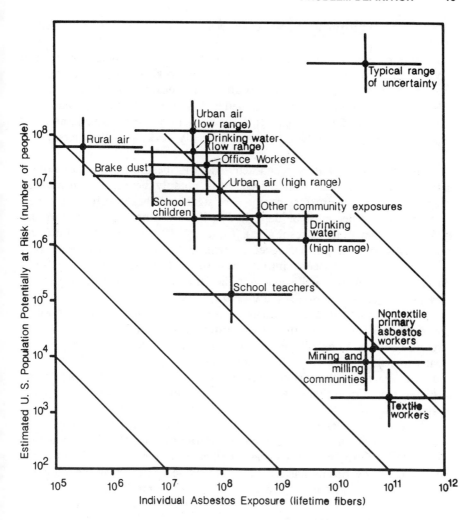

Figure 1.5.   Asbestos exposure estimates.[39]

Asbestosis is characterized by slowly progressing diffuse interstitial fibrosis which results in (1) an impaired ability to exchange gases between the blood and lung tissue, (2) restricted breathing, and (3) increased resistance in small airways. Asbestosis patients, as a consequence, have difficulty in breathing.[39] Symptoms of pleural disease are similar to asbestosis. They may include shortness of breath, chest tightness, and pulmonary restrictions. This disease is slow to progress and affected individuals may not experience any functional impairment.[39] Asbestosis and pleural disease are typically associated with heavy exposures to asbestos. The risk of either from nonoccupational exposures is expected to be very small. Mesothelioma is rare except in individuals exposed to asbestos. Approximately 1600 cases were estimated to have occurred in 1980.[55]

Asbestos-related lung abnormalities and mesothelioma have been reported for members of households of asbestos workers, suggesting that even asbestos brought home on work clothes poses a health risk. In New Jersey, 35% of 628 household contacts examined had asbestos-related lung abnormalities compared to 5% of controls.[56] Elevated rates of mesothelioma were observed among household contacts of asbestos workers in New Jersey and New York.[57]

The physical and chemical characteristics of asbestiform fibers are important in causing observed health effects. Most notably these include respirability, size, number, and surface charge. For significant health effects to occur, asbestiform particles must be deposited in the lower part of the respiratory system (i.e., they must be respirable). Typically, RSPs do not exceed aerodynamic diameters of 3 μm. However, particles that are much longer than they are wide can penetrate deeply into the lung since the aerodynamic diameter of fibers depends on the width and not the length. When asbestos fibers are inhaled, they align themselves parallel to respiratory airways.[58]

The size or dimensional characteristics of asbestos fibers determine where deposition will take place in the respiratory tract and how lung cells will respond to them. Long fibers are apparently more toxic or hazardous. This greater toxicity may be due to the inability of macrophages to engulf the fibers and/or to inactivate the various sites on long fibers.[59] There appears to be an increased risk of mesothelioma after exposure to long, thin fibers as compared to short, thick ones.[39]

Asbestos particles are exceptionally durable in biological systems. This durability is one of the basic factors contributing to observed biological effects. On autopsy or biopsy, asbestos fibers are found to be present in the lungs long after exposure has ceased. This may be significant in the induction of latent diseases such as lung cancer and mesothelioma.[39]

Health Risks from Indoor Air

A broad-based scientific and regulatory consensus has developed over the past decade that asbestos exposure, particularly in workers directly exposed in the extraction, manufacture, and application of asbestos and asbestos products, poses very significant health risks and that even low exposure levels that occur in nonoccupational environments carry some health risk. Because there is no apparent threshold for asbestos-related carcinogenicity, public health concern related to nonoccupational asbestos exposures is focused on lung cancer and mesothelioma.

Risk estimates for schoolchildren exposed to asbestos fibers associated with friable asbestos building materials have been recently made by Hughes and Weil.[60] They assumed an average enrollment period of six years and a fiber concentration of 0.001 f/cm³. Exposure to mixed asbestos fiber types under the assumed exposure conditions was projected to result in five lifetime excess cancer deaths per million exposed. For an exposure to chrysotile alone the estimated risk was considerably less (1.5 excess cancers/10^6 exposed), since chrysotile is less toxic than other forms. Breaking these down for lung cancer and mesothelioma, lifetime risks for mixed fiber exposures were projected to be 0.6×10^{-6} for lung cancer and 4.4×10^{-6} for mesothelioma. For chrysotile asbestos, the most common asbestos fibers found in schools and therefore the most likely exposures, the projected lifetime risk for lung cancer associated with classroom exposures was 0.6×10^{-6} and for mesothelioma 0.9×10^{-6}.

The risk of contracting mesothelioma is considered to be greater because the probability of contracting it, unlike lung cancer, is related to age at first exposure. From occupational studies, it is evident that mesothelioma usually appears 20 years after the initiation of exposure and the number of cases increases rapidly thereafter. As a consequence, asbestos exposure in childhood greatly increases the lifetime risk of mesothelioma compared to an equivalent dose later in life.[39]

Risk estimates for exposures to asbestos fibers in schools and other non-occupational environments have also been made by the Committee on Nonoc-cupational Exposures to Asbestiform Fibers of the NRC.[39] Using lifetime exposures to environmental levels of 0.002 f/cm³, plus an additional risk from a 10-year exposure of 0.02 f/cm³ in asbestos-contaminated schoolrooms, they estimated that the lifetime cancer risk for a male who eventually becomes a smoker is 66 in a million. For mesothelioma, the additional lifetime risk for such a 10-year schoolroom exposure would be 21×10^{-6}, compared to a background risk of mesothelioma of 46×10^{-6}.

Levels of exposure used by the NRC committee described above reflect those which the committee considered to be high; they are above the 90th percentile for both environmental (mostly ambient air) and schoolroom exposures. Thus, these risk estimates are at the high end of the risk spectrum.

COMBUSTION-GENERATED POLLUTANTS

Indoor air pollution associated with combustion has a long history, one that goes back to the first human dwellings and use of fire. Twentieth-century advancements in technology have provided advanced societies with vented gas and oil space heating and electric heating and cooking. Modern technology has provided us the opportunity to live in "smoke-free" homes and buildings.

Modern (or not-so-modern) technology also provides us with choices. We may choose to heat our homes with a variety of appliances, some of which may cause significant indoor contamination. These include wood stoves, furnaces, and fireplaces whose emissions are vented to the ambient environment through flues and chimneys. They also may include unvented gas and kerosene space heaters and gas cooking stoves and ovens. Improper appliance and flue gas system installation, lack of maintenance, and the vagaries of nature can cause combustion by-product contamination from appliances that should pose very low potential for contaminating indoor spaces. Significant contamination of indoor spaces by combustion by-products may also occur from personal habits such as smoking tobacco products (and possibly other weeds as well).

A large variety of gaseous and particulate-phase materials are produced as a result of tobacco smoking and combusting fuels such as oil, gas, wood, and coal. Those products, which are easily measured and produced in high concentrations, have received the most attention in monitoring and public health studies, and include CO, CO_2, NO_x, SO_2, formaldehyde, and RSPs.

Wood-Burning Appliances

In the past 10 years, the use of wood-burning stoves and furnaces for residential space heating has become very popular. This popularity is directly related to homeowner concerns about high utility costs of central oil and gas space heating systems and heating with electricity. An estimated six million wood-burning stoves were in use in 1985. Fireplaces, primarily for aesthetic reasons, are popular in middle-and upper-income households.

A large variety of wood-burning stoves are sold. Typically, they fall into two basic types: conventional and airtight. Conventional stoves have low combustion efficiencies, commonly in the range of 25–50%. Combustion efficiencies in airtight stoves exceed 50%.

The impact of wood appliance operation on indoor air quality has been studied by a number of investigators. Traynor et al.,[61] in field tests on wood-burning stoves and furnaces in three occupied houses, reported elevated levels of CO, NO, NO_2, and SO_2 during the period of appliance operation. Elevated levels of NO and NO_2 associated with wood-burning appliance operation were also reported by Knight et al.[62] Moschandreas et al.[63] observed elevated levels of CO and particulate matter, but not of NO, NO_2, or SO_2, associated with wood-burning stove operation. In comparisons of wood-burning stove houses with all-electric houses, no differences in NO_2 levels were observed.[64]

More consistent are reports of elevated levels of CO and suspended or respirable particles associated with wood-burning appliance operation. All studies that have measured CO levels report elevated concentrations. These typically are in the range of a few parts per million to approximately 30 ppm, the latter in non-airtight stoves under worst-case operating conditions.[62]

Wood-burning appliances are a significant source of particulate matter. Moschandreas[63] reported indoor/outdoor ratios for total suspended particles of 1.8, 6.1, and 8.5 in a residence heated with a wood stove and two residences with operating fireplaces, compared to nonwood-burning ratios of 1.0, 2.1, and 1.5 respectively. Knight et al.[62] reported indoor/outdoor suspended particulate matter ratios for five airtight stoves that varied from 1.2 to 1.3 and for three non-airtight wood stoves from 4.0 to 7.5.

Associated with increased particulate matter levels in homes heated with wood is an increase in carcinogenic polynuclear aromatic hydrocarbons

(PAH). Lowest and highest PAH indoor concentrations from a series of measurements for a variety of wood heater types, as well as outdoor concentrations, are illustrated in Figure 1.6. In studies conducted in Boston, Moschandreas et al.[63] reported benzo-α-pyrene levels in wood-burning residences which were four to five times higher than in those using all-gas or all-electric utilities.

Unvented Kerosene and Gas Space Heaters

For reasons similar to those for wood-burning appliances explained previously, unvented space heaters for residential heating have also become very popular in the last decade. This has been particularly the case for kerosene heaters, of which over ten million units were sold in the United States by 1985. Kerosene heaters are used to "spot-heat" homes which are usually heated centrally by gas, oil, or electricity. By spot-heating, comfortable temperatures are maintained in occupied rooms while the central thermostat is turned down. Kerosene heaters are typically used in northern climates where winter space heating needs are significant. Unvented gas-fired space heaters are commonly used in southern states where heating needs during the winter season are more limited.

Kerosene heaters are of three basic designs: convective, radiant and two-stage. Convective heaters employ a cylindrical wick and operate at relatively high combustion temperatures. Radiant heaters have a cylindrical wick whose flames extend up into a perforated metal baffle, which glows red hot, releasing infrared or radiant heat. Radiant heaters operate at lower combustion temperatures than convective ones. The two-stage heater is similar in design to the radiant type except that there is a second chamber above the radiant element. This second chamber is intended to further combust CO and unburned hydrocarbons.

Laboratory studies of both unvented kerosene and gas space heaters indicate that they can emit significant quantities of CO, CO_2, NO, NO_2, RSPs, SO_2, and formaldehyde into indoor spaces.[65-70] The magnitude of emissions depends on heater type, heater operation parameters, and the type of fuel (relative to SO_2 emissions). Specific heater emission characteristics are discussed in Chapter 2. In addition to emissions, indoor air concentrations of heater combustion by-products are also determined by such parameters as extent of use and house volume and ventilation characteristics.

The air contamination potential for both unvented gas and kerosene space heaters is considerable, as can be seen from data collected in chamber studies by Leaderer et al.,[65] illustrated in Figure 1.7. Under ventilation rates which are typical of residences ($<$ 1 air change per hour [ACH]), kerosene heater operation in a room similar in size to a moderate bedroom is observed to produce SO_2 levels in excess of 1 ppm, NO_2 levels in the range of 0.5 to 5 ppm, CO levels in the range of 5-50 ppm, and CO_2 levels in the range of 0.1-1% (1000-10,000 ppm). With reference to population exposed, exposure duration,

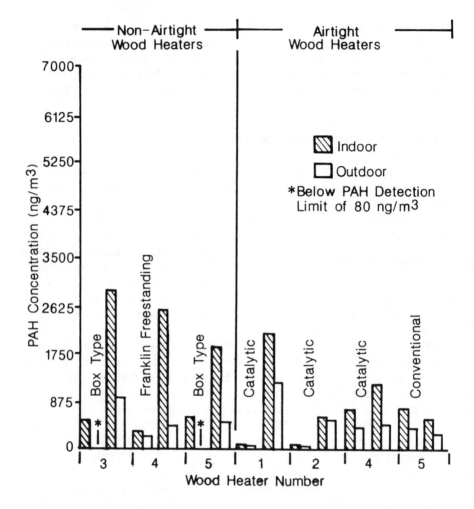

Figure 1.6. Indoor PAH concentrations (high and low 12-hr concentrations) associated with wood heater use compared to ambient levels.[62]

and various health guidelines, these worst-case chamber results appear to be potentially very significant.

Unvented kerosene and gas-fired space heaters are used by homeowners and apartment dwellers under widely varying conditions of home air space volume, air exchange rate, number of heaters used, and daily and seasonal operating hours. Therefore, exposures to contaminant emissions under real-world conditions can also be expected to vary widely. A limited number of investigations have been conducted in residential environments where unvented space heaters are routinely used to spot-heat homes.

Cooper and Alberti[71] monitored 14 homes with kerosene heaters in subur-

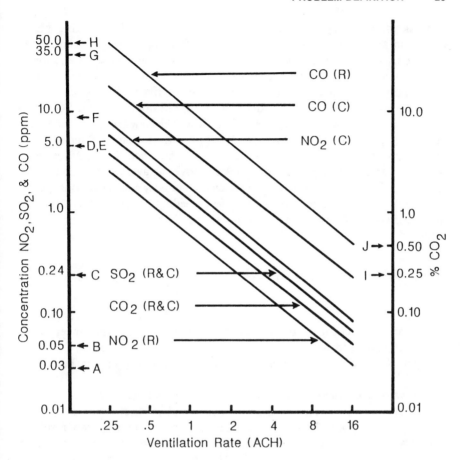

A   NAAQS SO2 Annual Average
B   NAAQS NO2 Annual Average
C   NAAQS SO2 24–Hour Average
D   OSHA NO2 Ceiling Value
E   OSHA SO2 8–Hour Time Weighted Average
F   NAAQS CO 8–Hour Average
G   NAAQS CO 1–Hour Average
H   OSHA CO 8–Hour Time Weighted Average
I   ASHRAE CO2 Guideline
J   OSHA CO2 8–Hour Time Weighted Average
(R) Radiant
(C) Convective

Figure 1.7.   Steady-state concentrations of contaminants associated with kerosene
heater operation.[65] (NAAQS: National Ambient Air Quality Standards.
ASHRAE: American Society of Heating, Refrigerating and Air-Conditioning Engineers.)

ban Richmond, Virginia. Carbon monoxide concentrations during heater operation averaged 6.8 ppm with a range of 0–14 ppm; SO_2 concentrations averaged 0.4 ppm with a range of 0–1.0 ppm. Traynor and Nitschke[72] reported NO_2, CO, and RSP levels in three kerosene-heated homes that ranged from 25 to 117 ppb, 1 to 5 ppm, and 10 to 88 $\mu g/m^3$, respectively.[3] In one of the most extensive studies conducted, Leaderer et al.[73] monitored NO_2 and SO_2 levels during one two-week period in over 100 houses in the New Haven, Connecticut area. Nitrogen dioxide levels averaged 19.6 ppb in those homes with one heater and 37 ppb in those with two. As a point of reference, in residences with neither a kerosene heater nor a gas cooking stove the average NO_2 concentration was 3.7 ppb. Over 49% of the residences had NO_2 concentrations greater than 53 ppb during heater use, with 8.4% exceeding 255 ppb. Over 21% had average SO_2 levels exceeding 0.24 ppm, the 24-hour ambient air quality standard.

Gas Cooking Stoves and Ovens

Although hoods are often installed with them, gas cooking ranges and ovens for all practical purposes are unvented combustion appliances. They have been shown to be significant sources or potential sources of CO, CO_2, NO, and NO_2. They may also release aldehydes, a variety of organic gases, and RSPs.[68,74,75]

A variety of factors affect contaminant emissions from gas cooking ranges and ovens. These are described in detail in discussions of control measures in Chapter 2. The use of gas ranges for cooking can result in significant levels of CO in the kitchen. Sterling and Sterling[75] reported kitchen CO levels in the range of 11–41 ppm depending on the number of burners in operation and whether the burner was covered by a pan. In studies of the effectiveness of local and general ventilation on contaminant levels associated with a gas range, Traynor et al.[76] have reported CO levels in the kitchen of an unoccupied research house to rise to approximately 20–25 ppm in the first 15–30 minutes of operation, with NO_x levels rising in the first 45 minutes to 1.2 ppm.

Nitrogen dioxide exposures of residents of homes with gas cooking stoves have been studied extensively. In Boston-area studies, indoor NO_2 concentrations in 38 homes averaged 27 ppb as compared to 5 ppb in 50 electric stove homes during the winter period.[77] In New Haven–area studies, NO_2 levels in 42 gas stove homes averaged 18 ppb.[73] Average NO_2 concentrations in 10 Indiana gas range homes were 35 ppb.[65]

The use of gas ranges and ovens for cooking represents contaminant exposures which are episodic. However, in ghetto dwellings these appliances are often used continuously during the winter period for supplemental heating. Such use would be expected to result in exposures to higher concentrations and for longer durations and would therefore have higher health risks.

Table 1.9   Contributing Factors to Household Carbon Monoxide Poisonings[79][a]

	%
1. Equipment problems due to defects, poor maintenance, damaged heat exchangers	46
2. Collapsed or blocked chimneys or flues, dislodged or damaged vents	31
3. Downdraft in chimneys or flues	25
4. Improper installation of equipment, chimneys, vents—lack of understanding of operation	24
5. Episodes in recreational settings (e.g. cottages, RVs, etc.)	18
6. Airtightness of building envelope/inadequate combustion air	17
7. Inadequate exhaust of space heaters, appliances	12
8. Exhaust ventilation/fireplace competing for air supply	11
9. Weather conditions	1

[a]More than one factor can be implicated in an episode.

Flue Gas Spillage/Backdrafting

Under normal operating conditions, vented gas-fired forced-air furnaces and gas-fired water heaters should not result in measurable effects on indoor air quality.[65,78] (This is fortunate; in North America the use of gas-fired furnaces is the predominant means of providing space heat during the colder months.) However, as a result of a variety of operating and climatic circumstances, substantial spillage of flue gases can occur with gas furnaces, gas water heaters, and wood-burning appliances.[62,79-83] Under the most extreme circumstances, such flue gas spillage and backdrafting can cause carbon monoxide poisoning. Russel and Robinson[79] reviewed 293 episodes of CO poisoning associated with flue gas spillage in Canada in the 10-year period from 1973 to 1982 and Statistics Canada data for the eight-year period from 1973 to 1981. In the latter period there were 238 deaths associated with flue gas spillage, a rate of 26 per annum. Russel and Robinson suggested that these figures may underestimate the true number of CO poisonings. Contributing factors to reported CO poisoning are summarized in Table 1.9.

Flue gas spillage occurs when flue gas flows upward at too slow a rate to evacuate all combustion products from the furnace, stove, or fireplace. Under some circumstances, such as chimney blockage or other failure, upward flue gas flow can be stalled, with most combustion gases entering the living space. In backdrafting, outside air flows down the flue or chimney with combustion gases spilling out of the dilution port and the furnace air intake port.

Causes of flue gas spillage and backdrafting are summarized in the table of Russel and Robinson.[79] Other failure mechanisms include chimney design flaws and appliance overfiring.[83]

Backdrafting can occur when a house is depressurized by competing exhaust systems, when the chimney is cold and the appliance is not operating, and under certain meteorological conditions.[83]

Backdrafting caused by house depressurization can be a significant problem in energy-efficient houses where infiltration air is not adequate to supply the needs of mechanical exhaust systems (kitchen, bath) and for fireplace combustion and dilution air. Fireplaces can serve as high-efficiency exhausts.[80] Swinton and White[81] developed a model to assess the components of pressure loss that affect flue performance under dynamic conditions. Their model shows that the operation of exhaust fans and fireplaces, increased envelope air tightness, downdrafting winds caused by local topography or high adjacent structures, and high leakage sites (high vertical center of leakage) all contribute to the impedance of flue gas flow.

Tobacco Smoking

The smoking of tobacco products indoors is in all probability the major source of combustion-generated contaminants found in indoor air. The significance of this contamination is told in the fact that approximately 42 million individuals (about 30% of the adult population) smoke. These individuals subject themselves and many nonsmokers to a large variety of gas- and particulate-phase contaminants. Several thousand different compounds have been identified in tobacco smoke and approximately 400 have been quantitatively characterized. Some of the more significant compounds or materials associated with tobacco smoke are respirable particles, nicotine, nitrosamines, polycyclic aromatic hydrocarbons, CO, CO_2, NO_x, acrolein, formaldehyde, and hydrogen cyanide.[84]

Indoor air contamination from tobacco smoke results from sidestream (SS) smoke (smoke given off between puffs) and exhaled mainstream (MS) smoke. The combination of sidestream smoke and exhaled mainstream smoke is usually referred to as environmental tobacco smoke (ETS). Qualitatively, ETS contains the same constituents as does MS tobacco smoke. However, quantitative differences exist between SS and MS tobacco smoke. On a mass basis, SS comprises about 55% of the cigarette column consumed. The SS smoke is produced at lower temperatures and under strongly reducing conditions. Differences in production rates for both MS and SS tobacco smoke can be seen for a variety of gas-phase and particulate constituents in Table 1.10. These are expressed as SS/MS ratios, with ratios greater than 1 indicating quantitatively higher mass concentrations emitted in SS smoke.[84] With few exceptions, constituents listed in Table 1.10 are seen to have higher concentrations in SS than MS smoke. In some cases, levels in SS are an order of magnitude higher than in MS. Concentrations of the animal carcinogen N-nitrosodimethylamine in SS smoke may be as much as two orders of magnitude higher than in MS smoke.

ETS undergoes physical-chemical changes as it ages. Such changes may include conversion of NO to the more toxic NO_2 and vaporization of volatiles,

Table 1.10   Ratios of Selected Gas- and Particulate-Phase Components in SS and MS Tobacco Smoke[84]

Vapor Phase	SS/MS Ratios	Particulate Phase	SS/MS Ratios
Carbon monoxide	2.5–4.7	Particulate matter	1.3–1.9
Carbon dioxide	8–11	Nicotine	2.6–3.3
Benzene[a]	10	Phenol	1.6–3.0
Acrolein	8–15	2-Naphthylamine[a]	30
Hydrogen cyanide	0.1–0.25	Benzo-α-anthracene[a]	2.0–4.0
Nitrogen oxides	4–10	Benzo-α-pyrene[a]	2.5–3.5
Hydrazine[a]	3	N-Nitrosodiethanolamine[a]	1.2
N-Nitrosodimethylamine[a]	20–100	Cadmium	7.2
N-Nitrosopyrrolidine	6–30	Nickel[a]	13–30

[a]Animal, suspected, or human carcinogen.

reducing median diameter of smoke particles. Aged ETS differs, therefore, from relatively fresh ETS.

The concentration of ETS components to which an individual is exposed depends on factors such as the type and number of cigarettes consumed, the volume of the room, the ventilation rate, and proximity to the source. A variety of studies have been conducted to monitor the impact of tobacco smoking on levels of contaminants in indoor spaces. Results from selected studies are summarized in Table 1.11.[85-91] RSP and nicotine levels are seen to be most significantly affected. Tobacco smoking appears to be the dominant source of respirable particles in residential indoor environments.

Health Effects

Health concerns associated with combustion-related contaminants in indoor spaces center on those which are produced in relatively high concentra-

Table 1.11   Tobacco-Related Contaminant Levels in Indoor Spaces

Contaminant	Type of Environment	Levels	Nonsmoking Controls	Reference No.
CO	Room (18 smokers)	50 ppm	0.0 ppm	85
	15 Restaurants	4 ppm	2.5 ppm	86
	Arena (11,806 people)	9 ppm	3.0 ppm	87
RSP	Bar and grill	589 μg/m^3	63 μg/m^3	88
	Bingo hall	1140 μg/m^3	40 μg/m^3	88
	Fast food restaurant	109 μg/m^3	24 μg/m^3	87,89
NO$_2$	Restaurant	63 ppb	50 ppb	90
	Bar	21 ppb	48 ppb	90
Nicotine	Room (18 smokers)	500 μg/m^3	—	85
	Restaurant	5.2 μg/m^3	—	91
Benzo-α-pyrene	Arena	9.9 ng/m^3	0.69 ng/m^3	87
Benzene	Room (18 smokers)	0.11 mg/m^3	—	85

tions and for which health standards or guidelines have been in place for other types of exposures, most notably occupational and ambient air quality standards. These include CO, NO_2, SO_2, particulate matter, formaldehyde, and CO_2. The health effects of CO, NO_2, SO_2, and particulate matter have been extensively reviewed by committees of the National Research Council[92-94] and by the EPA.[95-97] The reader is referred to those documents for an expanded treatment.

A variety of health effects have been associated with CO. These range from neuromotor effects such as decreased attention span and reaction time, to headaches, nausea, and extreme drowsiness, to death by asphyxiation. The response depends on the level of exposure. Carbon monoxide's toxicity is related to the fact that it binds with blood hemoglobin, thus decreasing hemoglobin's capacity to transport oxygen to body tissues. Except in cases of extensive flue gas spillage, which are life-threatening, CO exposures from indoor sources are quite low. Carbon monoxide levels in indoor spaces associated with the use of wood-burning appliances, unvented space heaters, gas cooking stoves and ovens, and tobacco smoking are usually below threshold levels implicit in air quality guidelines. Health risks associated with such exposures and even those somewhat above CO guidelines are unknown.

Potential health effects associated with indoor exposures to NO_2 have received considerable attention. Studies by Melia et al.[98] in the United Kingdom indicated a significantly increased incidence of respiratory illness in children in homes with gas cooking ranges as compared to those cooking with electricity, suggesting that the cause for the association was indoor air pollution. Initial results reported for the Harvard Six Cities studies indicated that gas ranges were associated with respiratory illness in children before the age of two and that lung function was lower in gas range homes than in those with electric cooking ranges.[99] In an update of the Harvard study, no significant relationship was observed between gas cooking and children's respiratory symptoms or pulmonary function.[100]

Melia et al.[98] surmised that if the relationship between childhood respiratory symptoms and gas cooking were real, then the most likely cause would be exposure to NO_2, based on NO_2's known clinical and toxicological effects. Further studies by Melia and colleagues[101] indicated that the presence of respiratory illness was not related to average NO_2 levels in the kitchen, although they did show a statistically weak association with NO_2 levels in the bedroom. Lung function in children was apparently not associated with average NO_2 levels in the kitchen or bedroom. In additional studies,[102] Melia et al. reported a tendency for the incidence of one or more respiratory conditions to be higher in homes with relatively high levels of NO_2 and lower in homes with lower levels in the bedroom and living room. They suggested that weekly average NO_2 levels used in previous epidemiological studies were not accurate estimates of exposures and that very high peak levels during gas range operation may be more harmful to health.

Because of the economic implications of a link between gas cooking and the respiratory health of children, this issue has been hotly debated. The evidence to date is suggestive of such a link, but is not definitive.

Health concerns associated with elevated NO_2 levels are not limited to gas cooking ranges and ovens. As previously described, NO_2 levels in laboratory chambers (32 m^3) in which a kerosene heater was being operated had high NO_2 levels, in excess of 0.5 ppm.[65] Residential monitoring studies indicate that NO_2 levels in homes with kerosene heaters are in the same range as those that use gas cooking.[73]

Contamination of residences by elevated levels of SO_2 is, for the most part, limited to the use of 2K kerosene in kerosene space heaters. Levels of SO_2 in poorly ventilated residences with an operating kerosene heater may on occasion be sufficient to induce asthmatic attacks in individuals with hypersensitive airways.[71]

Aside from health effects associated with specific contaminants, several studies[103,104] indicate a causal relationship between respiratory health in children and the use of wood-burning appliances. Honicky et al.[104] observed significant differences in the incidence of coughing and wheezing in wood-burning stove homes as compared to controls.

The health risks associated with passive or involuntary smoking have been extensively reviewed by the Surgeon General.[84] Most notably, passive smoking has been shown to be causally associated with lung cancer in healthy nonsmokers and to increase the frequency of respiratory infections and respiratory symptoms in children of smoking parents. Exposure to ETS is also a cause of sensory irritation and annoyance.

Passive smoking, in addition to its own risk, increases the risk of exposure to radon. Tobacco smoke is a vehicle for the transport of radon progeny to the lungs and thus increases exposures to radiation associated with radon decay products.[105]

RADON

As an indoor contaminant, radon is unique in two ways. First, it is the only naturally occurring indoor contaminant (other than biological contaminants); it is not a product of modern technology or even some ancient one. It is also unique in that it poses lifetime lung cancer risks orders of magnitude higher than those of commercial chemicals considered by EPA to pose significant cancer risks to the population.

Radon is the environmental quality problem of the moment if not the decade. It has received, and is continuing to receive, an enormous amount of media attention and, to a lesser degree, official public health concern. It has spawned a whole new air testing and radon mitigation services industry.

Radon is a noble gas produced in the radioactive decay of radium, found in uranium ores, phosphate rock, and a number of common minerals such as granite, schist, gneiss, and even limestone to some degree. Because it is inert,

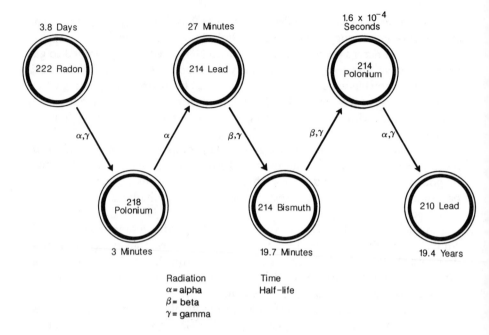

Figure 1.8. Radon decay series.

radon in itself poses no hazard. However, it undergoes radioactive decay, producing a series of short-lived progeny (Figure 1.8). The half-life of radon is 3.8 days. The decay products are solid elements, two of which emit alpha particles. Radon progeny are electrically charged and readily attach to airborne particles.

Concern related to radon contamination of indoor air was first expressed over a decade ago. It arose from the recognition that waste rock or tailings associated with the mining and processing of uranium and phosphate ores had significant concentrations of radium-226,[106] and that the decay of this radium to radon and its progeny meant individuals whose homes had been constructed on or with uranium or phosphate tailing materials could be at significant risk of radiation exposures. In response to this potential problem, Congress passed the Uranium Mill Tailings Act in 1978.

As a consequence of research conducted in conjunction with the uranium mill tailings problem, it was discovered that radon levels in houses far removed from uranium tailings or phosphate ore areas were often as high as or higher than levels in houses close to uranium or phosphate sites. Numerous subsequent studies have documented the widespread nature of radon contamination of indoor air.[14,107-109]

This indoor contamination was placed in an alarming context with the discovery of extremely high levels in a home in the Boyertown area of southeastern Pennsylvania.[109] Boyertown is located on a geological formation called the Reading Prong, which extends east of Reading, Pennsylvania through three counties in Pennsylvania and portions of north central New Jersey, southeastern New York, and parts of New England. Bedrock minerals in the Reading Prong contain elevated levels of uranium and thorium.

Concentrations

Radon and radon progeny concentrations are expressed in a variety of units. In the United States radon levels are most commonly expressed in pico-curies per liter (pCi/L). One picocurie is equivalent to 10^{-12} curies (Ci), where one Ci = 3.7×10^{10} becquerels (Bq). In the International System of Units (Systeme International, SI), radon concentrations are expressed as becquerels/ m^3. Under this system, 1 pCi/L = 37 Bq/m^3. Radon progeny concentrations are expressed in units called working levels (WL), a measure of the potential alpha energy concentration (PAEC). The PAEC for one WL is 1.3×10^5 million electron volts per liter. This would be equivalent to the PAEC/unit volume that would be associated with air containing 100 pCi/L each of the short-lived progeny. In the real world, where the equilibrium ratio between radon and radon progeny is approximately 2:1, one working level would be equal to 200 pCi/L.

Because of the common occurrence of radium-226 in the earth's crust, radon is ubiquitous in ambient air, with average concentrations in the range of 0.20–0.25 pCi/L.[110] Due to the effect of a variety of factors, radon levels in buildings may be several times, or even several orders of magnitude, higher than ambient levels. This is particularly the case in single-family dwellings.

In the Boyertown, Pennsylvania home, radon levels were as high as 2600 pCi/L or 13 WL. This is in comparison to the 4-pCi/L (0.02-WL) EPA guideline for remedial action[111] and the 8-pCi/L National Council on Radiation Protection (NCRP) guideline for unacceptably high levels.[112] Subsequent to this discovery, 18,000 homes were monitored in a cooperative program by the Pennsylvania Department of Health, the EPA, and the local utilities. Radon levels in excess of the EPA remedial action guideline were observed in 59% of the homes.[109]

Nero et al.[108] analyzed indoor monitoring data (primarily single-family dwellings) from 32 U.S. areas, concentrating primarily on urban areas. Radon levels for U.S. homes averaged 1.5 pCi/L, with a geometric mean of 0.9 pCi/L. (Radon levels were lognormally distributed.) The researchers estimated that 2% of U.S. homes have radon levels in excess of 8 pCi/L, the NCRP's guideline for unacceptability.

Terradex, Inc.[14] has one of the largest data bases for radon levels in U.S. homes, containing data for 30,000 nonrandomly distributed homes representing approximately 0.5% of the U.S. single-family housing stock. The highest three-month average was 4000 pCi/L. The average was 4.26 pCi/L, with median and geometric mean values of 1.65 and 1.74 pCi/L. Five percent of the

Table 1.12   Radon Levels (pCi/L) in U.S. Homes Summarized on a Regional Basis[14]

Region	Median	Geometric Mean	Arithmetic Mean
Northeast	3.24	3.43	11.13
Midwest	2.32	2.36	5.08
Northwest	0.64	0.64	1.20
Mountain states	2.78	2.90	6.47
Southeast	1.41	1.43	3.09

radon values were less than 0.2 pCi/L; 0.5% exceeded 100 pCi/L. Mean, median, and average values for five regions in the U.S. are summarized in Table 1.12. Highest radon levels were seen in the northeast, lowest in the northwest.

Data reported by Alter and Oswald (Terradex, Inc.[14]) were based on purchased tests and therefore are highly unlikely to be representative of U.S. housing stock. In a survey of 453 middle class homes in 42 states, Cohen[113] reported mean and geometric mean radon levels of 1.47 and 1.03 pCi/L, respectively.

Factors Affecting Radon Levels

The dominant source of radon in almost all instances is soil gas, which is transported into buildings by pressure-induced convective flows. Soil gas is responsible for 90+% of all radon in indoor air.[13] Other major sources of radon in indoor air include well water and masonry materials.

A variety of factors contribute to differences in radon levels found in buildings. The major cause of such differences is source strength.[107] Those buildings constructed on substrates with high radon release and transport potential will have the highest radon levels. Radon levels in houses and other buildings are increased by factors that enhance radon transport, including cracks in slabs and basement floors and walls, water sumps, and building pressure conditions.[114] Radon transport is significantly enhanced when a house is under negative pressure. This depressurization is associated with indoor/outdoor temperature differentials, excessive exhaust, and changes in barometric pressure. The effect of indoor/outdoor temperature and pressure differences on indoor radon levels can be seen in Figure 1.9. Note the significant diurnal variation.[115]

Other factors that may significantly affect indoor radon concentrations include the type of building substructure and natural or mechanically-induced ventilation. These are extensively treated in Chapter 2.

Health Effects

The only known health effect associated with exposure to elevated levels of radon is lung cancer. The lung cancer risk to uranium miners and other hard

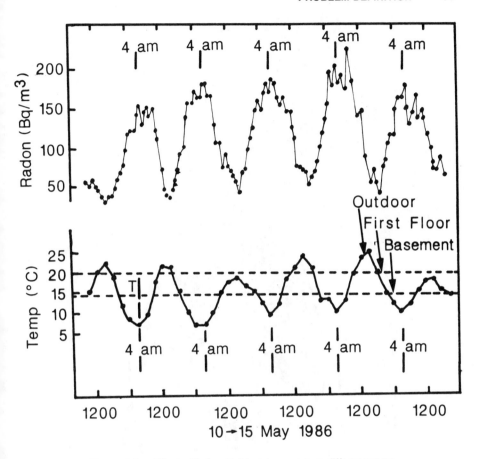

Figure 1.9. Effect of indoor/outdoor temperature differences on indoor radon concentrations.[115]

rock miners has been well documented.[116] The electrically charged radon progeny become attached to airborne particles, which upon inhalation are deposited in bifurcations of the respiratory airways (Figure 1.10). Alpha particle radiation released from radon progeny decay irradiates bronchial epithelium, producing lung cancer after long-term exposure. Budnitz et al.[117] have estimated that 10% of the lung cancer rate in the United States may be due to radon exposures in indoor air. A similar risk has been estimated for lung cancer in Norway.[118] Eidling et al.[119] in Sweden have attributed approximately 30% of lung cancer to indoor radon exposures. Radford and St. Clair Renard[120] estimated that, in Sweden, 30% of the lung cancer rate in nonsmokers and 5% of the rate in smokers (both aged 60) is due to radon exposures.

Nero et al.[108] have projected that the 1.5-pCi/L average concentration assumed for U.S. homes would represent a lifetime risk of 0.3%. This individual risk would correspond to about 10,000 annual cases of lung cancer in the

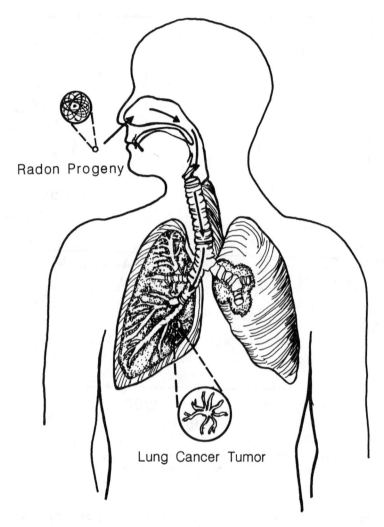

Figure 1.10.   Inhalation and deposition in the lungs of radon progeny attached to aerosol particles.

United States. A significant part of this incidence (about 15%) would be associated with homes exceeding 8 pCi/L. In these latter homes the lifetime risk would be greater than 2% for long-term occupants. Lifetime risk for the Boyertown home was estimated to be 13% per year, an extraordinarily high lung cancer risk.

Beyond risk assessments based on extrapolations from occupationally-exposed miners, a causal role for radon exposures in lung cancer is suggested from epidemiological studies. In Sweden, lung cancer was shown to be related to radon and radon progeny exposures, with 30% of lung cancers attributed to elevated radon levels.[121] In a study in Maine, Hess et al.[122] reported a significant correlation between tracheal, bronchial, and lung cancers with radon levels in well water and, presumably, indoor air. Lung cancer rates in central Florida (where radium-bearing phosphate ores are located) are reported to be several times higher than the national incidence of lung cancer on a county-by-county basis.[123]

Synergistic effects associated with exposures to both radon and tobacco smoke have been postulated. Studies of Bergman et al.[124] indicate that indoor radon exposure increases considerably in the presence of cigarette smoke, presumably because newly-formed, electrically-charged radon progeny tend to attach to smoke particles rather than walls and other surfaces. They have suggested that observations of a lung cancer hazard associated with passive smoking may be more consistent with the effect of alpha radiation associated with radon progeny than with chemical components of smoke. Martell[125] suggests that radon progeny exposure and cigarette smoking may explain the distribution of lung cancer in the general population. Results of his experimental studies indicate that radon progeny make a substantial contribution to alpha radiation at bronchial bifurcations in smokers' lungs. Cigarette smoke particles are of low mobility and persist in air; they are selectively deposited at bifurcations and highly resistant to dissolution. Because of these facts, a substantial amount of radiation exposure occurs to a small volume of bronchial tissue before clearance.

FORMALDEHYDE

Because of the extensive use of building materials and furnishings which release it, formaldehyde is a ubiquitous contaminant of indoor air. Elevated levels in indoor environments, including mobile homes, homes constructed with particleboard subflooring, and homes insulated with urea-formaldehyde foam (UFFI), have resulted in numerous odor and health complaints associated with formaldehyde contamination of indoor spaces, most notably in the late '70s and the early '80s. The formaldehyde indoor air quality problem has received considerable public and research attention. In the past decade it has been the major indoor air quality issue and public health concern.

Sources

Though many products have the potential for releasing formaldehyde into indoor air, relatively few are responsible for causing significant levels of contamination. Emission potentials for a variety of construction materials, furnishings, etc. are summarized in Table 1.13.[126-128] Pressed wood products (medium-density fiberboard, particleboard, and hardwood plywood paneling) and urea-formaldehyde foam release formaldehyde at rates which are one to two orders of magnitude greater than other products tested. Major formalde-

Table 1.13 Formaldehyde Emissions from a Variety of Construction Materials, Furnishings, and Consumer Products

	Range of Formaldehyde Emission Rates $\mu g/m^2/day$
Medium-density fiberboard[126]	17,600–55,000
Hardwood plywood paneling[127]	1,500–34,000
Particleboard[127,128]	2,000–25,000
Urea-formaldehyde foam insulation[128]	1,200–19,200
Softwood plywood[128]	240–720
Paper products[127]	260–680
Fiberglass products[127]	400–470
Clothing[127]	35–570
Resilient flooring[127]	<240
Carpeting[127]	NP[a]–65
Upholstery fabric[127]	NP–7

[a]NP = none present.

hyde sources are formulated, in whole or in part, from urea-formaldehyde (U-F) resins. Excess or residual formaldehyde trapped in the resin is later released; significant formaldehyde release also occurs when the resin polymer undergoes hydrolytic decomposition.

Urea-formaldehyde foam insulation has received the most media notoriety and regulatory attention. The walls of conventional homes in the United States and Canada were retrofitted with UFFI on a widespread basis. It was subsequently banned in Canada, and a ban was unsuccessfully attempted in the United States.

Despite the attention directed to UFFI, wood products are the major sources of indoor formaldehyde contamination. These products are bonded or finished with U-F resins. It is the U-F resins which are responsible for free formaldehyde liberation into indoor air.

Particleboard is a wood product manufactured from fine (about 1 mm) wood particles which are mixed with 6- to 8-wt% U-F resins and pressed into panels. Because of its relatively low cost, it has seen widespread use as subflooring in conventional houses, decking in mobile homes, low-grade decorative wall paneling, shelving, and components of cabinets and furniture. Because of the diversity of its use, most residences contain one or more products manufactured from particleboard.

The plys of hardwood (but not softwood) plywood are bonded with U-F resin. Common usages are as a decorative wall paneling and structural/decorative components of furniture and cabinetry.

Medium-density fiberboard (MDF) contains wood particles that are much smaller than those in particleboard. It is darkish in color and has the texture and appearance of a bonded paper–like product. It is often used in table tops and as a core material in the manufacture of wood members for cabinets and furniture. MDF is a potent source of formaldehyde, with a potency that has historically been 2–4 times that of particleboard.[129]

Table 1.14   Formaldehyde Levels in Residential Environments

Study	Number	Range	Mean	Median
		Concentration (ppm)		
Mobile Homes				
Minnesota[130]	397	0.02–3.69	0.42	—
Texas[131]	159	<0.02–0.78	0.15	—
Wisconsin				
Complaint[132]	65	<0.10–3.68	—	0.47
Noncomplaint[133]	137	<0.10–2.84	0.46	0.39
California[134]	663		0.09	0.07
UFFI Homes				
Ontario[135]	450		0.045	—
New York State[136]	1954	0.00–0.49	0.06	—
New Hampshire[137]	71	0.01–0.17	0.06	0.05
Conventional Homes				
Ontario[135]	225		0.03	—
CPSC[138]	41	0.01–0.08	0.03	—
Indiana[139]				
PB subflooring	30	0.01–0.46	0.11	0.09
Non-PB subflooring	58	0.00–0.14	0.06	0.06
Minnesota[130]	489	0.01–5.52	0.14	—
California[134]	51	0.01–0.09	0.04	—

Relatively less known as a source of indoor formaldehyde contamination are varnish-like surface coatings formulated from U-F or U-F/melamine-formaldehyde resin mixtures. These materials are widely used as base and finish coatings for hardwood cabinetry and furniture. They are also applied to hardwood floors and may even be applied to concrete. Immediately after application, they are significant sources of free formaldehyde. They may even be the most potent source of formaldehyde of any product intended for indoor use on the market today.

Indoor Concentrations

A variety of investigators have reported levels of formaldehyde in residential environments. Results of selected studies are reported for UFFI houses, mobile homes, and conventional houses in Table 1.14.[130-139] Highest average concentrations are reported for mobile homes. The high concentrations of formaldehyde in mobile homes are primarily caused by the large quantity of potent formaldehyde sources used in mobile home manufacture. These include the mobile home decking, decorative hardwood plywood wall covering, particleboard shelving, and cabinetry made from a variety of pressed wood products.

Despite the enormous publicity that has been associated with UFFI houses, their average formaldehyde levels are but a fraction of those reported in mobile homes. With few exceptions, average formaldehyde levels in UFFI houses are typically in the range of 0.5-0.06 ppm.

Conventional homes vary in kinds and quantities of formaldehyde-releasing products present. If conventional residences are classified by source type (note Indiana data in Table 1.14), formaldehyde levels are seen to be higher in residences that contain particleboard subflooring. Conventional homes without particleboard subflooring appear to have average formaldehyde concentrations in the range of 0.03-0.06 ppm.

Source Factors

Though it has been common to blame indoor air quality problems (such as formaldehyde) on building tightness and energy conservation practices, the reality is that formaldehyde levels in any given structure are primarily determined by both environmental and source factors. The latter includes source strength, source loading, and the presence of source combinations.

The more potent the source, the higher the levels of formaldehyde will be. Differences in the potency of major formaldehyde-releasing products can be seen in Table 1.13. In addition to source strength, formaldehyde levels are also affected by the volume of source materials used. The ratio of formaldehyde-releasing surface area to building air volume is called the loading factor. High loading factors for particleboard and hardwood plywood paneling common in mobile home manufacture are likely the primary reason that mobile homes have such high formaldehyde concentrations. The effect of loading factor on formaldehyde emissions and air concentrations has been reviewed by Myers.[140]

When multiple sources of formaldehyde are present in the same indoor space, such sources interact, resulting in formaldehyde concentrations which are not additive. Typically, source combination concentrations reflect those that would occur if the most potent source were present singly, or would be slightly augmented above these values.[141-144]

Environmental Factors

Formaldehyde concentrations reported in Table 1.14 are based on one-time samples collected over a period of an hour or two. These concentrations describe the level of contamination which occurred at the time of testing only, since the concentration at any particular time will be greatly influenced by environmental factors such as temperature, relative humidity and indoor/outdoor temperature differences.

Formaldehyde concentrations increase with increasing temperatures and decrease with decreasing temperatures. An approximate 5-6°C increase in temperature is seen to double the formaldehyde concentration.[145] Formaldehyde levels are also sensitive to both absolute and relative humidity. An increase of relative humidity from 30% to 70% is associated with an approximate 40% rise in indoor formaldehyde levels. Temperature and humidity changes in indoor environments cause formaldehyde levels to vary by as much as a factor of five.[145] The effect of temperature and humidity on emissions from source materials and indoor levels has been reviewed by Myers.[146]

Formaldehyde also varies as a function of season. The highest levels occur during the summer months, the lowest under winter conditions.[147] Under controlled indoor environmental conditions, formaldehyde levels show a strong linear correlation with the average outdoor temperature.[132] Changes in outdoor temperature change the pressure difference between the inside and outside of the structure. These pressure differences are a major factor in air infiltration, which serves to dilute formaldehyde concentrations.

Formaldehyde Level Decay

Formaldehyde levels decrease significantly with time. The nature of this decrease is described in Chapter 3.

Health Effects

Health concerns related to formaldehyde exposures in indoor spaces fall into three basic categories: irritant effects, ability to cause sensitization and asthma, and carcinogenicity.

Formaldehyde's irritant properties are most likely the cause of the large number of complaints that have been associated with formaldehyde exposures in indoor environments. Formaldehyde is a potent eye, upper respiratory, and skin irritant. In complaint investigations and epidemiological studies, mucous membrane irritation symptoms are most frequently reported.[132,135,139,148-150] These include eye irritation, nose and sinus irritation, sore throat, runny nose, sinus congestion, and cough. Pulmonary symptoms including difficulty in breathing, chest pain, and wheezing are also frequently reported. In addition to respiratory problems, formaldehyde-related symptoms (based on epidemiological correlations) may also be neurological (headaches, fatigue, nausea, difficulty sleeping), gastrointestinal (vomiting, diarrhea), and reproductive (menstrual irregularities).

Epidemiological studies have shown a dose-response relationship between formaldehyde exposures in indoor spaces and a variety of symptoms. Significant dose-response relationships have been observed between eight symptoms and formaldehyde concentrations (0.045 ppm average) in UFFI houses.[135,151] Other epidemiological studies[149,152] have shown that exposure to formaldehyde concentrations at levels of less than 0.10 ppm are sufficient to cause symptoms in sensitive individuals.

Menstrual and reproductive disorders are among the least recognized health problems that have been related to formaldehyde exposures. Menstrual and reproductive problems among permanent-press factory workers in Russia,[153] menstrual irregularities in Danish day care center workers,[150] and increased severity of menstrual problems among women living in mobile and other residential environments in the United States[139] have been reported.

Formaldehyde can induce asthma[154,155] and has the potential to induce asthmatic attacks as a nonspecific irritant. However, instances of formaldehyde-induced asthma appear to be rare.

One of the major questions associated with formaldehyde is its potential to cause sensitization. Dermal, as well as whole-body, sensitization has been well documented in dialysis patients.[156] Formaldehyde-induced asthma has been recently shown to be associated with formaldehyde-specific antibodies.[157] Formaldehyde-specific antibodies, which are indicative of an allergic-type sensitization, have also been reported in nonasthmatic subjects exposed to elevated formaldehyde levels in mobile homes.[158] The significance of the latter finding is that formaldehyde has the potential for causing allergy-type symptoms in sensitized individuals independent of its irritant effects and at much lower concentrations.

The carcinogenicity of formaldehyde has received considerable attention and been a source of controversy. Formaldehyde has been conclusively demonstrated to be an animal carcinogen.[159] A significant relationship between formaldehyde exposures and buccal cavity cancer has been reported for garment workers.[160] A significant association between formaldehyde exposures and nasopharyngeal cancer has been reported for residents of mobile homes.[161] The greatest cancer risk was observed for those individuals who lived in mobile homes for more than 10 years. The EPA has projected an upper bound risk of developing cancer among residents of mobile homes, exposed for more than 10 years to an average level of 0.10 ppm, to be 2 in 10,000; for residents exposed to an average level of 0.07 ppm in some conventional homes the risk is 1 in 10,000.[162]

VOLATILE ORGANIC COMPOUNDS

In addition to the special case of formaldehyde, a significant variety of organic contaminants are known to contaminate indoor air. Organic contaminants may also be present as major constituents of indoor aerosols. Because of the large number of chemical species present in indoor air, and the inherent difficulty and cost associated with the identification and quantification of organic chemicals in mixtures, the nature of VOC contamination of indoor air has not been well characterized to this date. The studies that have been conducted indicate that indoor air is contaminated to various degrees by a wide variety of hydrocarbons and hydrocarbon derivatives including aliphatics, aromatics, alkylbenzenes, ketones, polycyclic aromatics, and chlorinated hydrocarbons. Sources include combustion by-products, cooking, construction materials, furnishings, paints, varnishes and solvents, adhesives and caulks, gasoline and motor vehicle emissions, office equipment, home and personal care products, bioeffluents, and pesticides. Like formaldehyde, pesticides represent a special case of organic chemical contamination and as such will be treated in detail in a later section. They would be best described as being semivolatile, in contrast to the volatile compounds which are discussed here.

Contaminants and Indoor Levels

Public Access Buildings

One of the first studies attempting to characterize indoor air contamination by VOCs in public access buildings was by Molhave in Denmark.[163] Molhave identified and quantified a variety of VOCs in 14 office buildings characterized as being problem environments because of occupant complaints. Twenty-nine different compounds were identified, with an average concentration of 0.95 mg/m³ and a range of 0.03–2.8 mg/m³. Fifty percent of these were alkylbenzenes, 20% were alkanes, 9% were terpenes and 16% were miscellaneous.

Berglund et al.[164] studied VOC contamination in a newly built Swedish preschool. Over 160 different compounds were identified in school air samples. These included aliphatic and aromatic hydrocarbons, aldehydes, alcohols, and terpenes. Of those with the highest concentration selected for further study, 15 were observed to have concentrations which were usually less than 0.30 μg/m³, with maximum values of 160–190 μg/m³.

In California, Schmidt et al.[165] sampled air in four office buildings. Concentrations of organics identified were very low, ranging from 3–319 μg/m³. Compounds identified included n-hexane, n-heptane, n-octane, n-nonane, n-undecane, 2-methylpentene, 2,5-dimethylheptane, methylcyclopentane, ethylcyclohexane, and pentamethyl heptane. Additional studies of California office buildings were conducted by Miksch et al.[166] The largest groups of compounds consisted of aliphatic hydrocarbons followed by aromatic hydrocarbons, with toluene being the predominant chemical in this group. The third group consisted of chlorinated hydrocarbons, primarily tetrachloroethylene, 1,1,1-trichloroethane, and trichloroethylene.

EPA monitored volatile organic compounds in four public buildings including two homes for the elderly, a school, and a newly constructed office building.[167] Organic compounds found to be significantly higher in indoor as compared to outdoor air included trichloroethylene, styrene, m,p-dichlorobenzene, n-decane, n-undecane, and n-dodecane. In another EPA study, Wallace et al.[168] conducted qualitative and quantitative analyses of air samples in 10 public buildings. They identified over 200 aromatics, organic halogens, esters, alcohols, phenols, ethers, ketones, aldehydes, and epoxides, in addition to several hundred aliphatic hydrocarbons. Air samples from four of these buildings were given a full qualitative analysis. Table 1.15 lists 24 of the most common compounds (excluding aliphatic hydrocarbons). With few exceptions, VOCs were observed to decrease significantly in concentration over time. Increased levels following occupancy were observed for trichlo-

Table 1.15   Organic Compounds (Nonaliphatic) Most Commonly Found in Four Public Access Buildings[167]

Acetone	1,1,1-Trichloroethane[a]
Benzene[a]	Trichloroethylene[a]
Toluene	Propylmethylbenzene
Xylenes	Dichlorobenzenes[a]
Styrene[a]	Propylbenzene
Ethylbenzene	Nononal
Ethylmethylbenzenes	Trichlorofluoromethane
Trimethylbenzenes	Diethylbenzenes
Tetrachloroethylene[a]	Methylene chloride[a]
Naphthalene	Chloroform[a]
Methylnaphthalenes	Decanal
Dimethylbenzenes	Acetic acid

[a]Mutagnic or carcinogenic properties.

roethylene, p-dichlorobenzene, and 1,1,1-trichloroethane in one or more of the buildings sampled. The study suggests that these increases were associated with building occupant activities.

One factor which affects levels of organics in a building space is effluents from humans. Such effluents are released from body openings and surfaces. Wang[169] studied bioeffluents in a college classroom. Average concentrations for 12 organic compounds are summarized in Table 1.16. Significant effects on organic compound concentrations associated with human occupancy have been reported for two classrooms.[170] Both the number and concentration of organic compounds were observed to increase in the presence of humans. Significant increases were observed for acetone and ethanol, both of which are known to be exhaled in human breath.

Several very specific problems of organic chemical contamination of indoor air in office buildings have been investigated. The most noteworthy of these is polychlorinated biphenyls (PCBs) associated with transformers. In

Table 1.16   Average Concentrations (ppb) of Organic Bioeffluents in a College Classroom[168]

Compound	Population Present/Time		
	389/9:30 am	225/6:30 pm	0/—
Acetone	20.6	22.9	0.2
Acetaldehyde	4.2	3.1	0.1
Acetic acid	9.9	8.5	1.9
Alkyl alcohol	1.7	3.9	0.1
Amyl alcohol	7.6	3.7	0.1
Butyric acid	15.1	11.6	1.7
Diethyl ketone	5.7	2.2	0.8
Ethyl acetate	8.6	2.5	0.5
Ethyl alcohol	22.8	44.5	2.2
Methyl alcohol	54.8	28.4	1.0
Phenol	4.6	4.0	1.0
Toluene	1.8	1.5	0.3

such cases, PCB levels were twice as high as in buildings without PCB transformers. Levels in the former ranged from 191 to 881 ng/m^3.[171] Phthalate esters, ranging in concentration from 0.11 to 0.23 mg/m^3, have also been reported in an office building, apparently associated with polyvinyl wall covering.[172]

In most studies the identification and quantification of VOCs and semivolatile hydrocarbons has been based on the capture and analysis of vapor-phase materials. Organic contaminants are also found in the particulate phase. Numerous organic substances are found on indoor aerosols. Weschler[173] analyzed samples of airborne dust from two telephone buildings. Organic compounds associated with indoor particles included n-alkanes; branched alkanes; phthalate, phosphate, and azelate esters; chlorofluorocarbons; and nicotine. These organic contaminants were suggested to have become attached to aerosol particles by adsorption.

Residences

Molhave and Moller[174] studied air quality in seven unoccupied new and 39 occupied older apartments. Over 40 different compounds were identified, consisting mainly of C^{8-13} alkanes, C^{6-10} alkylbenzenes, and terpenes. Concentrations ranged from 0.02 to 18.7 mg/m^3, with an average of 6.2 mg/m^3 in new units and 0.4 mg/m^3 in older units. Alkylbenzenes such as toluene and the xylenes accounted for over 43% of the total hydrocarbon concentration.

Organic compounds identified in the air of 40 eastern Tennessee residences included toluene, ethyl benzene, n-xylene, p-xylene, nonane, cumene, benzaldehyde, mesitylene, decane, limonene, undecane, naphthalene, dodecane, tridecane, tetradecane, pentadecane, hexadecane, and 2,2-methylnapthalene.[175] With the exception of toluene, xylene, and benzaldehyde, average concentrations were less than 20 μg/m^3. Average toluene levels on a per-home basis ranged from 45 to 160 μg/m^3.

Organic chemical contamination of 10 Canadian homes was characterized by Otsin.[176] Fifty-six hydrocarbon compounds were detected, including 21 polynuclear aromatics, 15 aliphatics and aromatics, 7 halogenated compounds, 5 alcohols, 3 phthalates, and 2 organophosphates. No concentration exceeded 100 μg/m^3. Volatile organic compounds have been characterized in hundreds of randomly selected German houses.[177] Average concentrations and the concentration ranges for 57 VOCs are summarized in Table 1.17. Ranges varied by up to three orders of magnitude. With the exception of toluene, mean concentrations were below 25 μg/m^3 with the majority being lower than 10 μg/m^3. The average sum of identified VOCs was approximately 0.4 mg/m^3.

Table 1.17   Volatile Organic Compounds in the Air of 230 West German Residences[177]

Compound	Mean Concentration ($\mu g/m^3$)	Range ($\mu g/m^3$)
n-Alkanes	67.0	8.7– 432
n-Hexane	9.0	2.0– 46
n-Heptane	7.0	1.5– 82
n-Octane	4.9	<1.0– 64
n-Nonane	9.7	<1.0– 101
n-Decane	14.0	<1.0– 136
n-Undecane	10.0	<1.0– 88
n-Dodecane	5.6	<1.0– 35
n-Tridecane	6.3	<1.0– 79
i-Alkanes (C_6–C_9)	30.0	5.2– 231
Cycloalkanes	15.0	2.1– 79
Methylcyclopentane	2.8	<1.0– 15
Cyclohexane	7.4	<1.0– 29
Methylcyclohexane	5.5	1.1– 51
Aromatics	166.0	16.0–1260
Benzene	9.3	<1.0– 39
Toluene	76.0	11.0– 578
Ethylbenzene	10.0	1.5– 161
m + p-Xylene	23.0	3.3– 304
o-Xylene	7.0	1.2– 45
i + n-Propylbenzene	4.5	<1.0– 89
Styrene	2.5	<1.0– 41
1-Ethyl-2-methylbenzene	4.4	<1.0– 103
1-Ethyl-(3 + 4)methylbenzene	8.9	<1.0– 229
1,2,3-Trimethylbenzene	3.5	<1.0– 83
1,2,4-Trimethylbenzene	11.0	<1.0– 312
1,3,5-Trimethylbenzene	3.9	<1.0– 111
Naphthalene	2.3	<1.0– 14
Chlorinated Compounds	48.0	3.5–1630
1,1,1-Trichloroethane	7.9	<1.0– 264
Trichloroethylene	13.0	<1.0–1200
Tetrachloroethylene	12.0	<1.0– 617
1,4-Dichlorobenzene	22.0	<1.0–1270
Terpenes	42.0	2.1– 362
α-Pinene	11.0	<1.0– 97
β-Pinene	1.3	<1.0– 11
α-Terpinene	4.1	<1.0– 37
Limonene	25.0	<1.0– 315
Carbonyl Compounds	29.0	5.2– 347
Ethylacetate	12.0	1.0– 204
n-Butylacetate	6.1	<1.0– 125
i-Butylacetate	2.0	<1.0– 33
Methylethylketone	6.2	<1.0– 25
4-Methyl-2-pentanone	1.0	<1.0– 12
Hexanal	2.0	<1.0– 11
Alcohols	7.0	<1.0– 25
n-Butanol	1.0	<1.0– 11
i-Butanol	3.0	<1.0– 20
i-Amylalcohol	1.0	<1.0– 10
2-Ethylhexanol	2.0	<1.0– 10
All Calibrated Compounds	394.0	72.0–2670

Table 1.18 Volatile Organic Compounds in Building Materials and Consumer Products[179]

Material/Product	Major VOCs Identified
Latex caulk	Methylethylketone, butyl propionate, 2-butoxyethanol, butanol, benzene, toluene
Floor adhesive	Nonane, decane, undecane, dimethyloctane, 2-methylnonane, dimethylbenzene
Particleboard	Formaldehyde, acetone, hexanal, propanol, butanone, benzaldehyde, benzene
Moth crystals	p-Dichlorobenzene
Floor wax	Nonane, decane, undecane, dimethyloctane, trimethylcyclohexane, ethylmethylbenzene
Wood stain	Nonane, decane, undecane, methyloctane, dimethylnonane, trimethylbenzene
Latex paint	2-Propanol, butanone, ethylbenzene, propylbenzene, 1,1'-oxybisbutane, butylpropionate, toluene
Furniture polish	Trimethylpentane, dimethylhexane, trimethylhexane, trimethylheptane, ethylbenzene, limonene
Polyurethane floor finish	Nonane, decane, undecane, butanone, ethylbenzene, dimethylbenzene
Room freshener	Nonane, decane, undecane, ethylheptane, limonene, substituted aromatics (fragrances)

Sources

A variety of investigators have conducted laboratory studies of emission characteristics of potential VOC source materials. Molhave[163] identified an average of 13 (range: 4–22) VOCs associated with each of 38 different building materials evaluated. Toluene and xylene were identified in over 50% of the materials tested. In an expansion of this study, Molhave[178] identified 62 different chemical species in the emissions from 42 building materials. These VOCs were primarily aliphatic and aromatic hydrocarbons, their oxygen derivatives, and terpenes. The compounds with the highest steady-state emission concentrations, in decreasing order, were toluene, m-xylene, terpene, n-butylacetate, n-butanol, n-hexane, p-xylene, ethoxyethylacetate, n-heptane, and o-xylene.

Materials commonly used indoors emit a variety of VOCs. Major organic compounds identified in 10 building materials or consumer products are summarized in Table 1.18.[179] In addition to laboratory chamber studies, a variety of investigations designed to monitor VOCs have related elevated concentrations to specific sources. In the TEAM study, Wallace et al.[180] related personal pollution exposure concentrations to a variety of source-related activities. Examples included exposures to 1,1,1-trichloroethane from paneling; benzene from smoking; trichloroethylene from furniture stripping; p-dichlorobenzene from room fresheners and toilet bowl deodorizers; and ethylbenzene, o-xylene, m-xylene, p-xylene, decane, and undecane from painting. An association between tobacco smoking and residential VOC levels has been reported,[176] as well as VOC levels related to human bioeffluents.[168,169]

Health Effects

In most instances, concentrations of individual VOCs in indoor air are several orders of magnitude lower than Threshhold Limit Values (TLVs) set by the American Conference of Governmental Industrial Hygienists and recommended to provide a measure of health protection for occupational exposures. Because of this, it is unlikely that any specific VOC in public access buildings or residences is responsible for building-related health complaints.

Health concerns associated with VOCs center on the additive or additive/synergistic effects of the sum total of VOCs present and the carcinogenicity or potential carcinogenicity of specific VOCs. Relative to the former, Molhave et al.[181] exposed human volunteers to a mixture of 22 VOCs at three different exposure concentrations. Subjects reported irritation of the eyes, nose, and upper airways and signification subjective responses to questions related to air quality, odor intensity, and powers of concentration at total VOC concentrations (expressed as hexane) of 5 and 25 mg/m^3. VOC levels in the range of 5 mg/m^3 are not uncommon in new buildings. Organic compounds found in indoor air that are suspected to be human carcinogens include (in addition to formaldehyde) benzene, p-dichlorobenzene, chloroform, tetrachloroethylene, toluene, xylene, and styrene.

PESTICIDES

Though pesticides are organic compounds, they represent a distinct indoor air quality concern. They differ from the organic compounds discussed in the previous section in that most pesticides are semivolatile. They also differ in the sense that their introduction into indoor spaces is intended and not inadvertent as is the case for many VOCs. They are used because they are toxic to pest species. There is at least some minimal understanding that a potential toxic hazard is being introduced. It has been estimated that over 91% of U.S. households use pesticides.[182] Indoor applications include the control of insects, rodents, and microbial problems such as mold and bacteria. Rodenticides generally would not be expected to cause any significant contamination of indoor air. Because of the limited extent of their use at a given time, most antimicrobial agents would be expected to cause very limited indoor contamination as well. Antimicrobial agents used commonly in U.S. homes include sodium hypochlorite, ethanol, pine oil, hydrochloric acid, 2-benzyl-4-chlorophenol, isopropanol, phosphoric acid, calcium hypochlorite, glycolic acid, and 2-phenylphenol.[183] A variety of insecticides are used in U.S. homes (or enter them in some way), causing indoor air contamination. These include diazinon, paradichlorobenzene, pentachlorophenol, chlordane, chlorpyrifos,

Table 1.19   Air Concentrations of Pesticides Measured in 50 Jacksonville, Florida Homes[184]

| Pesticide | Air Concentration (μg/m³) | | | |
| | Indoor | | Outdoor | |
	Mean	Range	Mean	Range
Chlorpyrifos	0.47	0.02–2.2	0.06	0.02–0.21
Diazinon	1.10	0.06–13.7	0.15	0.07–0.29
Propoxur	0.70	0.05–7.9	0.07	0.05–0.09
Chlordane	1.60	0.23–5.6	0.76	0.19–2.30
Heptachlor	0.40	0.016–1.6	0.28	0.06–0.63

ronnel, dichlorvos (DDVP), malathion, heptachlor, lindane, naphthalene, coal tar, creosote oil, metaldehyde, boric acid, methyl demeton, propoxur, aldrin, dieldrin, folpet, bendiocarb, and methoxychlor.[183,184] Insecticides are primarily used indoors to control cockroaches, termites, flies, fleas, ants, etc. They may be applied in a variety of ways: as an emulsion spray, by fogging devices (bug bombs), impregnated in pest strips or pet flea collars, as a poison bait, etc. They may be applied directly to indoor air (fly spray, bug bombs) or surfaces (cockroach control); to the substructure and/or surrounding ground surfaces (termiticides); or as solids on materials to be protected (moth balls/flakes).

Indoor Concentrations

Since pesticides used indoors are of such relatively low volatilities (vapor pressures in the range of 10^{-4} to 10^{-2} mm Hg), indoor air concentrations typically are very low. Concentration ranges for five of the most prevalent pesticides found in the indoor air of 50 Jacksonville, Florida homes are summarized in Table 1.19. All are in the low microgram-per-cubic-meter range.

The highest reported indoor insecticide concentrations are reported for home foggers or bug bombs. These are pressurized insecticide canisters used to control fleas, flies, mosquitoes, cockroaches, ants, spiders, and moths. They use a variety of active ingredients. One of the more common ingredients used in such bug bombs is DDVP. DDVP is an organophosphorus insecticide with a relatively high volatility (1.2×10^{-2} mm Hg). Within the first hour after application, concentrations of DDVP are in the range of 4–6 mg/m³; they drop rapidly afterward.[185] The relatively high volatility causes initial levels to be very high, but it also results in a rapid dissipation of the product.

Termiticides

Indoor air contamination by termiticides, particularly chlordane, has received a significant amount of public attention. This attention has arisen out of occupant complaints associated with misapplication of chlordane by sub-slab pressure injection, most notably in U.S. Air Force (USAF) and Army housing. Air contamination in living spaces in such circumstances occurred as heating ducts were penetrated or as chlordane moved through leaky ducts located in slab and crawl space substructures.

Termiticide contamination of indoor air has been studied by a number of investigators. Indoor contamination associated with sub-slab pressure injection was first reported by Callahan in 1970.[186] Other incidents in service bases occurred at various times in the 1970s, prompting the USAF to conduct extensive chlordane sampling of base housing. Livingston et al.[187] sampled 498 ground floor units at a midwestern base. The average concentration was 1.9 $\mu g/m^3$; concentrations ranged from not detectible to 37.8 $\mu g/m^3$. The units had intra- or sub-slab ducts and many were treated by sub-slab injection. Lillie[188] measured chlordane levels in 474 family housing units on seven USAF installations that had been treated with chlordane by sub-slab injection or exterior ditching at some time after construction. Eighty-six percent had chlordane levels below 3.5 $\mu g/m^3$; 12% had levels from 3.5 to 6.5 $\mu g/m^3$; and 2% had levels above 6.5 $\mu g/m^3$. In another study, Lillie[189] reported chlordane levels for 2113 family housing units (with slab or sub-slab heating ducts) on four USAF bases treated with chlordane prior to construction. Chlordane concentrations were said to be extremely low; most houses had concentrations of < 1 $\mu g/m^3$, and only two houses had concentrations in excess of 5 $\mu g/m^3$, the action level used by the Air Force for remedial measures.

Fenske and Sternbach[190] reported chlordane levels in 157 homes tested in response to citizen complaints. Living area samples averaged 1.72 $\mu g/m^3$. One-half of these complaint homes exceeded the National Academy of Science (NAS) interim guideline of 5 $\mu g/m^3$. The authors concluded that misapplications were the cause of significantly elevated chlordane levels. The relatively high percentage of houses tested (50%) that exceeded the NAS guideline most likely reflects the bias toward cases of misapplication inherent in consumer complaint–based data. In a study of the effect of "proper" application procedures in 12 homes, Louis and Kisselbach[191] reported very low living area concentrations (about 0.14 $\mu g/m^3$).

Chlordane formulations are available as a technical product containing 72% chlordane and 7–13% heptachlor and as a mixture containing 39.2% chlordane and 19.6% heptachlor.[191] As a consequence, residences contaminated with chlordane will also be contaminated by heptachlor. Though heptachlor concentrations in termiticide formulations are significantly less than chlordane, heptachlor has a higher vapor pressure; consequently, heptachlor concentrations in indoor air are often in the same range as those of chlordane and in some cases are even higher.[192,193] Chlorpyrifos, more commonly known by its primary trade name, Dursban, has become increasingly popular as a termiticide. As is the case with chlordane, it too is applied to the ground around residences. Its use has grown more popular as health concerns have been raised about chlordane's carcinogenicity and its tendency to contaminate indoor air from termiticide applications. Monitoring studies of chlorpyrifos levels in residences following termiticidal treatments have been very limited.

Wright[194] reported detectable levels of chlorpyrifos in living spaces in 7 of 16 houses where chlorpyrifos was applied to surrounding soil for termite control. Levels of chlorpyrifos detected in postapplication air sampling ranged from 0.10 to 8.54 μg/m^3. Concentrations were significantly higher in houses built on sandy soils compared to those on clay.

Wood Preservatives

Wood used for a variety of indoor and outdoor purposes is treated to prevent decay by mold. Structural timbers for home construction are usually treated with a nonvolatile inorganic substance such as copper chromium arsenate. It is unlikely to contaminate indoor air unless such treated wood is burned in a fireplace or a leaky, non-airtight, wood-burning stove or furnace. Such burning could result in the release of very toxic arsenate products.

Pentachlorophenol (PCP) and lindane have been the two most widely used wood preservatives. PCP is commonly used to treat foundation lumber and structural members exposed to the weather and in coatings. It is used in paints, wood stains, and sealers.[195] PCP is used on outdoor timbers and lumber applications. PCP-treated timbers may be introduced indoors by design or when materials treated for outdoor use are brought indoors. Logs used in log home construction have been treated with PCP on the assumption that such use is an external application.

PCP levels in complaint log homes in California ranged from 10 to 30 μg/ m^3. PCP levels in an office building constructed with PCP-treated timbers were reported to range from 27.2 to 30.7 μg/m^3.[195] PCP is a chlorinated hydrocarbon and will therefore accumulate in fatty body tissues. Investigations conducted by the Centers for Disease Control indicated that PCP blood serum levels were seven times higher in residents of log homes than in individuals living in conventional houses.[196]

A variety of investigators have reported levels of PCP, lindane, and dieldrin in households in European countries where these biocides were applied to wood members for remedial treatment. In sampling conducted in houses (up to 10 years after treatment), Dobbs and Williams[197] reported living space concentrations of lindane and dieldrin (applied for insect control) in the range of 0.01–2.2 and 0.01–0.51 μg/m^3, respectively. PCP concentrations in houses treated for dry rot were significantly lower than in houses treated for insect control, presumably because dry rot treatments were less extensive in surface area.

West German investigators have reported on PCP and lindane levels in timber samples, household dust, and blood samples of occupants of wood preservative–treated houses. The content of PCP and lindane in house dust correlated with concentrations in timber samples. Elevated PCP (and, to a lesser extent, lindane) levels were found in the blood of occupants.[197] Other studies conducted in Germany have shown elevated levels of PCP (average 4.9 μg/m^3) in the air of three households that used wood preservatives; elevated levels of PCP and lindane were in the house dust.[198] House dust PCP concentrations were strongly correlated with concentrations in occupant urine samples. Volatilization of PCP from treated surfaces was observed to contaminate dust, curtains, furniture, carpets, clothing, and food.[199]

In addition to pesticidal contamination of indoor air associated with the application of wood preservatives, significant VOC contamination can occur as well. Van der Kolk[200] reported on measurements made in several hundred houses in which ground floors were treated with solvent-based PCP. VOC levels were on the order of 10 mg/m^3, with PCP levels ranging up to 10 μg/m^3.

Health Effects

Health concerns about pesticide use indoors focus on acute effects occurring immediately after exposures to high concentrations and on the carcinogenic potential of such pesticides as chlordane, heptachlor, and pentachlorophenol. Acute effects have been reported with the use of home foggers. There have been numerous reports to poison control centers in California concerning exposures and illness associated with such devices.[201] DDVP, with its relatively high volatility, is one of the pesticides most commonly involved in such incidents.[202] Principal complaints include headaches, nausea, dizziness, and eye and skin irritation. It has been estimated that such pesticide-related symptoms occurred in occupants of some 2.5 million households in 1976–1977.[203] Routes of exposure include both inhalation and skin absorption.

Acute effects have also been reported for misapplication of termiticides[186] and in the use of wood preservatives.[200] In the latter case, acute effects may have been due to high concentrations of the VOC solvent vehicle rather than the active pesticide ingredient itself. This may also be the case in a portion of other pesticide-related illness complaints.

Chlordane, heptachlor, lindane, dieldrin, and pentachlorophenol are chlorinated hydrocarbons, a class of chemicals which has a high probability of being carcinogenic. Because they are suspected human carcinogens, they have been restricted in their permissible uses in the United States. EPA permitted the use of chlordane as a termiticide because it had felt that there was a low probability of exposure when chlordane was applied for subterranean termite control. Data presented previously in this section indicate that chlordane and heptachlor are mobile and enter the living spaces of houses around which they are applied. Although the concentrations are low, they represent some cancer risk to occupants. This is similarly the case for PCP-containing wood preservatives applied to indoor surfaces.

In response to concerns associated with indoor contamination by termiticides such as chlordane, a committee of the National Research Council conducted a risk assessment of seven pesticides used in termite control.[204] Based on this assessment, they proposed interim guideline exposure levels for chlordane (5 μg/m^3), heptachlor (2 μg/m^3), aldrin/dieldrin (1 μg/m^3), and chlorpyrifos (10 $\mu g/m^3$). These values were set using occupational TLVs and, for chlordane, heptachlor, and aldrin/dieldrin, relative tumor incidence.

BIOGENIC PARTICLES

Biogenic particles are particles of biological origin. They include viable entities such as bacteria, fungi, viruses, amoebae, algae, and pollen grains and the no-longer-viable forms of the same. They also include plant parts; insect parts and wastes; animal saliva, urine, and dander; human dander; and a variety of organic dusts.

Exposure to airborne biogenic particles can result in a number of disease or illness syndromes. The relationship between pathogenic disease caused by bacteria and viruses and the airborne transmission of such disease has been known for over a half century. Many communicable diseases are transmitted from one individual to another by the suspension of infective particles in air and their subsequent inhalation by previously unaffected individuals. Influenza, tuberculosis, and colds are notable examples. The close proximity of individuals to each other in buildings results in an increased risk of airborne transmission of infective particles that cause disease.

In general, infectious diseases do not fall within the scope of indoor air quality or building-related health concerns. An exception is Legionnaires' disease, which has been linked to contamination of indoor air by building systems such as cooling towers and evaporative condensers. Airborne biogenic particles are notable because they have been known to cause allergies, asthma, and hypersensitivity pneumonitis/humidifier fever in addition to Legionnaires' disease, and are suspected by some investigators[205] to cause subjective symptoms associated with sick building syndrome.

The concept of biogenic particles as an indoor air quality concern has a history different from other contaminants described previously in this chapter. Biogenic particles have historically been the domain of physicians, most notably allergists and pulmonary specialists. Interest among nonmedical investigators in indoor air quality has lagged behind interest in other contaminant concerns. This lag belies the considerable health significance that indoor air contamination by biogenic particles has.

Allergies/Asthma

Allergies and asthma are the two most common disease syndromes associated with airborne particles of biological origin. The prevalence of each in the United States is considerable. Allergy (allergic rhinitis) probably afflicts 8% of the population, or about 20 million individuals, with characteristic symptoms of sinus inflammation, runny nose, eye irritation, etc.[206] It is caused by an immunological sensitization to antigens found in house dust; mold spore and hyphal fragments; animal danders, saliva, and urine; insect excreta and parts;

and pollen. The sensitization is described as a type I reaction involving IgE-mediated responses.[207] Type I reactors to skin prick and other allergy tests are described as being atopic.

Asthma is a more severe respiratory disease than allergies, which are limited to the upper respiratory system. Asthma is a disease of the pulmonary or lower respiratory system. It is characterized by episodic bronchial constriction resulting in severe shortness of breath and wheezing. It varies widely in severity. The airway obstruction resolves spontaneously or following treatment. Asthma affects about 10 million individuals in the United States. A variety of factors are involved in causing or contributing to asthma, including chemical irritants, exercise, cold air, respiratory infections, aspirin, and allergens.[208]

A strong link between asthma and allergens has been reported. A large percentage of asthmatic patients are also atopic for common allergens such as house dust, mold, and pollen.[208] Asthma in many of these cases appears to be an immunological response to allergens in airborne biogenic particles. For allergens to be of much importance in inducing sensitivity, they must be reasonably abundant in the air for fairly long periods of time. Once sensitized, individuals may develop symptoms after the most minute exposure.[208] The two most important allergen/particle contaminants found in indoor air that are known to cause both allergies and asthma are produced by dust mites and fungi (mold). Allergens are typically proteins with molecular weights of 16,000–50,000 daltons.

Dust Mites

One of the most strongly allergenic materials found indoors is house dust, which is heavily contaminated with the fecal pellets of dust mites, the two most common species of which are *Dermatophagoides pteronyssinus,* the European dust mite, and *D. farinae,* more commonly found in North America.[209] Dust mite allergen is the most important allergen associated with asthma in European countries.[206,210] It has been estimated to be the cause of asthma in over 50% of asthmatic patients in Denmark.[206] It is also probably one of the most important causes of asthma in North America, as well as the major cause of common allergies.

Dust mites are arachnids (members of the spider family) whose ecological niche is to consume the skin scales of humans and other animals. As a consequence, they are commonly found associated with humans and their dwellings. They are frequently recovered from house dust. Because of their small size (250–300 μm length) and translucent bodies, they are not visible to the unaided eye. We therefore are not aware of their presence. A dust mite can be seen in the photomicrograph in Figure 1.11. In Europe, highest populations are found in beds,[211] whereas in the United States bed populations are very low, with highest numbers found in bedroom carpeting and household upholstery.[212]

One of the major limiting factors in mite survival and population develop-

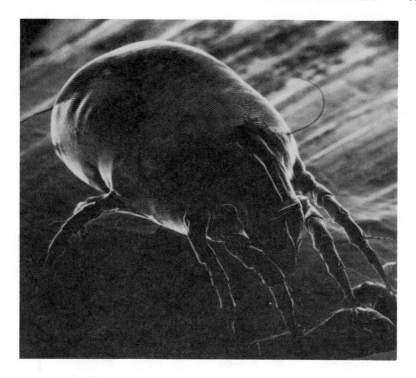

Figure 1.11.   Electron photomicrograph of a dust mite.

ment is the availability of water for sorption. The lowest relative humidity at which a fasting mite can sorb sufficient water to maintain water balance, described as the critical equilibrium activity (CEA), is 70–73% for the two most common species.[207] These relative humidity requirements must be met during a short time period or the population cannot survive. A strong dependence on indoor relative humidity levels has been observed for dust mite populations.[211-213] This can be seen in Figure 1.12. Mite populations show seasonal fluctuations which are related to relative humidity. Highest mite densities occur in the humid summer months and lowest densities occur in drier winter periods (Figure 1.12). Household relative humidities are typically lower than the CEA, suggesting that higher moisture levels must occur in the mites' microenvironment. Dust mite populations show a strong geographic dependence, with highest populations in humid regions and lowest densities in areas of high altitude and/or dry climate.[214,215]

Because of the large quantity of skin scales sloughed off daily by humans, mites have an abundant supply of food. They cannot, however, eat "fresh" skin scales. Mites thrive best on skin scales that have been defatted.[209] Some level of decomposition is essential before human skin scales can serve as mite

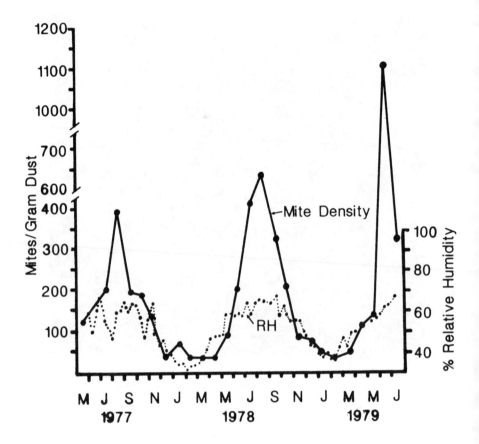

Figure 1.12.   Seasonal changes in dust mite populations and relative humidity levels.[213]

food. This decomposition or processing of skin scales into mite food is accomplished by mold species such as *Aspergillus anastelodami*.[216]

Dust mite antigen levels have been measured in bed and floor dust samples and in room air. Both bed and floor dust samples have been shown to vary widely in levels of antigen P(1) (100–100,000 ng/g of dust).[217] These antigen levels were strongly correlated to mite bodies. Antigen P(1) could only be detected in room air disturbed by bed-making and cleaning activities. Over 80% of airborne dust mite antigen was associated with particles greater than 10 μm in diameter. These particles had the physical appearance of mite feces.

Mold

Mold is also a significant cause of common allergies and asthma. Mold spores and hyphal fragments are typically dispersed by air currents; consequently, air contamination is common and the probability of inhalation exposure is high. In Western Europe, 5–10% of asthma patients test positive to fungi in skin prick tests.[218]

Mold spores range in size from 3 to 200 μm with most about 10 μm. They are often liberated in large concentrations and remain airborne for a long period of time. Spores are likely the major cause of mold allergy, since they can be inhaled and deposited on sensitive respiratory tissue.[219] The most important fungal groups involved in indoor air contamination are found among the so-called imperfect fungi, including *Penicillium, Aspergillus, Cladosporium,* and *Alternaria.* Most of these are saprophytic fungi that colonize nonviable organic matter; they are fungi of decay.

Studies of mold in indoor air show strong seasonal variations in levels and species composition. During the summer or non-heating-season months, indoor mold levels and species reflect those of the ambient environment.[220,221] Fungal genera most commonly reported indoors include *Cladosporium, Alternaria, Epicoccum, Penicillium, Aspergillus,* and *Rhodotorula.*[222-224] The number of molds isolated, expressed as colony-forming units per cubic meter (CFU/m^3), varies widely. Kozak et al.[223] reported mold levels to vary from 36 to 5984 CFU/m^3 in homes of asthmatic patients. The range of mold levels in control homes varied from 106 to 1923 CFU/m^3. High mold levels are usually associated with the presence of mold-damaged or -infested materials.

Solomon[222] conducted volumetric studies of mold prevalence in 150 houses in southeastern Michigan during the winter period. Mold recoveries from air samples varied from less than 10 to 20,000 CFU/m^3. Concurrent outdoor levels never exceeded 230 CFU/m^3. A positive association was suggested between fungal prevalence and bedroom relative humidity. Higher levels were observed for well-humidified homes. In other studies,[225] Solomon reported that cool mist humidifiers were a significant source of indoor mold contamination.

Outdoor levels of mold during summer months are usually significantly higher than those indoors. Indoor levels range from 20 to 300% of those outdoors and average about 40–50%.[226] Although Tyndall et al.[227] reported that with the exception of *Penicillium,* the average concentration of major mold genera was higher in indoor air than ambient air during the summer months, mold levels dropped dramatically during the winter period, indicating that the major source of mold spores was ambient air.

The prevalence of mold in indoor environments can be related to building moisture levels and to water-damaged materials. Mold spores require a high relative humidity to germinate. Minimum levels are on the order of 70%.[228] Studies conducted in the Netherlands[229] indicate that homes with at least two moisture characteristics (damp stains, mold growth, condensation, musty smell, and insects such as silverfish and saw bugs) had higher average mold counts and respiratory complaints. Although moisture problems increase fungal prevalence, the effect of increasing relative humidity in buildings (at least over the short term) is to reduce viable spore levels.[230] Low relative humidities cause increased spore release. These two moisture factors can confound the interpretation of mold counts determined from viable samples.

High mold levels in office buildings can result from mold growth in cooling coil drip pans found in HVAC systems, in evaporative condensers, and on water-damaged materials.[228]

In addition to allergenicity, other mechanisms that may cause mold/building-related health complaints have been proposed. These include the release of volatile irritant chemicals and mycotoxins.[231]

Organic Dust

Organic dust associated with carpeting in schools has been suggested to be the cause of symptoms in asthmatic children[232] as well as symptoms characteristic of sick building syndrome.[25,37]

Hypersensitivity Pneumonitis

Hypersensitivity pneumonitis is a form of respiratory disease caused by an immunological response to the inhalation of a variety of organic aerosols. In its acute form it is characterized by respiratory and systemic symptoms which occur 4–6 hours after inhalation exposure. Common symptoms include cough, shortness of breath, fever, chills, myalgia, and malaise. Recovery is spontaneous from symptoms which may last 18 hours. Attacks occur on each exposure, with severity depending on individual sensitivity and the degree of exposure. Pathological effects occur in lung tissue, with infiltration of lung interstitia and formation of fibroblasts. In the chronic form, irreversible lung damage may occur with an associated gas exchange impairment.[233]

Hypersensitivity pneumonitis is the disease syndrome known as farmers' lung, bagassosis (a sugar cane workers' disease), pigeon fanciers' disease, brown lung (cotton workers), etc. Commonly associated with organic dusts, it has been reported in office buildings where ventilation systems and office materials have been the source of the exposure. In Great Britain, it has been associated with humidification systems and units, and is referred to as humidifier fever. Humidifier fever and hypersensitivity pneumonitis appear to be the same disease syndrome. In Great Britain, however, the cause appears to be exposure to nonpathogenic amoebae.[234] Outbreaks of hypersensitivity pneumonitis can be caused by sensitization and exposure to thermophilic actinomycetes,[235,236] mold,[237,238] and *Flavobacterium* and its endotoxins,[239] as well as amoebae. A variety of components of building HVAC systems, including ductwork,[235,236] humidifiers and air washers,[236] and fan coil unit drip pans,[228,237] have been associated with such outbreaks. Outbreaks have also been attributed to moldy file folders and materials damaged by water intrusions.[228] Hypersensitivity pneumonitis has accounted for approximately 3% of NIOSH building health hazard investigations.[17]

Legionnaires' Disease

Legionnaires' disease was first identified after an extensive investigation of deaths and illness among attendees of an American Legion convention held in a Philadelphia hotel in 1976. The pneumonia-like symptoms were shown to be caused by a bacterium, *Legionella pneumophila,* which became entrained in an air handling system of the hotel, affecting primarily individuals who spent a relatively large amount of time in the hotel lobby. The disease has a relatively high mortality rate, but a low attack rate.[240]

The incubation period for Legionnaires' disease is 2–10 days, with initial symptoms consisting of malaise and headache followed by high fever. Dry cough and gastrointestinal problems may also occur in infected individuals. If untreated, the disease may progress to lung consolidation, respiratory failure, and death. Age of infected individuals varies widely, but they are most often in middle age. Predisposing factors include cigarette smoking, alcohol abuse, existing respiratory disease, malignancy, and immunosuppressing drugs.[241]

L. pneumophila also causes a nonpneumonic disease called Pontiac Fever, which has a high attack rate (95%) with no apparent fatalities. Pontiac Fever is a distinct syndrome characterized by fever, malaise, myalgia, and headaches.

L. pneumophila is found in stream water and therefore is widely distributed. It is commonly found in water samples in plumbing systems and cooling towers. Though it is relatively common in the environment, outbreaks of the disease are not common. Since the bacterium is present in the absence of disease, it is likely that the infective dose must be high and the means of disseminating the organism must be critical.[241]

Outbreaks of Legionnaires' disease have been linked with aerosol drift from cooling towers and evaporative condensors where water temperatures are very favorable to the growth of the disease bacterium. The source of bacterial introduction is believed to be potable water.[240,241]

Cooling towers are used to dissipate heat from air-conditioning systems of hospitals, hotels, and large office buildings.[242] In operation (Figure 1.13), warm water is forced through specially designed nozzles at the top of the tower. As the liquid is atomized, maximum contact occurs between air and water. As the droplets fall, they coalesce into large droplets that break up as they strike a wet deck surface, which is usually in the form of splash bars. The water is cooled by evaporation. Cooled water is collected in a basin at the bottom, from which it pumped back to the original heat source. To maximize cooling, large volumes of air are moved through the tower by large fans. Droplets entrained in air are removed by mist eliminators. However, some drift will be lost, and this drift has the potential for carrying infective *L. pneumophila* particles into building supply air.

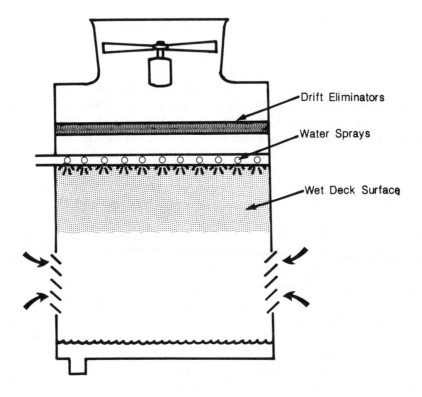

Figure 1.13. Diagrammatic representation of a cooling tower used with air conditioning systems.[242]

REFERENCES

1. Ferris, B.G., Jr. et al. 1978. "Relationship Between Outdoor and Indoor Air Pollution and Implications on Health." 25–37. In: *Proceedings of the First International Indoor Climate Symposium.* Danish Building Research Institute. Copenhagen.
2. Spengler, J.D. et al. 1979. "Sulfur Dioxide and Nitrogen Dioxide Levels Inside and Outside of Homes and the Implications on Health Effects Research." *Environ. Sci. Technol.* 13:1276–1280.
3. Berk, J.V. et al. 1981. "Impact of Energy-Conserving Retrofits on Indoor Air Quality in Residential Housing." LBL-12189. Lawrence Berkeley Laboratory, Berkeley, CA.
4. Sabersky, R.H. et al. 1973. "Concentrations, Decay Rates, and Removal of Ozone and Their Relationship to Establishing Clean Indoor Air." *Environ. Sci. Technol.* 7:347–351.
5. Godish, T. 1988. Unpublished data.
6. Yocum, J.E. 1982. "Indoor-Outdoor Air Quality Relationships. A Critical Review." *JAPCA* 35:500–520.

7. Yocum, J.E. et al. 1971. "Indoor/Outdoor Air Quality Relationships." *JAPCA* 21:251–259.

8. Spengler, J.D. et al. 1981. "Long-Term Measurements of Respirable Sulfates and Particles Inside and Outside of Homes." *Atmos. Environ.* 15:23–30.

9. Moschandreas, D.J. et al. 1981. "Comparison of Outdoor and Indoor Air Quality." Electric Power Research Institute Report EA-1733. Palo Alto, CA.

10. Wallace, L.A. et al. 1987. "The TEAM Study: Personal Exposures to Toxic Substances in Air, Drinking Water, and Breath of 400 Residents of New Jersey, North Carolina and North Dakota." *Env. Res.* 43:290–307.

11. Wallace, L.A. et al. 1986. "Concentrations of 20 Volatile Organic Compounds in the Air and Drinking Water of 350 Residents of New Jersey Compared with Concentrations in Their Exhaled Breath." *J. Occ. Med.* 28:603–608.

12. Reinert, J.C. 1984. "Pesticides in the Indoor Environment." 233–238. In: *Proceedings of the Third International Conference on Indoor Air Quality and Climate.* Swedish Council for Building Research. Stockholm. Vol. 1.

13. Bruno, R.C. 1983. "Sources of Indoor Radon in Houses: A Review." *JAPCA* 33:105–109.

14. Alter, W.H. and R.A. Oswald. 1987. "Nationwide Distribution of Indoor Radon Measurements: A Preliminary Data Base." *JAPCA* 37:227–231.

15. Solomon, W.R. 1975. "Assessing Fungus Prevalence in Domestic Interiors." *J. Allergy Clin. Immunol.* 56:235–242.

16. Moschandreas, D.J. 1981. "Exposure to Pollutants and Daily Time Budgets of People." In: Symposium on Health Aspects of Indoor Air Pollution. *Bull. N.Y. Acad. Med.* 57:845–860.

17. Budansky, S. 1980. "Indoor Air Pollution." *Environ. Sci. Technol.* 14:1023–1027.

18. Wallingford, K.M. and J. Carpenter. 1986. "Field Experience Overview: Investigating Sources of Indoor Air Quality Problems in Office Buildings." 448–453. In: *Proceedings of IAQ '86: Managing Indoor Air for Health and Energy Conservation.* American Society of Heating, Refrigerating and Air-Conditioning Engineers. Atlanta.

19. Melius, J. et al. 1984. "Indoor Air Quality—The NIOSH Experience." *Ann. ACGIH* 10:3–8.

20. Hodgson, M.J. et al. 1987. "Vibration as a Cause of 'Tight Building Syndrome' Symptoms." 449–453. In: *Proceedings of the Fourth International Conference on Indoor Air Quality and Climate.* Institute for Water, Soil and Air Hygiene. West Berlin. Vol. 2.

21. Rittfeldt, L. et al. 1984. "Indoor Air Pollutants Due to Vinyl Floor Tiles." 297–303. In: *Proceedings of the Third International Conference on Indoor Air Quality and Climate.* Swedish Council for Building Research. Stockholm. Vol. 3.

22. Breysse, P.A. 1984. "The Office Environment—How Dangerous?" 315–320. In: *Proceedings of the Third International Conference on Indoor Air Quality and Climate.* Swedish Council for Building Research. Stockholm. Vol. 3.

23. Molhave, L. et al. 1984. "Human Reactions During Controlled Exposures to Low Concentrations of Organic Gases and Vapor Known as Normal Indoor Air Pollutants." 431–436. In: *Proceedings of the Third International Conference on Indoor Air Quality and Climate.* Swedish Council for Building Research. Stockholm. Vol. 3.

24. Lauderdale, J. et al. 1984. "A Comprehensive Study of Adverse Health Effects in

Tight Buildings." American Industrial Hygiene Association Conference Abstract. Dallas.

25. Norback, D. and M. Torgen. 1987. "A Longitudinal Study of Symptoms Associated with Wall-to-Wall Carpets and Electrostatic Charge in Swedish School Buildings." 572–576. In: *Proceedings of the Fourth International Conference on Indoor Air Quality and Climate.* Institute for Water, Soil and Air Hygiene. West Berlin. Vol. 2.

26. Hijazi, N.R. 1983. "Indoor Organic Contaminants in Energy-Efficient Buildings." 471–477. In: *Proceedings: Measurements and Monitoring of Noncriteria (Toxic) Contaminants in Air.* Air Pollution Control Association. Pittsburgh.

27. Hawkins, L.H. 1981. "The Influence of Air Ions, Temperature and Humidity on Subjective Wellbeing and Comfort." *J. Environ. Psychol.* 1:279–292.

28. Hawkins, L.H. and L. Morriss. 1984. "Air Ions and the Sick Building Syndrome." 197–200. In: *Proceedings of the Third International Conference on Indoor Air Quality and Climate.* Swedish Council for Building Research. Stockholm. Vol. 3.

29. Nylen, P. et al. 1984. "Physical and Chemical Environment at VDT Work-Stations: Air Ions, Electrostatic Fields and PCBs." 163–167. In: *Proceedings of the Third International Conference on Indoor Air Quality and Climate.* Swedish Council for Building Research. Stockholm. Vol. 3.

30. Wallach, C. 1984. "Video Display Health Hazard Safeguards." 169–174. In: *Proceedings of the Third International Conference on Indoor Air Quality and Climate.* Swedish Council for Building Research. Stockholm. Vol. 3.

31. Nielsen, P.A. 1987. "Potential Pollutants—Their Importance to the Sick Building Syndrome—And Their Release Mechanism." In: *Proceedings of the Fourth International Conference on Indoor Air Quality and Climate.* Institute for Water, Soil and Air Hygiene. West Berlin. Vol. 2.

32. Hanssen, S.O. and E. Rodahl. 1984. "An Office Environment—Problems and Improvements." 303–308. In: *Proceedings of the Third International Conference on Indoor Air Quality and Climate.* Swedish Council for Building Research. Stockholm. Vol. 3.

33. Colligan, M.J. 1981. "The Psychological Effects of Indoor Air Pollution." *Bull. N.Y. Acad. Med.* 57:1014–1025.

34. Morriss, L. and R.L. Hawkins. 1987. "The Role of Stress in the Sick Building Syndrome." 566–571. In: *Proceedings of the Fourth International Conference on Indoor Air Quality and Climate.* Institute for Water, Soil and Air Hygiene. West Berlin. Vol. 2.

35. Skov, P. and O. Valbjorn. 1987. "The Sick Building Syndrome in the Office Environment—the Danish Town Hall Study." 439–443. In: *Proceedings of the Fourth International Conference on Indoor Air Quality and Climate.* Institute for Water, Soil and Air Hygiene. West Berlin. Vol. 2.

36. Woods, J.E. et al. 1987. "Office Worker Perceptions of Indoor Air Quality Effects on Discomfort and Performance." 464–468. In: *Proceedings of the Fourth International Conference on Indoor Air Quality and Climate.* Institute for Water, Soil and Air Hygiene. West Berlin. Vol. 2.

37. Valbjorn, O. and P. Skov. 1987. "Influence of Indoor Climate on the Sick Building Syndrome Prevalence." 593–597. In: *Proceedings of the Fourth International Conference on Indoor Air Quality and Climate.* Institute for Water, Soil and Air Hygiene. West Berlin. Vol. 2.

38. U.S. EPA. 1984. "National Emission Standards for Hazardous Air Pollutants — Asbestos Regulations." *Fed. Reg.* 49:13661. April 5.

39. National Research Council. 1984. *Asbestiform Fibers — Nonoccupational Health Risks.* National Academy Press. Washington, D.C.

40. U.S. EPA. 1985. "Guidance for Controlling Asbestos-Containing Materials in Buildings." EPA-560/5-85-024.

41. U.S. EPA. 1987. "Asbestos-Containing Materials in Schools — Final Rule." *Fed. Reg.* 52:41825. October 30.

42. Sebastien, P. et al. 1982. "Indoor Asbestos Pollution: From the Ceiling and the Floor." *Science.* 216:1410.

43. Ryan, D. 1986. "Advice on Asbestos in the Home." *EPA Journal,* August.

44. Constant, P.C. et al. 1982. "Airborne Asbestos Levels in Schools." Midwest Research Inst. Final Report to U.S. EPA. Contract #68-01-5915.

45. Nicholson, W.J. 1971. "Measurement of Asbestos in Ambient Air." NAPCA Contract CPA 70-92.

46. Nicholson, W.J. et al. 1975. "Asbestos Contamination of the Air in Public Buildings." EPA-450/3-76-004.

47. Nicholson, W.J. et al. 1976. "Asbestos Contamination of Building Air Supply Systems." In: *Proceedings of the International Conference on Environmental Sensing and Assessment.* Vol. II. Institute of Electrical and Electronic Engineers. New York.

48. Nicholson, W.J. 1978. "Control of Sprayed Asbestos Surfaces in School Buildings: A Feasibility Study." Final Report to NIEHS Contract #I-ES-2113.

49. Chadwick, D.A. et al. 1982. "Airborne Asbestos in Colorado Public Schools." *Env. Res.* 36:1-13.

50. Doll, R. 1955. "Mortality from Lung Cancer in Asbestos Workers." *Brit. J. Indust. Med.* 12:81-86.

51. McDonald, J.C. 1980. "Asbestos-Related Disease: An Epidemiological Review." 587-601. In: J.C. Wagner (Ed.). *Biological Effects of Mineral Fibers.* Vol. 2. IARC Scientific Pub. 30.

52. Selikoff, I.J. et al. 1979. "Mortality Experience of Insulation Workers in the United States and Canada, 1943-1976." *Ann. N.Y. Acad. Sci.* 330:91-116.

53. Wagner, J.C. et al. 1974. "The Effects of the Inhalation of Asbestos in Rats." *Brit. J. Cancer.* 29:252-269.

54. Hammond, E.C. et al. 1979. "Asbestos Exposure, Cigarette Smoking and Death Rates." *Ann. N.Y. Acad. Sci.* 330:473-490.

55. Connelly, P.R. and M.H. Myers. 1982. "The Incidence of Mesothelioma in the United States." In: *Proceedings of the Thirteenth International Cancer Congress.* Seattle, WA.

56. Anderson, H.A. et al. 1979. "Asbestosis Among Household Contacts of Asbestos Factory Workers." *Ann. N.Y. Acad. Sci.* 330:387-399.

57. Vianna, N.J. and A.K. Polan. 1978. "Non-Occupational Exposure to Asbestos and Malignant Mesothelioma in Females." *Lancet* 2:1061-1063.

58. Gross, P. 1981. "Consideration of the Aerodynamic Equivalent Diameter of Respirable Mineral Fibers." *Amer. Ind. Hyg. Assoc. J.* 42:449-452.

59. Morgan, A. 1979. "Fiber Dimensions: Their Significance in the Deposition and Clearance of Inhaled Fiber Dusts." 87-96. In: R. Lemen and J.M. Dement (Eds.). *Dusts and Disease.* Pathotox Pub., Inc., Park Forest South, IL.

60. Hughes, J.M. and H. Weil. 1986. "Asbestos Exposure—Quantitative Assessment of Risk." *Amer. Rev. Respir. Dis.* 133:5–13.
61. Traynor, G.W. et al. 1985. "Indoor Air Pollution Due to Emissions from Wood-Burning Stoves." LBL-17854. Lawrence Berkeley Laboratory, Berkeley, CA.
62. Knight, C.V. et al. 1985. "Indoor Air Quality Related to Wood Heaters." 430–447. In: *Proceedings of IAQ '86: Managing Indoor Air for Health and Energy Conservation.* American Society of Heating, Refrigerating and Air-Conditioning Engineers. Atlanta.
63. Moschandreas, D.J. et al. 1981. "Residential Indoor Air Quality and Wood Combustion." 99–108. In: Proceedings of the Conference on Wood Combustion and Environmental Assessment. New Orleans.
64. Godish, T. and I. Ritchie. 1985. "Nitrogen Dioxide in Residences—Effects of Source Type." 261–266. In: D.S. Walkinshaw (Ed.). *Transactions: Indoor Air Quality in Cold Climates.* Air Pollution Control Association. Pittsburgh.
65. Leaderer, B.P. 1982. "Air Pollutant Emissions from Kerosene Space Heaters." *Science.* 218:1113–1115.
66. Woodring, J.L. et al. 1985. "Measurement of Combustion Product Emission Factors of Unvented Kerosene Heaters." *Amer. Ind. Hyg. Assoc. J.* 46:350–356.
67. Ritchie, I.M. and L.A. Oatman. 1983. "Residential Air Pollution from Kerosene Heaters." *JAPCA* 33:879–881.
68. Girman, J.R. et al. "Pollutant Emission Rates from Indoor Combustion Appliances and Sidestream Cigarette Smoke." *Env. Int.* 8:213–221.
69. Traynor, G.W. et al. 1985. "Indoor Air Pollution Due to Emissions from Unvented Gas-Fired Space Heaters." *JAPCA* 35:231–416.
70. Apte, M.G. and G.W. Traynor. 1986. "Comparison of Pollutant Emission Rates from Unvented Kerosene and Gas Space Heaters." 405–416. In: *Proceedings of IAQ '86: Managing Indoor Air for Health and Energy Conservation.* American Society of Heating, Refrigerating and Air-Conditioning Engineers. Atlanta.
71. Cooper, K.R. and R.R. Alberti. 1984. "Effect of Kerosene Heater Emissions on Indoor Air Quality and Pulmonary Function." *Amer. Rev. Respir. Dis.* 129:629–631.
72. Traynor, G.W. and I. Nitschke. 1984. "Field Survey of Indoor Air Pollution in Residences with Suspected Combustion-Related Sources." 343–348. In : *Proceedings of the Third International Conference on Indoor Air Quality and Climate.* Swedish Council for Building Research. Stockholm. Vol. 4.
73. Leaderer, B.P. et al. 1984. "Residential Exposures to NO_2, SO_3, and HCHO Associated with Unvented Kerosene Space Heaters, Gas Appliances and Sidestream Tobacco Smoke." 151–156. In: *Proceedings of the Third International Conference on Indoor Air Quality and Climate.* Swedish Council for Building Research. Stockholm. Vol. 4.
74. Relwani, S.M. et al. 1986. "Effects of Operational Factors on Pollutant Emission Rates from Residential Gas Appliances." *JAPCA* 36:1233–1237.
75. Sterling, T.D. and E. Sterling. "Carbon Monoxide Levels in Kitchens and Homes with Gas Cookers." *JAPCA* 29:238–241.
76. Traynor, G.W. et al. 1982. "The Effects of Ventilation on Residential Air Pollution Due to Emissions from a Gas-Fired Range." *Env. Int.* 8:447-452.
77. Quackenboss, J.J. et al. 1986. "Personal Exposure to Nitrogen Dioxide: Relationship to Indoor/Outdoor Air Quality and Activity Patterns." *Env. Sci. Technol.* 20:775–783.

78. Fortmann, R.C. et al. 1984. "Characterization of Parameters Influencing Indoor Pollutant Concentrations." 259-264. In: *Proceedings of the Third International Conference on Indoor Air Quality and Climate.* Swedish Council for Building Research. Stockholm. Vol. 4.

79. Russel, P. and T. Robinson. "Carbon Monoxide Poisoning in Houses: Contributing Factors and Remedial Measures." Canada Mortgage and Housing Corp, Ottawa.

80. Wilson, A.G. et al. 1985. "House Depressurization and Flue Gas Spillage." 417-429. In: *Proceedings of IAQ '86: Managing Indoor Air for Health and Energy Conservation.* American Society of Heating, Refrigerating and Air-Conditioning Engineers. Atlanta.

81. Swinton, M.C. and J.H. White. 1985. "Avoidance of Chimney Backdrafting in Houses: Identifying the Critical Conditions." 413-424. In: D.S. Walkinshaw (Ed.). *Transactions: Indoor Air Quality in Cold Climates.* Air Pollution Control Association. Pittsburgh.

82. Moffatt, S. 1985. "Combustion Ventilation Hazards in Housing—Failure Mechanisms, Identification Technologies and Remedial Measures." 425-436. In: D.S. Walkinshaw (Ed.). *Transactions: Indoor Air Quality in Cold Climates.* Air Pollution Control Association. Pittsburgh.

83. American Society of Heating, Refrigerating and Air-Conditioning Engineers. 1983. "Chimney, Gas Vent and Fireplace Systems." 27.1-27.29. *Equipment Handbook.* ASHRAE, Atlanta.

84. U.S. Surgeon General. 1986. "The Health Consequences of Involuntary Smoking." DHHS (CDC) 87-8398.

85. Badre, R. et al. 1978. "Atmospheric Pollution by Smoking." *Ann. Pharm. Francaises.* 36:443-452.

86. Chappell, S.B. and R.J. Parker. 1977. "Smoking and Carbon Monoxide in Enclosed Public Places." *Can. J. Pub. Health.* 68:159-161.

87. Elliot, L.P. and D.R. Rowe. 1975. "Air Quality During Public Gatherings." *JAPCA* 25:635-636.

88. Repace, J.L. and A.H. Lowrey. 1982. "Tobacco Smoke Ventilation and Indoor Air Quality." 895-914. *ASHRAE Trans.* 88. Pt. 1.

89. Repace, J.L. and A.H. Lowrey. 1980. "Indoor Air Pollution, Tobacco Smoke and Public Health." *Science.* 208:464-472.

90. Weber, A. et al. 1979. "Passive Smoking in Experimental and Field Conditions." *Env. Res.* 20:205-216.

91. Hinds, W.C. and M.W. First. 1975. "Concentrations of Nicotine and Tobacco Smoke in Public Places." *New Eng. J. Med.* 292:844-845.

92. National Research Council. 1976. *Nitrogen Oxides.* National Academy Press. Washington, D.C.

93. National Research Council. 1977. *Carbon Monoxide.* National Academy Press. Washington, D.C.

94. National Research Council. 1979. *Airborne Particles.* National Academy Press. Washington, D.C.

95. U.S. EPA. 1982. "Review of the National Ambient Air Quality Standards for Nitrogen Oxides: Assessment of Technical and Scientific Information." EPA-450/5-82-002.

96. U.S. EPA. 1979. "Air Quality Criteria for Carbon Monoxide." EPA-600/8-79-022.

97. U.S. EPA. 1982. "Air Quality Criteria for Particulate Matter and Sulfur Oxides." Vols. I-III. EPA-600/8-82-029a, EPA-600/8-82-029b, EPA-600/8-82-029c.

98. Melia, R.J.W. et al. 1977. "Association Between Gas Cooking and Respiratory Disease in Children." *Brit. Med. J.* 2:149–152.

99. Speizer, F.E. et al. 1980. "Respiratory Disease Rates and Pulmonary Function in Children Associated with NO_2 Exposure." *Amer. Rev. Resp. Dis.* 121:3–10.

100. Ware, J.H. et al. 1984. "Passive Smoking, Gas Cooking, and Respiratory Health of Children Living in Six Cities." *Amer. Rev. Respir. Dis.* 129:236–374.

101. Florey, C.V. 1979. "The Relation Between Respiratory Illness in Primary School Children and the Use of Gas for Cooking. III. Nitrogen Dioxide, Respiratory Illness and Lung Infection." *Int. J. Epidemiol.* 8:347–353.

102. Melia, R.J.W. et al. 1982. "Childhood Respiratory Illness and the Home Environment: Association Between Respiratory Illness and Nitrogen Dioxide, Temperature and Relative Humidity." *Int. J. Epidemiol.* 11:164–168.

103. Honicky, R.E. et al. 1983. "Infant Respiratory Illness and Indoor Air Pollution from a Wood-Burning Stove." *Pediatrics.* 71:126–128.

104. Honicky, R.E. 1985. "Symptoms of Respiratory Illness in Young Children and the Use of Wood-Burning Stoves for Indoor Heating." *Pediatrics.* 75:587–593.

105. Moghissi, A.A. et al. 1987. "Enhancement of Exposure to Radon Progeny as a Consequence of Passive Smoking." 360–361. In: *Proceedings of the Fourth International Conference on Indoor Air Quality and Climate.* Institute for Water, Soil and Air Hygiene. West Berlin. Vol. 2.

106. Giumond, R.J. et al. 1979. "Indoor Radiation Exposure Due to Radium-226 in Florida Phosphate Lands." EPA-520/4-78-013.

107. Nero, A.V. 1983. "Indoor Radiation Exposures from Radon-222 and its Daughters: A View of the Issue." *Health Physics.* 45:277–288.

108. Nero, A.V. et al. 1986. "Distribution of Airborne Radon-222 Concentrations in U.S. Homes." *Science.* 234:992–997.

109. Gerusky, T.M. 1987. "The Pennsylvania Radon Story." *J. Env. Health.* 49:197–200.

110. Gesell, T.F. 1983. "Background Atmospheric Radon Concentrations Outdoors and Indoors: A Review." *Health Physics.* 45:284–302.

111. U.S. EPA. 1979. "Indoor Radiation Exposure Due to Radium-226 in Florida Phosphate Lands: Radiation Protection Recommendations and Request for Comment." *Fed. Reg.* 44:38644–38670.

112. National Council on Radiation Protection. 1984. "Exposures from the Uranium Series with Emphasis on Radon and Its Daughters." Report No. 77. Bethesda, MD.

113. Cohen, B.L. 1986. "A National Survey of Radon-222 in U.S. Homes and Correlating Factors." *Health Physics.* 51:175–183.

114. Nazaroff, W.W. and A.V. Nero. 1984. "Transport of Radon from Soil into Residences." 15–21. In: *Proceedings of the Third International Conference on Indoor Air Quality and Climate.* Swedish Council for Building Research. Stockholm. Vol. 2.

115. Osborne, M.C. 1987. "Four Common Diagnostic Problems that Inhibit Radon Mitigation." *JAPCA* 37:604–606.

116. Radford, E.P. 1985. "Potential Health Effects of Indoor Radon Exposure." *Env. Health Perspectives.* 62:281–287.

117. Budnitz, R.J. et al. 1978. "Human Disease from Exposure: The Impact of Energy

Conservation in Buildings." LBL-7809. Lawrence Berkeley Laboratory, Berkeley, CA.

118. Stranden, E. 1980. "Radon in Dwellings and Lung Cancer: A Discussion." *Health Physics.* 38:301-306.

119. Eidling, C. et al. 1982. "Effects of Low-Dose Radiation — A Correlation Study." *Scand. J. Work Environ. Health.* 8: suppl. 1:59-64.

120. Radford, E.P. and K.G. St. Clair Renard. 1984. "Application of Studies of Miners to Radon Problem in Homes." 93-96. In: *Proceedings of the Third International Conference on Indoor Air Quality and Climate.* Swedish Council for Building Research. Stockholm. Vol. 2.

121. Axelson, O.C. et al. 1979. "Lung Cancer and Residency — A Case Referent Study on the Possible Impact of Exposure to Radon and Its Daughters in Dwellings." *Scand. J. Work Environ. Health.* 5:10-15.

122. Hess, C.T. et al. 1983. "Environmental Radon and Cancer Correlation in Maine." *Health Physics.* 45:339-348.

123. Fleisher, R.L. 1981. "Lung Cancer and Phosphates." In: Proceedings of the International Symposium on Indoor Air Pollution, Health and Energy Conservation. Amherst, MA.

124. Bergman, H. et al. 1984. "Indoor Radon Daughter Concentrations and Passive Smoking." 79-84. In: *Proceedings of the Third International Conference on Indoor Air Quality and Climate.* Swedish Council for Building Research. Stockholm. Vol. 2.

125. Martell, E.A. 1984. "Aerosol Properties of Indoor Radon Decay Products." 161-165. In: *Proceedings of the Third International Conference on Indoor Air Quality and Climate.* Swedish Council for Building Research. Stockholm. Vol. 2.

126. Grot, R.A. et al. 1985. "Validity of Models for Predicting Formaldehyde Concentrations in Residences Due to Pressed Wood Products, Phase 1." National Bureau of Standards. NBSIR 85-3255.

127. Pickrell, J.A. et al. 1983. "Formaldehyde Release Coefficients from Selected Consumer Products: Influence of Chamber Loading, Multiple Products, Relative Humidity and Temperature." *Env. Sci. Technol.* 17:753-757.

128. Matthews, T.G. 1984. "Modeling and Testing Formaldehyde Emission Characteristics of Pressed Wood Products." CPSC-IAG-84-1103. Consumer Products Safety Commission.

129. Matthews, T.G. et al. 1983. "Formaldehyde Release from Pressed Wood Products." In: *Proceedings of the 17th International WSU Particleboard/Composite Materials Symposium.* Washington State University. Pullman, WA.

130. Ritchie, I. and R.E. Lehnen. 1985. "An Analysis of Formaldehyde Concentrations in Mobile and Conventional Homes." *J. Env. Health.* 47:300-305.

131. University of Texas School of Public Health. 1984. "Texas Indoor Air Quality Study. Vol. III: Environmental Monitoring."

132. Dally, K.A. et al. 1981. "Formaldehyde Exposure in Nonoccupational Environments." *Arch. Env. Health.* 33:277-284.

133. Hanrahan, L.P. et al. 1985. "Formaldehyde Concentrations in Wisconsin Mobile Homes." *JAPCA* 35:1164-1167.

134. Sexton, K. et al. 1986. "Formaldehyde Concentrations Inside Private Residences: A Mailout Approach to Indoor Air Monitoring." *JAPCA* 36:698-704.

135. Broder, I. et al. 1986. "Health Status of Residents in Homes Insulated with Urea-Formaldehyde Foam." 155-167. In: D.S. Walkinshaw (Ed.). *Transactions:*

Indoor Air Quality in Cold Climates. Air Pollution Control Association. Pittsburgh.

136. Syrotynski, S. 1985. "Prevalence of Formaldehyde Concentrations in Residential Settings." 127–136. In: D.S. Walkinshaw (Ed.). *Transactions: Indoor Air Quality in Cold Climates.* Air Pollution Control Association. Pittsburgh.

137. Godish, T. et al. 1984. "Formaldehyde Levels in New Hampshire Urea-Formaldehyde Foam Insulated Houses. Relationship to Outdoor Temperatures." *JAPCA* 34:1051–1052.

138. Ulsamer, A.G. et al. 1982. "Formaldehyde in Indoor Air: Toxicity and Risk." In: Proceedings of the 75th Annual Meeting of the Air Pollution Contol Association. New Orleans.

139. Godish, T. et al. 1986. "Relationship Between Residential Formaldehyde Exposures and Symptom Severity in a Self-Referred Population." In: Proceedings of the 79th Annual Meeting of the Air Pollution Control Association. Minneapolis.

140. Myers, G.E. 1984. "Effect of Ventilation Rate and Board Loading on Formaldehyde Concentration: A Critical Review of the Literature." *For. Prod. J.* 34:59–68.

141. Newton, L.R. 1982. "Formaldehyde Emission from Wood Products: Correlating Environmental Chamber Levels to Secondary Laboratory Test." 45–61. In: *Proceedings of the 16th Annual WSU Particleboard Symposium.* Washington State University. Pullman, WA.

142. Singh, J. et al. 1982. "Evaluation of the Relationship Between Formaldehyde Emissions from Particleboard Decking in Experimental Mobile Homes." U.S. Department of Housing and Urban Development Technical Report.

143. Godish, T. and B. Kanyer. 1985. "Formaldehyde Source Interaction Studies." *For. Prod. J.* 35:13–17.

144. Matthews, T.J. et al. 1984. "Environmental Dependence of Formaldehyde Emission from Pressed Wood Products: Experimental Studies and Modeling." In: *Proceedings of the 18th Annual WSU Particleboard/Composite Materials Symposium.* Washington State University. Pullman, WA.

145. Godish, T. and J. Rouch. 1986. "Mitigation of Residential Formaldehyde Contamination by Climate Control." *Amer. Ind. Hyg. Assoc. J.* 47:792–797.

146. Myers, G.E. 1985. "The Effects of Temperature and Humidity on Formaldehyde Emission from U-F Bonded Boards: A Literature Critique." *For. Prod. J.* 35:20–31.

147. Garry, V.F. 1980. "Formaldehyde in the Home. Environmental Disease Perspectives." *Minn. Med.* 63:107–111.

148. Ritchie, I.M. and R.H. Lehnen. 1987. "Formaldehyde-Related Health Complaints of Residents Living in Mobile and Conventional Homes." *Amer. J. Pub. Health.* 77:323–328.

149. Sterling, T.D. et al. 1986. "Dose-Response Effects of UFFI." In: Proceedings of the 79th Annual Meeting of the Air Pollution Control Association. Minneapolis.

150. Olsen, J.H. and M. Dossing. 1982. "Formaldehyde-Induced Symptoms in Day Care Centers." *Amer. Ind. Hyg. Assoc. J.* 43:366–370."

151. Broder, I. et al. 1987. "Comparison of Health of Occupants of Control Homes and Homes Insulated with Urea-Formaldehyde Foam Before and After Corrective Work." 605–609. In: *Proceedings of the Fourth International Conference on Indoor Air Quality and Climate.* Institute for Water, Soil and Air Hygiene. West Berlin. Vol. 2.

152. Liu, K.S. et al. 1987. "Irritant Effects of Formaldehyde in Mobile Homes." 610–614. In: *Proceedings of the Fourth International Conference on Indoor Air Quality and Climate.* Institute for Water, Soil and Air Hygiene. West Berlin. Vol. 2.

153. Shumilina, A.V. 1975. "Menstrual and Reproductive Functions in Workers with Occupational Exposure to Formaldehyde." *Gig. Truda. Prof. Zabol.* 12:18–21.

154. Hendrick, D.J. et al. 1982. "Formaldehyde Asthma: Challenge Exposure Levels and Fate after Five Years." *J. Occup. Med.* 24:893–897.

155. Nordman, H. et al. 1985. "Formaldehyde Asthma—Rare or Overlooked?" *J. Allergy Clin. Immunol.* 75:91–99.

156. Anonymous. 1984. "Report on the Consensus Workshop on Formaldehyde." *Env. Health Perspectives* 43:139–168.

157. Patterson, R. et al. 1986. "Human Antibodies Against Formaldehyde—Human Serum Albumin Conjugates." *Int. Arch. All Appl. Immunol.* 79:53–59.

158. Thrasher, J.D. et al. 1988. "Evidence for Formaldehyde Antibodies and Altered Cellular Immunity in Subjects Exposed to Formaldehyde in Mobile Homes." *Arch. Environ. Health.* 42:347–350.

159. Anonymous. 1982. "Report of the Federal Panel on Formaldehyde." *Env. Health Perspectives.* 43:139–168.

160. Stayner, L.T. et al. 1986. "A Retrospective Study of Workers Exposed to Formaldehyde in the Garment Industry." NIOSH. Washington, D.C.

161. Vaughan, T.L. et al. 1986. "Formaldehyde and Cancers of the Pharynx, Sinus and Naval Cavity: Residential Exposure." *Int. J. Cancer.* 38:685–688.

162. Hefter, R. et al. 1987. "Assessment of Health Risks to Garment Workers and Certain Home Residents from Exposure to Formaldehyde." U.S. EPA.

163. Molhave, L. 1979. "Indoor Air Pollution Due to Building Materials." 89–104. In: *Proceedings of the First International Indoor Climate Symposium.* Danish Building Research Institute. Copenhagen.

164. Berglund, B. et al. 1981. "A Longitudinal Study of Air Contaminants in a Newly Built Preschool." In: Proceedings of the International Symposium on Indoor Air Pollution, Health and Energy Conservation. Amherst, MA.

165. Schmidt, H.E. et al. 1980. "Trace Organics in Offices." LBL-11378. Lawrence Berkeley Laboratory, Berkeley, CA.

166. Miksch, R.R. et al. 1981. "Trace Organics in Office Spaces." In: Proceedings of the International Symposium on Indoor Air Pollution, Health and Energy Conservation. Amherst, MA.

167. Hartwell, T.D. et al. 1985. "Levels of Volatile Organics in Indoor Air." In: Proceedings of the 78th Annual Meeting of the Air Pollution Control Association. Detroit.

168. Wallace, L.A. et al. 1987. "Volatile Organic Chemicals in 10 Public Access Buildings." 188–192. In: *Proceedings of the Fourth International Conference on Indoor Air Quality and Climate.* Institute for Water, Soil and Air Hygiene. West Berlin. Vol. 2.

169. Wang, T.C. 1975. "A Study of Bioeffluents in a College Classroom." *ASHRAE Trans.* 81:32–44.

170. Johannson, I. 1978. "Determination of Organic Compounds Indoors with Potential Reference to Air Quality." *Atmos. Environ.* 12:1371–1377.

171. Oatman, L. and R. Ray. 1986. "Surface and Indoor Air Levels of Polychlorinated Biphenyls in Public Buildings." *Bull. Environ. Contam. Toxicol.* 37:461–466.

172. Vedel, A. and P.A. Nielson. 1984. "Phthalate Esters in the Indoor Environment." 309–314. In: *Proceedings of the Third International Conference on Indoor Air Quality and Climate*. Swedish Council for Building Research. Stockholm. Vol. 3.

173. Weschler, C.J. 1984. "Indoor-Outdoor Relationships for Nonpolar Organic Constituents of Aerosol Particles." *Environ. Sci. Technol.* 18:648–652.

174. Molhave, L. and J. Moller. 1979. "The Atmospheric Environment in Modern Danish Dwellings—Measurements in 39 Flats." 171–184. In: *Proceedings of the First International Indoor Climate Symposium*. Danish Building Research Institute. Copenhagen.

175. Hawthorne, A.R. et al. 1985. "Results of a Forty-Home Indoor Air Quality Monitoring Study." Oak Ridge National Laboratory, Oak Ridge, TN.

176. Otsin, R. 1985. "Surveys of Selected Organics in Residential Air." 224–236. In: D.S. Walkinshaw (Ed.). *Transactions: Indoor Air Quality in Cold Climates*. Air Pollution Control Association. Pittsburgh.

177. Krause, C. et al. 1987. "Occurrence of Volatile Organic Compounds in the Air of 500 Homes in the Federal Republic of Germany." 102–106. In: *Proceedings of the Fourth International Conference on Indoor Air Quality and Climate*. Institute for Water, Soil and Air Hygiene. West Berlin. Vol. 1.

178. Molhave, L. 1982. "Indoor Air Pollution Due to Organic Gases and Vapors of Solvents in Building Materials." *Env. Int.* 8:117–127.

179. Tichenor, B.A. 1987. "Organic Emission Measurements Via Small Chamber Testing." 8–15. In: *Proceedings of the Fourth International Conference on Indoor Air Quality and Climate*. Institute for Water, Soil and Air Hygiene. West Berlin. Vol. 1.

180. Wallace, L.A. 1987. "The Influence of Personal Activities on Exposure to Volatile Organic Compounds." 117–121. In: *Proceedings of the Fourth International Conference on Indoor Air Quality and Climate*. Institute for Water, Soil and Air Hygiene. West Berlin. Vol. 1.

181. Molhave, L. et al. 1986. "Human Reactions to Low Concentrations of Volatile Organic Compounds." *Environ. Int.* 12:167–175.

182. U.S. EPA. 1979. "National Household Pesticide Usage Study, 1976–1977." EPA-540/9-80-002.

183. Reinert, J.C. 1984. "Pesticides in the Indoor Environment." 233–238. In: *Proceedings of the Third International Conference on Indoor Air Quality and Climate*. Swedish Council for Building Research. Stockholm. Vol. 1.

184. Lewis, R.G. and A.E. Bond. 1987. "Non-Occupational Exposure to Household Pesticides." 195–199. In: *Proceedings of the Fourth International Conference on Indoor Air Quality and Climate*. Institute for Water, Soil and Air Hygiene. West Berlin. Vol. 1.

185. Goh, K.S. et al. 1987. "Dissipation of DDVP and Propoxur Following the Use of a Home Fogger: Implication for Safe Re-Entry." *Bull. Environ. Contam. Toxicol.* 39:762–768.

186. Callahan, R.A. 1970. "Chlordane Contamination of Government Quarters and Personal Property." USAF. OEHL. Tech. Report 70-7.

187. Livingston, J.M. et al. 1981. "Airborne Chlordane Contamination in Houses Treated for Termites at a Midwestern Air Force Base." USAF. OEHL. Tech. Report 81-11.

188. Lillie, T.H. 1981. "Chlordane in Air Force Family Housing: A Study of Houses Treated After Construction." USAF. OEHL. Tech. Report 81-45.

189. Lillie, T.H. 1982. "Chlordane in Air Force Family Housing: A Study of Houses Treated Prior to Construction." USAF. OEHL. Tech. Report 82-9.
190. Fenske, R.A. and T. Sternbach. 1987. "Indoor Levels of Chlordane in Residences in New Jersey." *Bull. Environ. Contam. Toxicol.* 39:903–910.
191. Louis, J.B. and K.C. Kisselbach. 1987. "Indoor Air Levels of Chlordane and Heptachlor Following Termiticide Applications." *Bull. Environ. Contam. Toxicol.* 39:911–918.
192. Wright, C.G. and R.B. Leidy. 1982. "Chlordane and Heptachlor in the Ambient Air of Houses Treated for Termites." *Bull. Environ. Contam. Toxicol.* 28:617–623.
193. Jurinski, N.B. 1984. "The Evaluation of Chlordane and Heptachlor Vapor Concentrations Within Buildings Treated for Insect Pest Control." 51–56. In: *Proceedings of the Third International Conference on Indoor Air Quality and Climate.* Swedish Council for Building Research. Stockholm. Vol. 4.
194. Wright, C.G. et al. 1988. "Chlorpyrifos in the Ambient Air of Houses Treated for Termites." *Bull. Environ. Contam. Toxicol.* 40:561–567.
195. Levin, H. and J. Hahn. 1984. "Pentachlorophenol in Indoor Air: The Effectiveness of Sealing Exposed Pressure-Treated Wood Beams and Improving Ventilation in Office Buildings to Address Public Health Concerns and Reduce Occupant Complaints." 123–130. In: *Proceedings of the Third International Conference on Indoor Air Quality and Climate.* Swedish Council for Building Research. Stockholm. Vol. 5.
196. Centers for Disease Control. 1980. "Pentachlorophenol in Log Homes—Kentucky." *Morbidity and Mortality Weekly Reports.* 29:431–432, 437.
197. Ruh, C. et al. 1984. "The Indoor Biocide Pollution: Occurrence of Pentachlorophenol and Lindane in Homes." 309–315. In: *Proceedings of the Third International Conference on Indoor Air Quality and Climate.* Swedish Council for Building Research. Stockholm. Vol. 4.
198. Krause, C. et al. 1987. "Pentachlorophenol-Containing Wood Preservatives—Analysis and Evaluation." 220–224. In: *Proceedings of the Fourth International Conference on Indoor Air Quality and Climate.* Institute for Water, Soil and Air Hygiene. West Berlin. Vol. 1.
199. Gebefugi, I. and F. Korte. 1984. "Indoor Contamination of Household Articles Through Pentachlorophenol and Lindane." 317–322. In: *Proceedings of the Third International Conference on Indoor Air Quality and Climate.* Swedish Council for Building Research. Stockholm. Vol. 4.
200. Van der Kolk, J. 1984. "Wood Preservatives and Indoor Air, Experiences in the Netherlands." 251–256. In: *Proceedings of the Third International Conference on Indoor Air Quality and Climate.* Swedish Council for Building Research. Stockholm. Vol. 1.
201. Maddy, K.T. and C.R. Smith. 1984. "Pesticide-Related Human Illnesses Reported as Occurring in California Between January 1 and December 31, 1983." California Department of Food and Agriculture. HS-1207.
202. Maddy, K.T. et al. 1981. "Monitoring Residues of DDVP in Room Air and on Horizontal Surfaces Following Use of a Room Fogger." California Department of Food and Agriculture. HS 897.
203. U.S. EPA. 1979. "National Household Pesticide Usage Study, 1976-1977." EPA-540/9-80-002.
204. National Research Council. 1982. "An Assessment of the Health Risks of Seven

Pesticides Used for Termite Control." National Academy of Sciences, Washington, D.C.

205. Holmberg, K. 1987. "Indoor Mold Exposure and Health Effects." 637–642. In: *Proceedings of the Fourth International Conference on Indoor Air Quality and Climate*. Institute for Water, Soil and Air Hygiene. West Berlin. Vol. 1.

206. Anderson, I. and J. Korsgaard. 1984. "Asthma and the Indoor Environment — Assessment of the Health Implications of High Indoor Air Humidity." 79–86. In: *Proceedings of the Third International Conference on Indoor Air Quality and Climate*. Swedish Council for Building Research. Stockholm. Vol. 1.

207. Lowenstein, H. et al. 1979. "Airborne Allergens — Identification of Problems and the Influence of Temperature, Humidity and Ventilation." 111–125. In: *Proceedings of the First International Indoor Climate Symposium*. Danish Building Research Institute. Copenhagen.

208. Reed, C.E. and R.G. Toconley. 1983. "Asthma — Classification and Pathogenesis." 811–831. In: E. Middleton et al. (Eds.). *Allergy: Principles and Practice*. 2nd Ed. C.V. Mosby, St. Louis.

209. Wharton, G.W. 1976. "House Dust Mites." *J. Med. Ent.* 12:577–621.

210. Voorhorst, R. et al. 1967. "The House Dust Mite and the Allergens It Produces. Identity with the House Dust Allergen." *J. Allergy* 39:325–339.

211. Korsgaard, J. 1982. "Preventative Measures in House Dust Allergy." *Amer. Rev. Respir. Dis.* 125:80–84.

212. Arlian, L.G. et al. 1982. "The Prevalence of House Dust Mites *Dermatophagoides* spp. and Associated Environmental Conditions in Ohio." *J. All. Clin. Immunol.* 69:527–532.

213. Woodford, P.J. et al. 1979. "Population Dynamics of *Dermatophagoides* spp in Southwest Ohio Homes." 197–204. In: J.G Rodriguez (Ed.). *Recent Advances in Acarology*. Academic Press. New York. Vol. II.

214. Murray, A.B. et al. 1985. "House Dust Mites in Different Climatic Areas." *J. All. Clin. Immunol.* 76:109–112.

215. Pauli, G. et al. 1987. "Indoor Mite Pollution and Mite Asthma." 738–741. In: *Proceedings of the Fourth International Conference on Indoor Air Quality and Climate*. Institute for Water, Soil and Air Hygiene. West Berlin. Vol. 2.

216. Bronswiik, J.E. et al. 1973. "Role of Fungi in the Survival of *Dermatophagoides (Acarina: Pyroglyphidae)* in House Dust Environment." *Environ. Ent.* 2:142–145.

217. Tovey, E.R. et al. 1981. "The Distribution of Dust Mite Allergen in the Houses of Patients with Asthma." *Amer. Rev. Respir. Dis.* 124:630–635.

218. Beaumont, F. et al. 1985. "Sequential Sampling of Fungal Air Spores Inside and Outside the Homes of Mold-Sensitive, Asthmatic Patients: A Search for a Relationship to Obstructive Reactions." *Ann. Allergy* 55:740–746.

219. Gravesen, S. 1979. "Fungi as a Cause of Allergenic Disease." *Allergy* 34:135–154.

220. Hirsch, S.R. and J.A. Sosman. 1976. "A One-Year Survey of Mold Growth Inside Twelve Homes." *Ann. Allergy* 36:30–38.

221. Colin, J. et al. 1961. "Molds of Allergenic Significance in the Puget Sound Area." *Ann. Allergy* 19:1399–1406.

222. Solomon, W.R. 1976. "A Volumetric Study of Winter Fungus Prevalence in the Air of Midwestern Homes." *J. Allergy Clin. Immunol.* 57:46–55.

223. Kozak, P.P. et al. 1979. "Factors of Importance in Determining the Prevalence of Indoor Molds." *Ann. Allergy* 43:88–94.

224. Kozak, P.P. et al. 1980. "Currently Available Methods for Home Mold Surveys: Examples of Problem Homes Surveyed." *Ann. Allergy* 45:167-176.
225. Solomon, W.R. 1974. "Fungus Aerosols from Cold-Mist Vaporizers." *J. Allergy and Clin. Immunol.* 54:222-227.
226. Solomon, W.R. and H.A. Burge. 1984. "Allergens and Pathogens." 174-175. In: Walsh et al. (Eds.). *Indoor Air Quality.* CRC Press, Boca Raton, FL.
227. Tyndall, R.L. et al. 1987. "Microflora of the Typical Home." 617-621. In: *Proceedings of the Fourth International Conference on Indoor Air Quality and Climate.* Institute for Water, Soil and Air Hygiene. West Berlin. Vol. 2.
228. Morey, P.R. et al. 1984. "Environmental Studies in Moldy Office Buildings: Biological Agents, Sources, Preventative Measures." *Ann. ACGIH* 10:21-36.
229. van Wageningen, N. et al. 1987. "Health Complaints and Indoor Molds in Relation to Moist Problems in Homes." 723-727. In: *Proceedings of the Fourth International Conference on Indoor Air Quality and Climate.* Institute for Water, Soil and Air Hygiene. West Berlin. Vol. 1.
230. Jontunen, M.J. and A. Nevalainen. 1987. "The Effect of Humidification on Indoor Air Fungal Spore Counts in Apartment Buildings." 643-647. In: *Proceedings of the Fourth International Conference on Indoor Air Quality and Climate.* Institute for Water, Soil and Air Hygiene. West Berlin. Vol. 1.
231. Tobin, R.S. et al. 1987. "Significance of Fungi in Indoor Air: Report of a Canadian Working Group." 718-722. In: *Proceedings of the Fourth International Conference on Indoor Air Quality and Climate.* Institute for Water, Soil and Air Hygiene. West Berlin. Vol. 1.
232. Hansen, L. et al. 1987. "Carpeting in Schools as an Indoor Pollutant." 727-731. In: *Proceedings of the Fourth International Conference on Indoor Air Quality and Climate.* Institute for Water, Soil and Air Hygiene. West Berlin. Vol. 2.
233. Fink, J. 1983. "Hypersensitivity Pneumonitis." 1085-1099. In: E. Middleton et al. (Eds.). *Allergy: Principles and Practice.* 2nd Ed. C.V. Mosby, St. Louis.
234. Edwards, J.H. 1980. "Microbial and Immunological Investigations and Remedial Action After an Outbreak of Humidifier Fever." *Brit. J. Ind. Med.* 37:55-62.
235. Patterson, R. et al. 1981. "Hypersensitivity Lung Disease Presumptively Due to *Cephalosporium* in Homes Contaminated by Sewage Flooding or by Humidifier Water." *J. Allergy Clin. Immunol.* 68:128-132.
236. Banaszak, E.F. et al. 1970. "Hypersensitivity Pneumonitis Due to Contamination of an Air Conditioner." *New Engl. J. Med.* 283:271-276.
237. Bernstein, R.S. et al. 1983. "Exposures to Respirable Airborne *Penicillium* from a Contaminated Ventilation System: Clinical, Environmental and Epidemiological Aspects." *AIHIJ* 44:161-169.
238. Solley, G.O. and R.E. Hyatt. 1980. "Hypersensitivity Pneumonitis Induced by *Penicillium* Species." *J. Allergy Clin. Immunol.* 65:65-70.
239. Rylander, R. et al. 1978. "Humidifier Fever and Endotoxin Exposure." *Clin. Allergy* 8:511-516.
240. Imperato, P.J. 1981. "Legionellosis and the Indoor Environment." In: Proc. Symp. Health Aspects of Indoor Air Pollution. *Bull. New York Acad. Med.* 57:922-935.
241. Ager, B.P. and J.A. Tickner. 1983. "The Control of Microbiological Hazards Associated with Air Conditioning and Ventilation Systems." *Ann. Occup. Hyg.* 27:341-358.
242. Miller, R.P. 1979. "Cooling Towers and Evaporative Condensers." *Ann. Int. Med.* 90:667-670.

2   SOURCE CONTROL—
INORGANIC CONTAMINANTS

In theory, it is technically desirable to control contaminants at their source before they are released to indoor air. Source control is potentially a simpler and more effective contaminant mitigation measure than those measures that focus on removing contaminants after they become airborne or become entrained in indoor air. (See Chapters 5 and 6.)

Source control comprises a variety of principles and applications based on the nature of particular contaminants and contamination problems. It includes actions that prevent or exclude the entry of contaminant-releasing building materials, furnishings, devices, consumer products, etc. into building spaces; various elements of building design and maintenance that prevent air contamination problems from occurring; use of treatment measures in which a physical or chemical barrier is placed between the source and the indoor environment; and physical removal of the source or source materials and replacement with nonemitting or low-emitting sources. It may also include control of indoor climate to reduce emission rates or prevent the occurrence of conditions which would lead to the generation of contaminants.

Source control often is described as an indoor mitigation measure in the generic sense. Its importance and the full dimension of its application and effectiveness, however, can best be described in the context of specific contamination problems. Source control will therefore be described and illustrated in this and the following two chapters on a contaminant or problem-by-problem basis. In this chapter, source control is discussed in the context of contaminants which are basically inorganic in nature. They include asbestos, combustion-generated pollutants, and radon.

ASBESTOS

As described in Chapter 1, friable asbestos-containing material (ACM) has been widely used as a fire-and heat-retardant material in schools and a variety of public access buildings. To a lesser extent, it has also been used in some residential applications. Because of its hazardous nature, most applications of unbound asbestos have been banned by regulatory action. Therefore, the prevention of entry, or exclusion, of friable asbestos products is something that building owners or prospective builders of new buildings no longer have to consider. Because of regulatory bans, asbestos is unlikely to become an indoor air quality concern in any building where it is not already present.

Asbestos is unique among indoor air quality problems in that source control is the only acceptable mitigation option. This reflects its hazardous nature and the compelling need to prevent asbestos fibers from becoming airborne, thereby preventing or greatly reducing human exposure.

A variety of source control options are available to reduce asbestos fiber exposures to occupants of buildings in which ACM is present. These include building operation and maintenance (O&M) practices to prevent or reduce the likelihood of asbestos fibers becoming airborne or resuspended; repair; enclosure; encapsulation; and removal.

Building Operation and Maintenance Practices

Building O&M practices can at best be described as an interim asbestos control or management effort. Their use is appropriate in the interval between the identification of a potential asbestos problem and application of a more permanent measure. Building O&M practices have considerable importance, since building owners or managers often temporize on the application of permanent measures because of the enormous costs often involved and physical dislocations of personnel necessary during removal operations.

A detailed guide to building operation and maintenance practices related to schools containing friable ACM has been developed by the EPA,[1] as has a guidance document for service and maintenance personnel.[2]

In conducting an operation and maintenance program, it is important that service and maintenance personnel know how to recognize friable ACM and/or know where it is located. To minimize release of asbestos fibers to the air, they must avoid damaging or disturbing it. Asbestos-disturbing activities to be avoided include (1) drilling holes in ACM; (2) hanging or attaching materials to walls or ceilings containing ACM; (3) causing physical abrasion by moving construction materials and furniture, replacing light bulbs, etc.; and (4) removing ceiling tiles below ACM.

Asbestos fibers can be resuspended by custodial activities such as dry sweeping and dusting. To minimize resuspension, custodians are advised to dust with a damp cloth and wet mop floors. Where asbestos debris has fallen on floors, it should be wetted down and removed with a dust pan and placed into a plastic bag. The pan should be rinsed with water. A high-efficiency particulate air (HEPA)–filtered vacuum cleaner should be used as a follow-up. All cloths, mopheads, and other asbestos-containing wastes should be disposed of using double-ply plastic bags. When service workers must do routine maintenance work around ACM, they are advised to wear a half-mask respirator (with disposable cartridges), clear the area of other people, use a drop cloth (or glove bag around boiler pipe insulation), and clean up carefully after each job. If HVAC filters are to be replaced, they should be misted with water, removed carefully, and disposed of in plastic bags.[2]

Asbestos-containing pipe and boiler insulation is not as significant a source of building asbestos contamination as is surfacing ACM. Protective jackets usually prevent fiber release from pipe and boiler insulation. Special operation and maintenance practices include alerting workers to ACM location, inspecting for damage, and taking precautions prior to any building construction or maintenance activities.

Although an O&M program is considered to be an interim asbestos control measure, it is likely to be used on a long-term basis when asbestos assessments indicate that asbestos-containing surface material is in good condition with low potential for future disturbance or erosion, and the likelihood of air contamination is very low. Cementitious ACM, for example, with a low asbestos fiber content and low friability, would be considered to have a low fiber release potential.

The application of more stringent asbestos control measures depends on the outcome of systematic assessments to evaluate the likelihood of fiber release. Assessment factors include evidence of deterioration or delamination from substrates, physical debris, and water damage. They also include factors indicating potential for future disturbance, such as (1) proximity to a direct airstream or plenum, (2) accessibility to building occupants or personnel and degree of activity (vibration, etc.), and (3) change in building use. Assessment may also include air monitoring for asbestos fibers.

Special asbestos abatement measures include repair, enclosure, encapsulation, and removal. The EPA considers removal to be the only permanent solution and recommends the use of enclosure or encapsulation only under limited circumstances.[1] If the building is to be demolished at some future date, the ACM will have to be removed anyway under requirements of the National Emission Standard for Hazardous Pollutants (NESHAP) for asbestos, which specifically addresses control measures for asbestos in building demolitions.[3]

Repair

In some instances, damaged ACM can be easily repaired. This is particularly the case with thermal insulation applied to steam lines and other boiler-related surfaces. Repair usually involves patching the cloth wrappings that normally enclose ACM used for thermal insulation.

Enclosure

Exposure to asbestos fibers released by friable asbestos material can be reduced or minimized by placement of physical barriers between ACM and the air space near it. In theory, any physical barrier that reduces the introduction of asbestos fibers into occupied spaces (e.g., a suspended ceiling) could be described as an enclosure. Because suspended ceilings are not airtight and are subject to intrusion by maintenance and service personnel, EPA does not recommend the use of a suspended ceiling as an enclosure for purposes of asbestos mitigation in schools.[1]

Outside of the school context, suspended ceilings, where they are already in place, may in some measure serve to reduce asbestos fiber exposure to building occupants, particularly where they do not serve as a base for a cold air return plenum. Where they do, they may increase exposures as a consequence of the high velocity air flows maintained in the suspended ceiling-based plenum. Removal of suspended ceiling tiles by service personnel may cause a substantial number of asbestos fibers to become airborne.

The use of enclosures as recommended by EPA[1] involves the construction of nearly airtight walls and ceilings around the ACM. The application of enclosures requires that construction material be impact-resistant and assembled to reduce air movement across the enclosure boundary.

Enclosures may be appropriate where ACM is located in a small area and where disturbance or entry into the enclosed area is unlikely. They are inappropriate where (1) water damage is evident, (2) damaged or deteriorating materials cause rapid fiber release, (3) damage or entry into the enclosure is likely, and (4) the ceiling to be enclosed is low.

Enclosures have the potential advantage of reducing exposures to individuals outside the area of enclosure at relatively low cost (unless utilities require relocation). However, the asbestos source remains and will eventually require removal. Additionally, enclosures require periodic reinspection to check for damage and a special operations program to control access for maintenance and renovation.

Encapsulation

Another approach to asbestos control is the use of sprayed-on encapsulants designed to bind asbestos fibers in the ACM matrix and/or achieve adherence of ACM to building substrates. Like enclosure, encapsulation serves to reduce asbestos fiber exposures to building occupants. The purpose of encapsulation is to prevent fiber release. It has the advantage of lowering initial costs (as compared to removal) and does not require replacement of insulating material.

Encapsulation is an appropriate application where ACM (1) retains its bound integrity, (2) is granular or cementitious in nature, (3) is unlikely to be damaged, and (4) is not highly accessible. If an encapsulant is applied to ACM that is in poor condition, the ACM may delaminate and fall off, thereby posing a risk of asbestos fiber exposure.

Because the asbestos source remains, encapsulation cannot be considered a long-term solution. The encapsulated ACM will require periodic reinspection and reapplication. Because of costs involved in the application of encapsulants and the fact that the asbestos will have to be removed sometime in the future, the long-term costs of encapsulation may be higher than removal. Additional costs of removal may be imposed since the encapsulated surface is often more difficult to remove. From a sociopolitical standpoint, use of encapsulation in school buildings has the advantage of deferring costs of removal to a different generation of taxpayers!

The use of encapsulation is advised where ACM is extremely difficult to remove—for example, where it has been sprayed on complex surfaces such as ceilings with numerous surface irregularities or on pipe or duct work. It is also advised on troweled-on or cementitious ACM.[4] The latter materials do not pose exposure problems as severe as the spongy/fluffy form of ACM.

Encapsulants are of two types. Bridging encapsulants are designed to form a tough membrane over the surface to prevent fiber release. Penetrating encapsulants are designed to enter the ACM material to bind fibers to one another and to the substrate. Penetrating encapsulants have lower viscosities and solids contents than do the bridging types.

Application of either of the two encapsulant types depends on characteristics of the ACM. If it is moderately friable and less than one inch thick, a penetrating encapsulant is advised.[4] A bridging encapsulant is recommended for moderately friable ACM found on complex surfaces such as pipes, ducts, and beams, and on cementitious materials.

Bridging encapsulants do not improve the ACM's adhesion to substrate materials. The additional weight may cause the ACM to fall in an improper application. Consequences of failure may be severe; the bridging compound covers but does not bind fibers. If ACM does fall, damaged or delaminated fibers behind the encapsulant can escape into building air. A bridging encapsulant is not suitable on large flat ceilings unless the ACM is cementitious and in good condition.

A penetrating encapsulant is suitable where the binding together of asbestos fibers and improved adhesion to surface substrates are desired. In tests conducted by the EPA, penetrating encapsulants were found to be slightly more susceptible to impact and abrasion than bridging compounds. They are not recommended for painted or previously encapsulated surfaces. Bridging encapsulants are recommended for recoating.

The performance of encapsulants varies with the type of material to which they are applied. Because of variability in performance, the EPA recommends that encapsulants be field-tested before a decision is made on using a particular one. The American Society for Testing and Materials (ASTM) has developed standards by which encapsulants can be judged. Desirable properties include (1) penetration of the material to bind fibers together or to form a tough membrane, (2) tolerance to some abuse without releasing fibers, (3) water-insolubility when cured, and (4) sufficient durability after six years to allow recoating. Additionally, encapsulants should not release toxic substances which may be dangerous to workers applying them in rooms isolated by plastic barriers. They should not be flammable.[1]

Battelle Columbus Laboratories, on contract to EPA, evaluated 100 commercially available encapsulants. Battelle judged 11 of 100 products to be acceptable and an additional 23 as marginally acceptable. Acceptability was based on ratings of fire resistance and smoke generation, toxic gas release on burning, and surface integrity capable of sealing or binding fibers together by penetrating 0.5 inches or more. The Battelle studies were done on a mineral wool matrix and may not duplicate results when the encapsulant is applied to ACM.[4]

Good application techniques are required to ensure high performance of the encapsulant and prevent exposure to workers and building occupants. Encapsulants should be applied with airless spray equipment to reduce the likelihood of dislodging fibers. Good performance requires that workers apply the encapsulant at the recommended rate and under relatively dry atmospheric conditions. Reduced performance can result from both under- and over-application. Special precautions are required to maintain worker safety, including use of plastic sheeting and duct tape to seal off the work space, half-face respirators and disposable clothing, portable shower facilities, etc.[1]

Encapsulation is only a temporary solution to a building asbestos problem. Its use requires periodic reinspection and probably recoating to maintain safe conditions for building occupants.

Removal

In theory, the best way to reduce exposure to a hazardous contaminant such as asbestos is to eliminate its source. By removing ACM from a building, the threat of asbestos fiber exposure can be eliminated on a permanent basis. Asbestos removal is applicable in most situations and eliminates the need for special operation and maintenance programs.

There are, however, hard realities associated with asbestos removal. A serious concern is that improper removal by contractors may introduce a major risk of asbestos exposure to building occupants that would not have occurred had the ACM remained in place. Anecdotal testimonials abound of gross asbestos removal practices by inexperienced and unscrupulous contractors. Choosing an experienced and reputable contractor is essential to make removal the safe and effective asbestos control measure that, in theory, it should be.

Asbestos removal is an extraordinarily expensive task. Specifications for asbestos removal procedures are stringent and labor-intensive. Costs for contractor liability insurance are high; in many cases such insurance may not be available. Asbestos removal costs for relatively large school buildings may be hundreds of thousands of dollars. It is these costs that cause building owners/operators to defer the removal of asbestos.

Asbestos removal must be conducted under the minimum guidelines set forth by EPA under the asbestos NESHAP[4] and meet Occupational Safety and Health Administration (OSHA) worker protection requirements.[5] Additional guidelines have been developed for the school asbestos program and are recommended by EPA for ACM removal projects.

The focus of asbestos removal is to accomplish the task with a minimum degree of risk to both asbestos removal workers and present and future building occupants. To achieve this, asbestos removal is conducted by isolating the working area in airtight plastic containment barriers. Negative pressure systems[1] with HEPA filtration are recommended by EPA to reduce the likelihood of fiber release to other parts of the building if the containment barrier is accidentally torn. ACM must be wetted with a solution of water and a wetting agent prior to removal. Friable ACM is disposed of in leak-tight containers, typically 6-mil polyethylene plastic bags. Rigorous post-removal cleanup is extremely important. EPA recommends wet mopping or HEPA vacuuming all horizontal and vertical surfaces in each asbestos removal area on two consecutive days to provide assurance of good fiber reduction.

After asbestos has been removed and before the contractor is released, the worksite should be visually inspected. On passing visual inspection, air quality monitoring should be conducted in the area of asbestos removal using test methods recommended by EPA.[6] The contractor can be released when the average fiber concentration from samples collected in the asbestos removal area is (1) not statistically larger than the average of outside samples when test samples are analyzed with the transmission electron microscopic method, or (2) when every sample is below the release quantification limit (0.01 f/cm^3 per 3000-L sample) determined by the phase contrast microscopic method.

COMBUSTION-GENERATED POLLUTANTS

As described in Chapter 1, elevated combusted-generated pollutants in indoor spaces usually result from unvented, improperly vented, or poorly maintained combustion appliances and the smoking of tobacco products. In many instances, entrance of combustion-generated contaminants into the air of building spaces reflects personal choices and lack of individual understanding of what the consequences of those choices may be.

Combustion Appliances

Avoidance

Most of the contamination of indoor air associated with combustion can be avoided. Avoidance is based on decision making that would choose the use

Table 2.1 Average Contaminant Emission Rates from Well-Tuned Kerosene Space Heaters[7]

Heater Type	No. of Heaters	Emission Rates (μg/kJ)				
		CO	NO	NO_2	HCHO	Particles
Convective	3	25.0	14.1	16.3	0.31	—
Radiant	4	64.0	1.3	4.7	0.29	0.49
Two-stage	2	9.2	4.3	2.2	0.20	1.70

of well-vented combustion devices over those that are unvented (e.g., kerosene and gas space heaters). It would include the choices of electric cooking ranges and ovens over gas-fueled ones; not using a wood-burning appliance (stove, furnace, fireplace); and cessation of smoking in indoor spaces.

These noncombustion choices, if implemented, would prevent combustion-generated indoor contamination from occurring in new and existing buildings and mitigate problems which already exist. In the latter case, sources are removed and excluded.

Selective Avoidance

Let us assume, however, that by individual choice, complete avoidance will not be practiced by many individuals. This is apparently the case for unvented kerosene and gas space heaters and wood-burning stoves and furnaces, as such devices are used to some degree to space heat 10 million or more households. It is likely that millions of new devices will be sold for space heating homes in the future as well.

In regard to future purchases, individuals can, in theory, minimize or at least reduce/alter the magnitude and nature of the contaminants that they and their families will be exposed to. They would in such instances practice selective avoidance.

The magnitude of emissions from kerosene space heaters may vary by an order of magnitude depending on heater type and design. In studies conducted at Lawrence Berkeley Laboratory of emissions from three types of kerosene heaters (radiant, convective, and two-stage), mean CO emission rates for well-tuned heaters ranged from 9.2 μg/kJ (0.021 lb/10^6 Btu) for two-stage heaters to 64 μg/kJ (0.15 lb/10^6 Btu) for radiant-type heaters (Table 2.1).[7] Mean emission rates for NO_2 ranged from 2.2 μg/kJ (0.005 lb/10^6 Btu) for the two-stage type to 16.3 μg/kJ (0.037 lb/10^6 Btu) for the convective type. Two-stage heaters also had the lowest emissions of formaldehyde.

In choosing a kerosene heater based on contaminant emissions, the two-stage heater is clearly the best choice. In selecting between radiant and convective types, one has a choice between lower emissions of NO_2/higher emissions of CO (radiant heaters) and lower emissions of CO/higher emissions of NO_2 (convective heaters).

Apte and Traynor[7] evaluated emissions from several types of unvented gas space heaters. They observed no significant differences in mean emission rates of CO, NO_2, formaldehyde, and submicron suspended particles between convective propane-fueled and natural gas–fueled unvented space heaters. Infrared unvented gas space heaters had significantly lower emissions of nitrogen oxides (1.6 $\mu g/kJ$; 0.0037 $lb/10^6$ Btu) than the convective propane-fueled gas space heaters. Infrared unvented space heaters therefore have a clear advantage over convective types.

Kerosene heater emissions can also be reduced by properly sizing the unit for the space to be served. An oversized unit will consume more fuel and therefore have higher total emissions. An oversized unit will also likely result in operating practices which increase emission rates.

Combustion-generated contaminant exposures increase with increasing duration of use and with each additional unit used.[8] Both indoor contamination and human exposures increase as more units are employed and as units are used for extended durations during a single day and the heating season. The use of multiple units and the use of any unvented device for extended periods is not advisable and should be selectively avoided.

Selective avoidance can also include the type of fuel used. The amount of SO_2 produced by a kerosene heater depends on the sulfur content of kerosene. Kerosene produced according to ASTM specification D3699 Grade No. 1-K has no more than 0.04% sulfur by weight, while Grade No. 2-K has a sulfur content of no more than 0.30%. The maximum sulfur content of the latter is 7.5 times the former. 1-K fuel produces less sulfur dioxide and does not tend to gum up the wick as do 2-K fuels.[9] Unfortunately, 1-K fuel has a more limited distribution than does 2-K kerosene.

Wood-burning appliances also differ in design and in their ability to cause indoor air contamination. Significantly higher indoor emissions of CO and respirable particles have been reported for nonairtight wood burning appliances than for those of airtight designs.[10-12] Airtight wood burners therefore are the clear choice. In theory, design factors will also have significant effects on contaminant emissions from fireplaces.[13] For example, a recirculating eddy tends to form inside the upper front half of the fireplace opening. The high velocity of air flow along the back (Figure 2.1) transfers momentum to the slow-moving mixture of air and smoke, creating the eddy. The tendency to eddy is increased by restrictions or poor design features in the throat area between the lintel and damper. The eddy causes smoke to move away from the area of maximum velocity and increases the likelihood that it will escape from the fireplace. Few studies have been conducted on fireplace emissions into indoor spaces and no comparative information is available to make recommendations relative to low-polluting choices.

Gas-fired cooking stoves and ranges have been reported to produce substantial emissions of CO and NO_x.[14,15] Recent research by the American Gas Association and Thermo Electron Corporation has indicated that technical and design modifications to gas appliance burners can result in lower emissions.

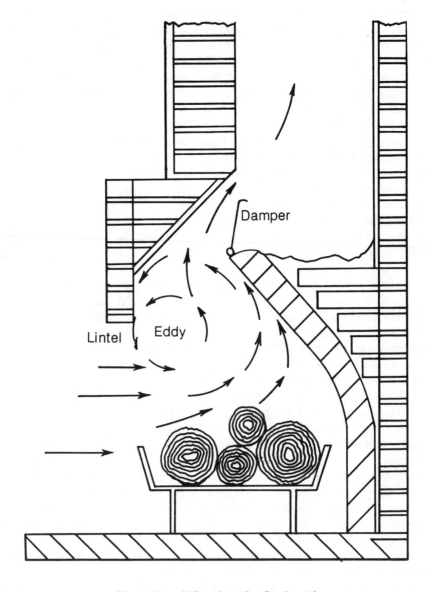

Figure 2.1. Airflow through a fireplace.[13]

Adapting technology developed to reduce NO_x emissions from residential gas furnaces and water heaters (to comply with California NO_x emission limitations), Dewerth and Sterbik[16] developed a production-type wire insert for range-top burners. The insert was able to reduce NO_x and NO_2 emissions by 35% and 26%, respectively. They also developed and evaluated a low NO_x

dual rod insert oven burner, which reduced NO_x by 63% and NO_2 by 25%. In addition, they evaluated an alternate low-NO_x range-top burner that employed high primary aeration to reduce NO_x. This burner significantly reduced NO_x and NO_2 emissions. However, it produced marginal combustion results.

Thermo Electron Corporation, in conjunction with the Gas Research Institute, has developed a gas range with perforated tile burners with glass ceramic plates above them. Heat is transferred radiantly from burner to pot through the glass ceramic plate. In addition to radiant heat, heat is also transferred to pots by convection as combustion gases pass through the holes in the glass plate, forming jets. Combustion air is delivered by a blower and mixed with gas prior to ignition. The burner design results in significantly lower emissions of CO, NO_x, and NO_2.[17]

Lower emissions can also be achieved by using gas ranges and ovens with electronic ignition. Baker et al.[18] report that homes with electronic ignition ranges have significantly lower NO_2 concentrations than homes having pilot light gas ranges/ovens.

Operation and Maintenance

The way that unvented kerosene and gas appliances and wood-burning appliances are operated and maintained can have significant effects on emissions. Good operation and maintenance in all instances results in lower emissions.

Emission rates of CO, NO_2, and formaldehyde are all substantially increased when wick lengths are reduced from those recommended by manufacturers.[7] Mean CO emission rates were increased by factors of 5.1, 2.5, and 5.9, respectively, for convective, radiant, and two-stage burner heaters when wick height was reduced. The effect of reducing wick height is to decrease fuel input and heat output. Because these are often desirable from a consumer perspective, they are likely to operate heaters at lower wick heights than recommended by manufacturers. As a consequence, emissions and exposures to CO are greatly increased. Operation at the manufacturer's recommended wick heights is therefore advisable.

Unvented gas space heaters have been tested under maltuned conditions by adjusting the heater's primary air shutter.[7] Carbon monoxide and formaldehyde emissions were observed to increase 20- and 30-fold, respectively.

Wood-burning stoves and furnaces can release contaminants to indoor spaces during starting, stoking, reloading, and ash removal operations. Decreasing the duration of stove door openings would be expected to reduce emissions, as would reductions in the severity and duration of disturbances to the fire box and during ash removal.

Operating and maintenance factors also affect emissions from range-top burners. Relwani et al.[19] reported that gas range burners operating with an improperly adjusted air shutter (33% of stoichiometric air) produced substantially higher emissions of CO with little change in NO_x emissions. In addition to the effect of primary aeration rate levels, significant effects of firing rate at burning settings of warm, low, medium, and high were observed. Carbon monoxide emissions were highest at the warm setting and were lower and similar at the other three settings. Total NO_x emissions were observed to be highest at the high fuel input rate or burner setting. The more significant NO_2 had its highest emission rate at the lowest burner setting. The lowest fuel input rate therefore resulted in the largest emissions of the two pollutants of prime concern. In field studies Moschandreas and Relwani[20] reported that in general, NO_2 emissions increased with range-top burner age, suggesting that maintenance be considered as a means of reducing NO_2 emission rates.

Reaction Decay Removal

Reactive gases such as NO_2 and SO_2 decrease in concentration in indoor spaces more rapidly than can be accounted for by air exchange. This decrease is associated with the reaction of these gases with interior surfaces and other airborne species. It has been observed in field studies of NO_2 and SO_2 emissions from unvented kerosene space heaters that reaction decay rates are at least as important in removing NO_2 and SO_2 as is building air exchange rate.[21] Decay rates vary greatly within a residence from time period to time period, as well as between residences. In chamber studies, a pronounced increase in SO_2 decay rates with increasing relative humidity has been observed.[21] Nitrogen dioxide decay was, however, relatively insensitive to humidity changes. This is in contrast to another study which reported strong positive correlations between NO_2 decay and relative humidities from 40% and 70%.[22] In chambers with carpeting, nitric oxide (NO) concentrations were observed to decrease more slowly than would be expected based on air exchange rates. Since outdoor NO concentrations were very low, it was suggested that NO might be supplied to room air by the catalytic reduction of NO_2 adsorbed on surfaces.

Decay rates for NO_2 and SO_2 are dependent on temperature and on the nature and extent of surfaces present. Renes et al.[21] observed a low negative correlation between SO_2 removal rates and temperature, which they suggested was due to the increased relative humidities associated with lower temperatures. Miyazaki[23] reported significant increases in NO_2 decay rates with increases in temperature or relative humidity. Miyazaki reported that NO_2 decay rates were significantly affected by the type of interior materials and their surface area; degree of air movement was also an important variable.

Nitrogen dioxide decay rates for a variety of materials (Table 2.2) have been reported from studies conducted on behalf of the Gas Research Institute.[24,25] Highest NO_2 removal rates were observed for unpainted gypsum board, cement block, and wool carpeting. The effects of relative humidity were mixed, with humidity-associated decay rates increasing with wool carpeting, decreasing for cement blocks, and unaffected by wallboard. It was also observed that the rate of NO_2 decay after the appliance is turned off decreases as the length of appliance operation is increased, suggesting a surface satura-

Table 2.2   Decay Rates of NO_2 in the Presence of Selected Residential Materials[24]

Highest Removal (≥ 6 hr^{-1})	Moderate Removal (1.9–2.7 hr^{-1})
Unpainted plasterboard (>8.4)	Painted plasterboard (2.6)
Cement block (8.4)	Ceiling tile (1.9)
Wood carpet (6.0)	Acrylic fiber or nylon carpeting (1.9–2.0)
	Low Removal (<1 hr^{-1})
High Removal (3.7–4.2 hr^{-1})	Particleboard (0.7)
Used bricks (4.2)	Ceramic tile (0.7)
Masonite (4.1)	Window glass (<0.1)

tion of decay sites. The effect of appliance operating time on decay rates is illustrated in Figure 2.2. The decay rate was seen to decrease as the NO_2 concentration decreased. There was a significant conversion of NO_2 to NO for some materials but not for others, and it was suggested that NO_2 may be removed by at least two mechanisms—reduction to NO and physical adsorption.[24]

The significance of pollutant decay is that it tends to reduce indoor contaminant concentrations and therefore human exposures. Evidence that decay rates vary with the type and surface area of materials present as well as environmental factors suggests that, at least in theory, strategies can be developed to use this phenomenon to control levels and exposures to reactive pollutants such as NO_2 and SO_2. This would require judicious selection of building mate-

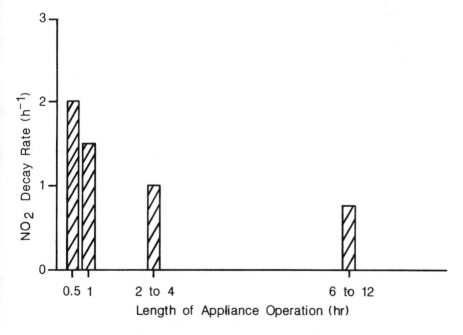

Figure 2.2.   Effect of appliance operating time on NO_2 decay rates.[25]

rials and furnishings to increase decay and therefore removal rates.[25] The likelihood of homeowners constructing and furnishing a house to take advantage of this phenomenon is very remote. This remains at best a novel but interesting approach for controlling combustion-generated pollutants in indoor air. For those residences and spaces where materials favorable to contaminant decay are present, they apparently provide an unintended side benefit.

Flue Systems

Design and Installation

Indoor contamination problems associated with flue gas spillage and their causes were described in Chapter 1. Avoidance and mitigation of such problems requires proper design, construction, and installation of flues and chimneys as well as proper operation and maintenance of combustion equipment, flues, and chimneys. Design principles for chimneys, gas vents, and fireplaces are extensively described in Chapter 27 of the ASHRAE *Equipment Handbook*.[13] Careful attention to ASHRAE chimney and gas vent design considerations by flue gas equipment manufacturers and installation contractors can avoid many problems of flue gas spillage and minimize safety hazards.

The provision of a suitable chimney for any type of combustion appliance requires a knowledge of draft needs. With respect to draft needs, combustion system equipment can be classified as follows:

1. equipment requiring negative pressure at the outlet to induce air flow into the combustion zone
2. equipment operating at neutral pressure without the need for chimney draft (combustion is isolated from chimney flow variations)
3. equipment with a forced draft combustion system operating above atmospheric pressure so that no chimney draft is required

For a given application, a proper type of chimney is determined from the type, size, and operating temperature of the appliance, construction of surroundings, National Fire Protection Association appliance usage classification, and height of the building. Acceptable chimney construction requires familiarity with applicable codes and standards.

In addition to chimney selection based on draft needs and building safety, additional attention must be given to air supply, draft hood dilution, flue gas temperature levels, oxygen depletion, external wind eddy zones, and air pollution considerations. In the operation of combustion equipment requiring negative overfire draft and negative static outlet draft, the chimney assists the supply of combustion air in the flow circuit from the ambient to the indoor environment. The system design may require an allowance for combustion air flow resistance. Failure to provide adequate combustion air can result in oxygen depletion and flue gas spillage in draft hood-type gas appliances. Excess capacity is also needed to draw in draft hood dilution air to prevent spillage. Selection of vent connectors consistent with operating conditions is essential to assure building safety and long-term system performance. Consideration should be given to flue gas temperatures, the propensity to form condensate, combustion air contaminants (e.g., sulfur and halogens) which lead to corrosion, and the operating cycle of appliances. Consideration of circulating and eddying wind currents external to buildings is necessary in order to avoid recirculation around the building and reentry through openings and fresh air intakes.

Inspection and Maintenance

Avoidance of flue gas spillage also requires periodic inspection and maintenance. In a study conducted in California, gas appliances were inspected in 50 homes which had previously been determined to have elevated NO_2 levels.[26] Inspection consisted of (1) visually checking the appliance, (2) checking for exhaust leaks, (3) monitoring the venting operation, and (4) metering fuel consumption. Seventeen of 40 furnaces were observed to be operating unsafely, with 10 additional ones spilling combustion products into the home interior. Spillage was also observed from 11 hot water heaters and two of 14 gas dryers.

Inspection and maintenance become particularly important with increasing age of gas appliances and flue systems. As with any system, aging increases the likelihood of failure. Failures may result from noncode vent connections, corrosion, vibration-induced joint separations, debris in chimneys, erosion of chimney mortar, cracks in heat exchangers, etc. Inspection and maintenance are particularly desirable when individuals plan on purchasing an older house.

House Depressurization

Flue gas spillage may occur when houses are depressurized by the simultaneous operation of vented gas furnaces, exhaust fans, or fireplaces. This may particularly be a problem in energy-efficient houses.[27-30] Special precautions should be taken in venting combustion appliances in tight houses. To prevent significant depressurization of the house when gravity flow venting is utilized for furnaces, outside air should be provided for both combustion appliances and flue gas systems. Various other approaches to prevent depressurization-induced flue gas spillage into energy-efficient houses have been suggested.[27] In addition to the provision of outside air for combustion devices and flue gas systems, these include (1) isolation of the combustion device from the living space, (2) incorporation of combustion air requirements into the design of mechanical ventilation systems, (3) installation of an induced-draft fan in the flue stack interlocked with burner controls to ensure that downdrafting cannot occur, (4) using only electric appliances for space and water heating, and (5) not installing fireplaces in energy-efficient houses.

Direct vent appliances, such as mobile home–type furnaces and side wall–vented, pulse-combustion furnaces, are one example of isolating combustion equipment from the living space. The former are relatively inefficient and the latter are expensive. The use of outdoor furnaces is another example. Induced draft fans and flow-improving devices are employed in certain models of high-efficiency furnaces. They may, however, cause unbalanced operation of heat recovery ventilation systems and excessive loss of house air when other air-consuming devices are not operating. Fireplaces are special problems because they can act as high-capacity exhausts and lead to draft failure when left smoldering.

Fail-Safe Systems

Another approach to the problem of flue gas spillage and its avoidance or minimization would be to install fail-safe systems on combustion appliances. Moffatt[29] has devised a fail-safe device for gas water heaters and appliances he considers to be susceptible to spillage and backdrafting. It consists of a reset snap therm-o-disc mounted on the draft hood and wired in series with a thermocouple. Spillage gases continuing for 30+ seconds will automatically cause the water heater to shut off. Moffatt[29] also devised an alarm system for gas furnaces.

RADON

Radon levels in residential structures can be reduced by a variety of source control techniques. Some are applied in the construction of new houses; some are used on a retrofit basis.

New House Construction

Potential exposures to elevated radon levels can be reduced by prospective new home owners and their contractors. With some foreknowledge, decisions can be made on whether to build on a particular site, construct the house using techniques that minimize radon entry, or install a proven radon mitigation system during the construction phase.

Site Selection

One might choose to build (or not build) on a site based on knowledge of geology or soil characteristics. High radon contamination potential, for example, can be expected in houses constructed on glacially deposited materials in geological features such as eskers and kames. Additionally, higher-than-normal radon contamination potential can be expected on soils of moderate to high permeability.[31] Such soils can be identified from soil maps prepared by the Soil Conservation Service. Soil maps are available locally to most U.S. citizens. Additionally, services of consulting soil scientists are available on a fee basis in many parts of the country. Building on a site with clay subsoil does not guarantee that radon levels will be within acceptable guidelines. It is, however, probable that clay subsoil building sites will have significantly lower radon emanation potentials than sandy/gravelly subsoil. Additionally, soils that are prone to frost heaving and/or having high shrink-swell capacities are likely to be unstable and may thus cause settling and cracking of subsurface structures. Unstable soils should be avoided.

Site Modification

Site modification applied prior to construction can serve to limit radon entry into new houses. Rogers et al.[32] and Fitzgerald et al.[33] have developed recommendations for building on uranium mining waste and reclaimed phosphate lands. These include covering high radium–content surface materials with soil that has a low radium content and high resistance to soil gas flow, or removing high-radium soil altogether. How effective removing/replacing and covering are depends, in part, on the depth of fill. The effect of fill depth on fractional reduction of radon on a reclaimed phosphate site can be seen in Figure 2.3. A fill depth of 10 feet, for example, would cause an 80% reduction in radon emanation rates.[34]

Construction

Radon exposures can be reduced by choice of substructure. In studies conducted on houses located on reclaimed Florida phosphate lands, Guimond et al.[35] reported radon progeny levels to average 0.02 WL in houses with basements, 0.014 WL in slab-on-grade, 0.010 WL in houses on crawl spaces and 0.008 WL in mobile homes. Though their data were not shown to be statistically significant (due probably to low numbers of basement, slab-on-grade, and mobile house types), there exist strong theoretical reasons why significant differences between substructure types can be expected. Basements, for example, have a very large amount of surface area in intimate contact with the soil. Free-standing mobile homes would have no substructure and minimal soil contact. In slab-on-grade houses, soil contact would be limited to the floor. Of the three substructure types, the crawl space has the least direct soil contact and, in theory, the lowest potential to conduct soil gas. This potential, however, depends on the degree to which natural ventilation occurs. When crawl spaces are completely enclosed, they restrict air flow and provide a pathway for radon entry.[35]

Crawl spaces have been recommended as a means of reducing radon entry. Their effectiveness depends on whether, and how well, they are naturally

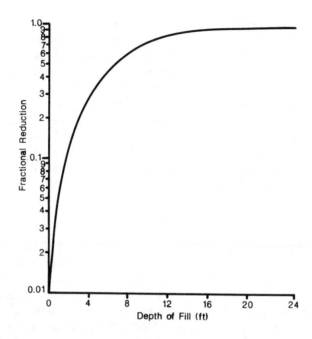

Figure 2.3. Effect of fill depth over a radium-bearing substrate on radon movement to soil surface.[33]

ventilated. Scott and Findlay[36] observed a fourfold decrease in indoor radon progeny concentrations when crawl space ventilation area was increased. Nazaroff et al.[37] observed that radon concentrations are decreased by approximately 50% when crawl space vents are opened. Houses in Sweden constructed on crawl spaces equipped with forced ventilation fans in areas of uranium-rich alumshale soils required no additional measures to maintain indoor radon at acceptable levels. By constructing houses on crawl spaces which were ventilated mechanically, unacceptably high radon levels were easily reduced at low cost.[38]

If, for whatever reason, a homeowner or contractor chooses to build a basement or slab-on-grade substructure, it can be designed and constructed to minimize radon entry points and to facilitate the installation of a subsurface ventilation system should it become necessary. Desirable design and construction practices include:

1. use of poured reinforced concrete for basement walls and slab
2. use of good quality concrete and mortar
3. use of adequate (depthwise) and good quality (low radon-releasing potential and absence of sharp edges) sub-slab aggregate
4. use of an impermeable plastic barrier between slab and aggregate
5. use of a complete system of drainage tile around basement perimeters

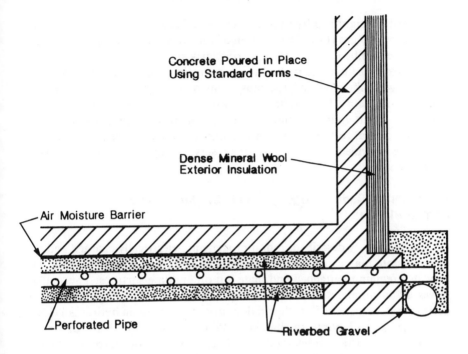

Figure 2.4.   Radon mitigation substructure design.[40]

6. maintenance of a monolithic structure, as much as possible, by avoidance of nonessential slab and wall penetrations and sealing all service entries and other penetrations of concrete structures
7. capping of all sumps
8. avoidance of heating duct installation in slabs
9. proper waterproofing of basement walls[33,39,40]

Harrje and Gadsby[40] have suggested the design approach illustrated in Figure 2.4. Poured concrete with wire mesh throughout is recommended for floor and wall construction, including corners between walls and floors. A low-permeability membrane is placed beneath the floor; beneath that is a layer of riverbed gravel. Lengths of perforated pipe running beyond the perimeter of the foundation are placed within the gravel to assure soil gas movement under the entire concrete floor. Mineral wool insulation on the exterior basement walls is used to divert water down to the gravel bed and relieve soil gas pressure, allowing soil gas to move freely to the ambient atmosphere. This design helps to mitigate potential water as well as radon problems. It has the advantage of lending itself to active mitigation techniques such as soil gas ventilation should postconstruction radon levels be unacceptable.

The choice of a heating system can also potentially affect radon levels and radon progeny exposures. Combustion appliances placed in building interiors can contribute significantly to house depressurization,[28] which is the cause of radon-laden soil gas flow. The location of vented combustion appliances in the basement could be expected to be particularly significant.

Depressurization occurs because combustion appliances consume air in supporting combustion and the movement of flue gases through vents and chimneys. In theory, lower radon levels would be expected in houses heated by electricity. An alternative would be to locate combustion appliances in a garage or space external to the living environment. If fireplaces are to be used, they should be provided with outside air to serve combustion air and flue gas exhaust needs.

Reduction of Radon Progeny Levels by Natural and Forced Plateout

As previously described, radon progeny are electrically charged and their charged state may cause them to become attached to aerosol particles. They may also remain unattached and plate out on building surfaces. This plateout serves as an unintended mechanism by which radon progeny exposures are decreased. George et al.[41] studied radon progeny plateout in both small (1.9-m) and large (20-m) laboratory chambers. Plateout was observed to be strongly dependent on particle concentrations. The proportion of plateout activity to total radon progeny activity varied from 4% at particle concentrations of $> 10^5/cm^3$ to 86% for concentrations of $< 10^3/cm^3$. The average attachment rate for radon progeny was a linear function of particle concentration.

Scott[42] observed that plateout activity was higher on walls than on floors or tables, suggesting that turbulence near the walls induced by convection was likely responsible for higher deposition velocities on wall surfaces. He also suggested that the strong air currents produced by forced-air heating systems can explain differences in radon progeny removal rates for summer and winter.

Holub et al.,[43] Rudnick et al.,[44] and Bigu[45] observed significant reductions in radon progeny activity when air-mixing fans were used. Holub et al. observed a 41% decrease in working levels when a mixing fan was operated in an environmental chamber, concluding that the observed decrease was due to plateout on the fan blades. An even larger effect was observed by Bigu with fan-induced reduction of WL by factors ranging from less than 4 to approximately 7. Rudnick et al., also working with environmental chambers, observed mixing fan-related reductions in WLs on the order of 40–70%, with greater than 90% of the plateout taking place on room surfaces. Rector et al.[46] found that constant operation of a central circulation fan in a house resulted in reduction in radon progeny levels of 40%, approximately equal to that achieved by using an air-air heat exchanger.

Practical applications of plateout phenomena would include (1) the use of ceiling and HVAC system fans to reduce radon progeny exposures; (2) choice of forced-air heating systems over radiant heat ones; and (3) actions that reduce particle generation, such as limitations on smoking and house dust control (e.g., use of high-efficiency vacuum cleaners).

Diffusion Barriers and Sealants

Radon entry into buildings occurs as a consequence of diffusion from building materials and pressure-driven flows through cracks in building substructures. Though the latter is the predominant source of radon in houses, under some circumstances building products may liberate sufficient radon by diffusion to warrant mitigation attempts. For example, in the southeastern United States, phosphate slag, a by-product of the fertilizer industry, has been incorporated into concrete blocks that exhibit significantly enhanced radium and radon levels, leading to higher-than-average exposure levels. Eichholz et al.[47] tested the effectiveness of a variety of sealants for the control of radon emanation from phosphate slag concrete block. They observed radon reduction percentages as follows: polyethylene sheet—78%, epoxy paint—49%, epoxy paint on filler—59%, latex paint—32%, Surewall, troweled—18%; Surewall plus wallpaper—51%.

Pohl-Ruling et al.[48] investigated a variety of plastic materials as radon diffusion barriers in radon-filled containers. Effectiveness was observed to decrease with storage time over a period of five days. Only the polyamide foil achieved no detectable loss in radon. Polyethylene and polyvinylchloride were approximately 80% effective over a period of five days. Thickness was a factor, with plastic films 0.2 mm thick significantly more effective than those 0.05 mm thick.

Auxier et al.[49] studied the effects of sealants on radon emissions from concrete. Epoxy paint was observed to reduce radon release by approximately 75%. Bedrosian et al.[50] evaluated a variety of gel seals, showing that those with thicknesses over 80 mm effectively reduced radon emissions by 99%. Such seals had limited applicability.

Low-permeability polymeric sealants such as epoxy have proven effective in reducing radon levels in houses when properly applied. Studies by Culot et al.[34] have shown that radon entry into a building could be reduced by approximately 50% by utilizing an epoxy sealant. Such effectiveness, however, was only achieved when building substructures were free of cracks, which serve as major pathways for radon entry. Because such cracks are the rule rather than the exception, sealants alone as a radon mitigation measure are of limited usefulness.

Sealing Substructure Openings

The entry of soil gas into buildings by pressure-driven convective flow occurs through cracks and other penetrations in the building substructure. In theory, significant reductions in radon levels can be achieved if these were identified and filled with nonpermeable material such as any of a variety of caulking and concrete patching compounds.[39] Sealing openings alone, without further mitigation actions, may produce inconsistent results. In some cases it may be partially effective in reducing radon levels,[51] and in other instances it may be totally ineffective.[52] This inconsistency is due to the difficulty of locating and filling all major cracks and openings that facilitate radon entry and the fact that significant radon movement may occur in cracks as small as 0.5 mm. It is not unusual for soil gas prevented from moving through its normal entry routes to flow through previously minor ones after normal routes are sealed.

The sealing of wall joints, substructure cracks, and penetrations is important and effective when it is coupled with measures such as subsurface ventilation. Sealants used in the EPA-sponsored Clinton, New Jersey radon development and demonstration project included Acryl 60 by Thoroseal and urethane caulk. Acryl 60 was used for large openings such as cavities around electrical outlets, for patching block and slab holes, and for dug-out perimeters. Urethane caulking was used for sealing around fans, ducts, pipes, and slab penetrations.[53]

Soil Gas Ventilation

Typically, ventilation is used to control an indoor contamination problem after the contaminant is released by a source. (See Chapter 5.) However, it can also be used to prevent the entry of contaminants into building air. This is the case with radon and other soil gases, which are drawn into indoor spaces by suction forces. By exhausting soil gases from beneath and (in the case of basements) around building substructures, radon concentrations in indoor air can be reduced to acceptable levels. Since ventilation is applied directly to the source, i.e., soil gas, it can be described as a source control method. It is within that context that soil gas ventilation is treated in this chapter.

Henschel and Scott[54] evaluated three different soil gas ventilation approaches to mitigating high radon levels in 18 concrete block–basement homes in the Reading Prong area of southeastern Pennsylvania. These were to apply suction on the (1) void network within the concrete basement block walls, (2) drain tile system, and (3) aggregate beneath the concrete floor system, and are illustrated diagrammatically in Figures 2.5, 2.6, and 2.7.

In Figure 2.5, pressure is reduced in walls by one or two suction points per wall (including interior walls). This system pulls basement air through unsealed cracks and block pores, preventing soil gas from entering basements through these routes. Since there is contact between block voids and sub-slab aggregate via the mortar joint in the wall, some portion of the sub-slab is also ventilated. For wall suction to be effective, all major openings in the block wall must be closed. Uncapped voids on the top of the wall are closed by mortaring them shut or filling them with polyurethane foam. Without closure of major openings, the fan's capacity would be spent on drawing basement or outdoor air through those openings and would therefore not be effective in

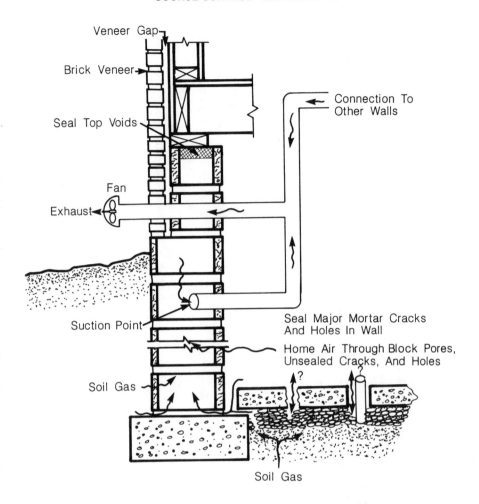

Figure 2.5.   Basement block ventilation for radon control.[54]

removing soil gas from the void network. Additionally, the resulting depressurization of the basement would cause radon inflow at points not associated with the wall (e.g., basement floor) and cause flue gas spillage from combustion appliances in the basement.

Test results for five homes where suction was applied to individual suction points are summarized in Table 2.3. Data are also presented for two houses in which suction was applied to a specially constructed baseboard duct. In the latter, a sheet-metal fabricated duct was sealed over the wall/joint around the entire perimeter of the basement. Holes were drilled into block voids and allowed to vent into the duct on which suction was drawn. This approach is recommended for houses with French drains. French drains consist of 1- to 2-

Table 2.3 Efficacy of Radon Reduction Associated with Block Wall Ventilation[54]

Home No.	Summer 1985 Performance					Winter 1985–86	Comments
	WL—System Off		WL—System On		% Reduction	WL—System On	
	Range	Mean	Range	Mean		Range	
Individual Suction Points							
3A	4.2–7.4	6.2	0.005–0.01	0.005	99+	0.005–0.01	Major wall openings amenable to effective closure. Includes sub-slab suction.
5	0–0.01	0.008	0.14–0.25				Voids in top blocks not effectively closed. Floor drain soil gas source untreated.
6	0.14–0.24	0.20	0.01–0.05	0.032	84	0.30–0.53	Top voids are not effectively closed. Brick veneer on 4 walls.
7	0.02–0.70	0.28	0.01–0.06	0.02	93	0.54–2.5	Veneer on 4 walls. Top voids in interior wall not closed.
8	0.26–0.80	0.44	0.005–0.02	0.010	98	0.01–0.02	Major wall openings amenable to effective closure.
Baseboard Duct							
9	1.2–2.4	1.8	0.01–0.04	0.036	98	0.05–0.15	Contains French drain. About 20% of drain not covered, in inaccessible locations.
11	0.30		0.005–0.01		97	0.07–0.14	Contains French drain, completely covered. Veneer on 3 walls.

inch (25–50 mm) gaps between walls and floor slabs. A baseboard duct is more expensive and difficult to install than the suction point technique described previously.

In the seven homes for which performance data are presented in Table 2.3, all but one achieved substantial reduction (84–99%) in radon levels. In one home no reduction occurred. This extremely poor performance was believed to be due to the fact that the top wall voids were not adequately closed, causing basement depressurization and inflow of radon-laden soil gas through an untrapped floor drain.

Features that were observed to make effective closure of major wall openings difficult included (1) inaccessible uncapped voids in a portion of the top course of concrete wall blocks, (2) exterior brick veneer, which causes air leakage through the gap between the brick and block, and (3) the fireplace incorporated into the block wall, serving as a conduit for air to enter the void network.

The application of suction to drain tile (Figure 2.6) is designed to draw soil gas away from the bottom mortar joint (suspected of being one of the major routes of soil gas entry into wall voids and the house). The suction is drawn on the basement sump. A fan, riser, and water trap are installed when drain tiles empty to above-grade soakaways. Radon reduction percentages ranged from 84–96% in three homes where drain tiles completely surrounded houses to 75–98% reductions where drain tiles did not completely surround the house. In general, higher performance was observed in the former cases.

In a subsequent study, Henschel and Scott[55] suggested that the presence of a complete loop of perimeter drain tile would make the drain tile approach very desirable. This is because drain tiles allow suction to be drawn where it is needed the most in houses with concrete block basement walls, making it a highly effective radon mitigation measure, particularly in houses with very high radon levels (> 100 pCi/L). In general, the drain tile approach is the least expensive of the soil ventilation techniques; however, complete drains are not common.

In the sub-slab approach, loose aggregate below the slab is used to collect soil gas (Figure 2.7). Two suction points are installed in each slab. As evidenced in Table 2.4, the performance of sub-slab ventilation was generally less effective and more variable than suction applied to drain tile. Best performance occurred in those houses with an adequate layer of aggregate. In one home the sub-slab system was unable to prevent soil gas from moving up through uncapped wall voids when the basement was depressurized during the winter.

The house that showed 0% reduction had no aggregate under the suction points. Sub-slab aggregate presence, characteristics, and uniformity vary considerably from builder to builder and from house to house. In some cases, slabs are poured directly on bedrock, while others are poured over compacted fill or disturbed soil.[56] At present the only approach to determining the nature of the aggregate is drilling test holes.

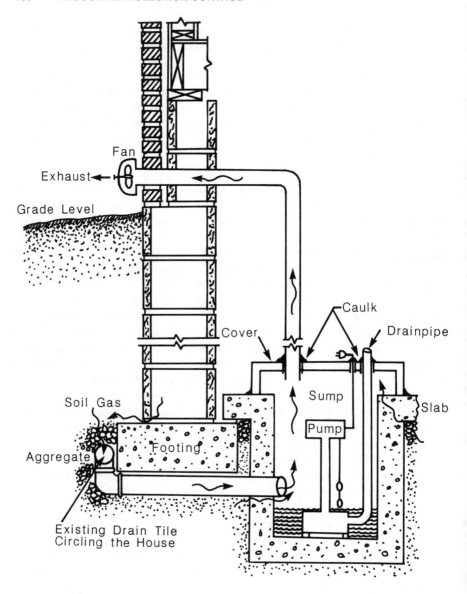

Figure 2.6.   Suction applied to slab drain tile system for radon control.[54]

In other sub-slab studies, Henschel and Scott[55] increased the number of suction points in basement slabs, placing them near the perimeter joint between the foundation wall and slab. They observed substantial reductions (usually greater than 90%) in radon levels in very high-radon houses. They concluded that with use of a suitable number of judiciously positioned suction

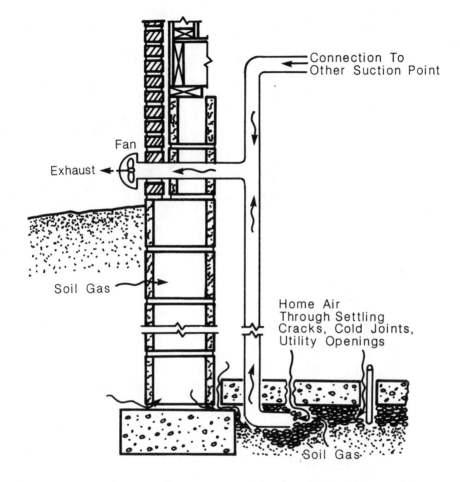

Figure 2.7. Suction applied to aggregate below floor slab for radon control.[54]

points there would be an increased likelihood that suction would be evenly distributed under the entire slab, treating all the major entry points and preventing the movement of soil gas up the void network. The number and location of suction points is a function of sub-slab permeability, location of major radon entry routes, and presence of openings which cannot be closed. When suction points are more limited and placed more remotely from the wall-floor joint, radon reduction is less predictable. Reasonable sub-slab permeability, however, is required to achieve good radon reduction performance.

With the application of soil ventilation techniques it is of critical importance that all major routes of radon entry into the wall and slab are closed. It is also necessary to trap or plug floor drains and cap sumps.[54,55] Best results are

Table 2.4 Efficacy of Radon Reduction Associated with Sub-Slab Ventilation[54]

Home No.	Summer 1985 Performance					Winter 1985–86		Comments
	WL—System Off		WL—System On		% Reduction	WL—System On	% Reduction	
	Range	Mean	Range	Mean		Range		
Individual Suction Points								
1	0.63–1.1	0.7	0.54–0.94	0.7	0			No aggregate under slab. System not operating in winter. Two suction points in slab.
2	0.44–2.5	1.2	0.15–0.29	0.17	86			Good aggregate; suction extends under entire slab. Two suction points in slab.
2A	0.44–2.5	1.2	0.06–0.23	0.15	88	1.8–7.4		Home 2 with suction on one wall in addition to sub-slab suction. Brick veneer on 1 wall. Has fireplace.
3	4.2–7.4	6.2	2.5–2.8	2.6	58			Reasonable aggregate. Home 3A but with sub-slab suction only (no wall suction). One suction point in slab.
4	0.2		0.03		85			

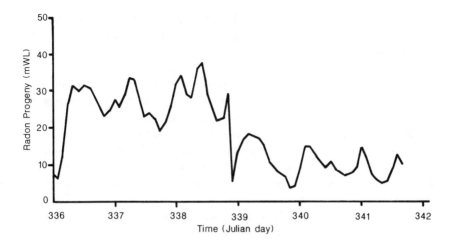

Figure 2.8.   Effect of sump venting on radon levels.[51]

achieved when suction fans used are capable of maintaining 170 pascals at the suction point, with soil gas flows in the range of 20–70 L/sec.

House-Specific Techniques

Evaluations of soil gas ventilation techniques on radon levels in residential housing structures have also been reported by Nitschke et al.,[51] Osborne,[57] and Turk et al.[52] Nitschke et al. also observed that sub-slab soil gas ventilation coupled with sealing openings was the most cost-effective technique applicable to houses with below-grade floors and walls with high to very high radon levels. The effect of sump venting, seen in Figure 2.8, decreased when houses were under negative pressure due to indoor/outdoor temperature differences and elevated wind speeds. Under such circumstances, competition for soil gas occurs between the sump and depressurized basement, resulting in reduced soil gas ventilation effectiveness. Sump ventilation was ineffective when it was not coupled with sealing cracks and other penetrations.

Osborne[57] applied a variety of house-specific mitigation techniques, including soil gas ventilation, to 10 high radon–concentration houses in Clinton, New Jersey. Four of the houses were split-level, with a combination of block wall basement and slab-on-grade substructure. The application of sub-slab ventilation was complicated by the fact that forced-air heating ducts were present beneath the slab. This required the sealing of ducts at the registers and installation of new heating ducts in attics. A plastic pipe with an in-line fan was then connected to the sub-slab heating ducts and vented to the outside. The application of soil gas ventilation, coupled with other measures such as sealing cracks, capping openings, relocating heating ducts, etc., resulted in radon mitigation performance levels on the order of 98.3–99.8% radon level control. Based on this 10-home demonstration project, EPA developed 20 house-specific radon reduction plans to assist homeowners with different floor plans.[57]

Although soil gas ventilation represents one of the most effective radon mitigation measures (particularly in high-radon houses), it has its disadvantages. For example, finished basements limit access to radon entry points and thus require considerable additional expense. The installation of pipes in concrete walls can be both physically difficult and costly. Performance of the system depends on the operation of blower fans. If homeowners turn these off, or fail to maintain and replace them when they periodically wear out, the system becomes inoperative.

Soil gas ventilation techniques can also be applied to crawl spaces. Turk et al.[52] applied mechanical exhaust ventilation to crawl spaces in two houses with both a basement and a crawl space. When it was applied alone to one house, the radon concentration was reduced from 23.3 to 10.8 pCi/L. Combined with other modifications to the crawl space in a second house, radon concentration was reduced from 49.9 to 19.9 pCi/L. Substantial additional reductions were achieved when mitigation measures were applied to basement areas.

EPA[58] has estimated installation costs for soil gas ventilation systems for concrete block and slab floor substructures. From EPA's experience, a pipe-wall ventilation system (assuming reasonably accessible top voids, no exterior veneer, and no fireplace structure) would cost approximately $2500 if installed by a contractor. For a baseboard duct system, the cost would be about $5000. Ventilation of the sub-slab with a piping network would cost anywhere from $2000 to $7500, depending on the extent of the piping network installed.

Pressurization

Positive pressure techniques can also provide effective radon control. Pressurization or overpressurization reverses the indoor-outdoor pressure differentials that cause soil gas flow into basements and other substructures. As can be seen in Figure 2.9, effectiveness increases as overpressurization increases. Pressurization was achieved by drawing air into basements from a heated upper floor.[52] Blower fan flow rates of 118–189 L/sec were required to develop a 3-pascal overpressurization. Effective use of this technique necessitated the sealing of holes and cracks in the basement walls and ceiling. It is only applicable to basements with tight shells and closeable doors. Overpressurization can only be effective if occupants keep windows and doors closed. A possible problem associated with the application of overpressurization is excessive upstairs depressurization, which can cause flue-gas spillage from combustion appliances and drafts from infiltrating air.

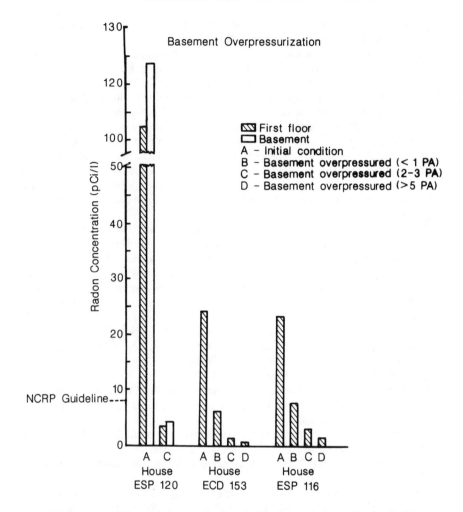

Figure 2.9. Effect of basement overpressurization on indoor radon levels.[52]

REFERENCES

1. EPA. 1985. "Guidance for Controlling Asbestos-Containing Materials in Buildings." EPA-560/5-85-024. Office of Pesticides & Toxic Substances, Washington, D.C.
2. EPA. 1985. "Asbestos in Buildings—Guidance for Service and Maintenance Personnel." EPA-560/15-85-018. Office of Pesticides & Toxic Substances, Washington, D.C.
3. EPA. 1984. "National Emission Standards for Hazardous Air Pollutants. Asbestos Regulations." *Fed. Reg.* 49:13661. April 5.

4. EPA. 1981. "Guidelines for the Use of Encapsulants on Asbestos-Containing Materials." Office of Pesticides and Toxic Substances.
5. 1976. "Asbestos Regulations." *Code of Federal Regulations.* 29 CFR 1910.1001.
6. EPA. 1985. "Measuring Airborne Asbestos Following an Abatement Action." Office of Research & Development/Office of Toxic Substances.
7. Apte, M.G. and G.W. Traynor. 1986. "Comparison of Pollutant Emission Rates from Unvented Kerosene and Gas Space Heaters." 405–414. In: *Proceedings of IAQ '86: Managing Indoor Air for Health and Energy Conservation.* American Society of Heating, Refrigerating and Air-Conditioning Engineers. Atlanta.
8. Leaderer, B.P. et al. 1984. "Residential Exposures to NO_2, SO_2 and HCHO Associated with Unvented Kerosene Space Heaters, Gas Appliances and Side-Stream Tobacco Smoke." 151–156. In: *Proceedings of the Third International Conference on Indoor Air Quality and Climate.* Swedish Council for Building Research. Stockholm. Vol. 4.
9. Anonymous. 1982. "Are Kerosene Heaters Safe?" *Cons. Rep.* 47:499–507.
10. Traynor, G.W. et al. 1985. "Indoor Air Pollution Due to Emissions from Wood-Burning Stoves." LBL-17854. Lawrence Berkeley Laboratory, Berkeley, CA.
11. Knight, C.V. et al. 1986. "Indoor Air Quality Related to Wood Heaters." 430–447. In: *Proceedings of IAQ '86: Managing Indoor Air for Health and Energy Conservation.* American Society of Heating, Refrigerating and Air-Conditioning Engineers. Atlanta.
12. Kaurakka, P. et al. 1987. "Assessment and Control of Indoor Air Pollution Resulting from Wood-Burning Appliance Use." 425–429. In: *Proceedings of the Fourth International Conference on Indoor Air Quality and Climate.* Institute for Water, Soil and Air Hygiene. West Berlin. Vol. 1.
13. American Society of Heating, Refrigerating and Air-Conditioning Engineers. 1983. "Chimney, Gas Vent, and Fireplace Systems." 27.1–27.29. *Equipment Handbook.* ASHRAE, Atlanta.
14. Traynor, G.W. et al. 1982. "The Effects of Ventilation on Residential Air Pollution Due to Emissions From a Gas-Fired Range." *Environ. Int.* 8:447–452.
15. Moschandreas, D.J. et al. 1986. "Emission Rates From Unvented Gas Appliances." *Environ. Int.* 12:247–253.
16. Dewerth, S.W. and W.G. Sterbik. 1985. "Development of Advanced Residential Range Burners with Low NO_x Characteristics." Gas Research Institute, Chicago. GRI-85/0204.
17. Shakula, K.C. et al. 1985. "Development of an Efficient, Low-NO_x Domestic Gas Range Cooktop. Phase II." Gas Research Institute, Chicago. GRI-85-0080.
18. Baker, P.E. et al. 1987. "An Overview of the Residential Indoor Air Quality Characterization Study of Nitrogen Dioxide." 395–399. In: *Proceedings of the Fourth International Conference on Indoor Air Quality and Climate.* Institute for Water, Soil and Air Hygiene. West Berlin. Vol. 1.
19. Relwani, S.M. et al. 1986. "Effects of Operational Factors on Pollutant Emission Rates from Residential Gas Appliances." *JAPCA* 36:1233–1237.
20. Moschandreas, D.J. and S.M. Relwani. 1987. "Field Measurements of NO_2 Gas Range-Top Burner Emission Rates." 343–348. In: *Proceedings of the Fourth International Conference on Indoor Air Quality and Climate.* Institute for Water, Soil and Air Hygiene. West Berlin. Vol. 1.
21. Renes, S. et al. 1985. "An Evaluation of Sink Terms in Removing NO_2 and SO_2 from Indoor Air." *Clima. 2000* 4:221–225.

22. Yamanaka, S. 1984. "Decay Rates of Nitrogen Oxides in a Typical Japanese Living Room." *Environ. Sci. Technol.* 18:566–570.
23. Miyazaki, T. 1984. "Adsorption Characteristics of NO_x by Several Kinds of Interior Materials." 103–110. In: *Proceedings of the Third International Conference on Indoor Air Quality and Climate.* Swedish Council for Building Research. Stockholm. Vol. 4.
24. Billick, I.H. and N.L. Nagda. 1987. "Reaction Decay of Nitrogen Dioxide." 311–315. In: *Proceedings of the Fourth International Conference on Indoor Air Quality and Climate.* Institute for Water, Soil and Air Hygiene. West Berlin. Vol. 1.
25. Spicer, C.W. et al. 1986. "Removal of Nitrogen Dioxide from Indoor Air by Residential Materials." 584–590. In: *Proceedings of IAQ '86: Managing Indoor Air for Health and Energy Conservation.* American Society of Heating, Refrigerating and Air-Conditioning Engineers. Atlanta.
26. Baker, P.E. et al. 1987. "Evaluation of Housing and Appliance Characteristics With Elevated Indoor Levels of Nitrogen Dioxide." 390–394. In: *Proceedings of the Fourth International Conference on Indoor Air Quality and Climate.* Institute for Water, Soil and Air Hygiene. West Berlin. Vol. 1.
27. Energy, Mines & Resources Canada. 1983. "Combustion Air Requirements for Well-Sealed Houses." Energy Conservation Techniques for Buildings. Ottawa, Canada. OR-82-3.1
28. Wilson, A.G. et al. 1986. "House Depressurization and Flue Gas Spillage." 417–429. In: *Proceedings of IAQ '86: Managing Indoor Air for Health and Energy Conservation.* American Society of Heating, Refrigerating and Air-Conditioning Engineers. Atlanta.
29. Moffatt, S. 1986. "Combustion Ventilation Hazards in Housing—Failure Mechanisms, Identification Technologies, and Remedial Measures." 425–436. In: D.S. Walkinshaw (Ed.). *Transactions: Indoor Air Quality in Cold Climates.* Air Pollution Control Association, Pittsburgh.
30. Swinton, M.C. and J.H. White. 1986. "Chimney Backdrafting in Houses: Identifying the Critical Conditions." 413–424. In: D.S. Walkinshaw (Ed.). *Transactions: Indoor Air Quality in Cold Climates.* Air Pollution Contol Association, Pittsburgh.
31. Tanner, A.B. 1986. "Geological Factors that Influence Radon Availability." 1–12. In: *Proceedings: Indoor Radon.* APCA Specialty Conference. Air Pollution Control Association, Pittsburgh.
32. Rogers, V.C. et al. 1984. *Radon Attenuation Handbook for Uranium Tailings Cover Design.* Rogers & Assoc. Engineering Corp. NUREG/CR-3533.
33. Fitzgerald, J.E. et al. 1976. "A Preliminary Evaluation of the Control of Indoor Radon Daughter Levels in New Structures." EPA-520/4-76-018.
34. Culot, M.V.J. et al. 1973. "Radon Progeny Control in Buildings." EPA-RO 1 EC00015.3.
35. Guimond, R.J. et al. 1979. "Indoor Radiation Exposure to Radium-226 in Florida Phosphate Lands." EPA-520/4-78-013.
36. Scott, A.G. and W.O. Findlay. 1983. "Demonstration of Remedial Techniques Against Radon in Houses on Florida Phosphate Lands." American Atcon, Inc.
37. Nazaroff, W.W. and S.M. Doyle. 1983. "Radon Entry into Houses Having Crawl Spaces." LBL-16637. Lawrence Berkeley Laboratory, Berkeley, CA.
38. Erickson, S.O. et al. 1984. "Modified Technology in New Constructions and Cost

Effective Remedial Action in Existing Structures to Prevent Infiltration of Soil Gas Carrying Radon." 153–158. In: *Proceedings of the Third International Conference on Indoor Air Quality and Climate*. Swedish Council for Building Research. Stockholm. Vol. 5.

39. Atomic Energy Board. 1979. "Report on Investigation and Implementation of Remedial Measures for the Radiation Reduction and Radioactive Decontamination of Elliot Lake, Ontario." Dilworth, Secord, Meagher & Assoc., Ltd. Toronto.

40. Harrje, D.T. and K.J. Gadsby. 1986. "Practical Engineering Solutions for Optimizing Energy Conservation and Indoor Air Quality in Residential Buildings." 333–341. In: *Proceedings of IAQ '86: Managing Indoor Air for Health and Energy Conservation*. American Society of Heating, Refrigerating and Air-Conditioning Engineers. Atlanta.

41. George, A.C. et al. 1983. "Radon Daughter Plateout. I. Measurements." *Health Physics* 45:439–444.

42. Scott, A.G. 1983. "Radon Daughter Deposition Velocities Estimated from Field Measurements." *Health Physics* 45:481–485.

43. Holub, R.F. et al. 1979. "The Reduction of Airborne Radon Daughter Concentration by Plate-Out on an Air-Mixing Fan." *Health Physics* 36:497–504.

44. Rudnick, S.N. et al. 1983. "Effect of Plateout, Air Motion and Dust Removal on Radon Decay Product Concentration in a Simulated Residence." *Health Physics* 45:463–470.

45. Bigu, J. 1983. "On the Effect of Negative Ion Generator and a Mixing Fan on the Plate-Out of Radon Decay Products in a Radon Box." *Health Physics* 44:259–266.

46. Rector, H.E. et al. 1985. "Impact of Energy Conservation Measures on Radon Progeny Concentrations: A Controlled Study." *ASHRAE Trans.* 91. Pt. 2. Atlanta.

47. Eichholz, G.G. et al. 1980. "Control of Radon Emanating From Building Materials by Surface Coating." *Health Physics* 39:301–304.

48. Pohl-Ruling, J. et al. 1980. "Investigation on the Suitability of Various Materials as (222) Radon Diffusion Barriers." *Health Physics* 39:299–301.

49. Auxier, J.A. et al. 1974. "Preliminary Studies of the Effects of Sealants on Radon Emanation from Concrete." *Health Physics* 27:390–392.

50. Bedrosian, P.H. et al. 1974. "Sealants for Sources Emanating Rn-222." *Health Physics* 27:387–389.

51. Nitschke, I.A. et al. 1985. "Indoor Air Quality, Infiltration and Ventilation in Residential Buildings. Final Report." Prepared for New York ERDA & Niagara Mohawk Power Corp. W.S. Fleming & Associates, Inc. Syracuse, NY.

52. Turk, B.H. et al. 1986. "Radon and Remedial Action in Spokane River Valley Residences: An Interim Report." LBL-21399. Lawrence Berkeley Laboratory, Berkeley, CA.

53. Michaels, L.D. et al. 1987. "Development and Demonstration of Indoor Radon Reduction Measures for 10 Homes in Clinton, New Jersey." EPA-600/18-87-027.

54. Henschel, D.B. and A.G. Scott. 1986. "The EPA Program to Demonstrate Mitigation Measures for Indoor Radon: Initial Results." 110–121. In: *Proceedings: Indoor Radon*. APCA Specialty Conference. Air Pollution Control Association, Pittsburgh.

55. Henschel, D.B. and A.G. Scott. 1987. "Some Results From the Demonstration of Indoor Radon Reduction Measures In Block Houses." 340–346. In: *Proceedings of*

the Fourth International Conference on Indoor Air Quality and Climate. Institute for Water, Soil and Air Hygiene. West Berlin. Vol. 2.

56. Osborne, M.C. 1987. "Four Common Diagnostic Problems that Inhibit Radon Mitigation." *JAPCA* 37:604–606.

57. Osborne, M.C. 1987. "Resolving the Radon Problem in Clinton, New Jersey Houses." 305–309. In: *Proceedings of the Fourth International Conference on Indoor Air Quality and Climate.* Institute for Water, Soil and Air Hygiene. West Berlin. Vol. 2.

58. EPA. 1986. "Radon Reduction Techniques for Detached Houses. Technical Guidance." EPA-625/5–86–019.

3	SOURCE CONTROL—
	ORGANIC CONTAMINANTS

In the previous chapter, source control was discussed within the context of those contaminants that are for the most part inorganic. Here the focus is placed on controlling organic contaminants at the source.

FORMALDEHYDE

Formaldehyde exposures in residential and other indoor environments can be reduced by a variety of source control techniques, including both those applied prior to the introduction of formaldehyde-releasing products into residential and other indoor environments and those applied on a retrofit basis. Preintroduction control techniques include avoidance (total, near total, and selective) and product improvement. Postintroduction source control techniques include source removal and substitution, source barriers, ammonia fumigation, and climate control. Source aging could, in theory, be applied prior to product introduction as well as afterward.

Avoidance

Total/Near Total Avoidance

In theory, the best way of minimizing exposures to formaldehyde in indoor air spaces is by avoiding or not using formaldehyde-releasing products with high emission rates, thereby avoiding the potential for elevated formaldehyde levels. Products with such potential include wood products bonded with urea-formaldehyde resins, urea-formaldehyde foam insulation (UFFI) and urea-formaldehyde wood finishes.

Wood products to avoid include hardwood plywood, particleboard, and medium-density fiberboard (MDF). Principal uses of hardwood plywood are

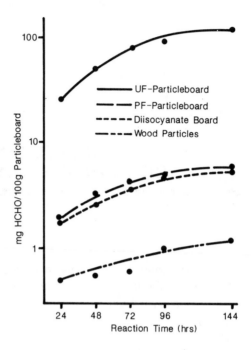

Figure 3.1. Formaldehyde emissions associated with urea-formaldehyde, phenol-formaldehyde and diisocyanate-bonded particleboards.[2]

decorative wall paneling, doors, cabinets, and furniture. Non-formaldehyde-emitting or low-emitting wall paneling substitutes include gypsum board, decorative gypsum board panels, and decorative hardboard panels. Solid wood or metal cabinets can be substituted for hardwood plywood cabinets. All-wood, metal, or plastic furniture may substitute for formaldehyde-releasing hardwood plywood. Doors may be solid wood or metal.

Uses of particleboard include subflooring, decorative wall paneling, cabinetry (core material and shelving), cabinet tops, closet shelving, and core material in doors and furniture. Because of high formaldehyde concentrations associated with particleboard as subflooring, the avoidance of such use is particularly advisable. Acceptable substitutes (relative to formaldehyde emissions) are softwood plywood, waferboard, oriented-strand board, iso-board, and phenol-formaldehyde (P-F) bonded particleboard. Formaldehyde emissions for iso-board and P-F particleboard are only a fraction of those associated with urea-formaldehyde bonded wood products.[1] Differences can be seen in Figure 3.1.[2] These materials can also serve as substitutes for particleboard ordinarily used for shelving and cabinet tops. Substitutes for particleboard used as decorative wall covering and components used in the manufacture of doors, cabinets, and furniture are the same as those listed for hardwood plywood above.

Table 3.1   Effect of Loading Factor on Formaldehyde Levels in a Conventional House[6]

Source Condition	Formaldehyde Concentration (ppm)[a]	% of Maximum
Baseline	0.03 a	
100% wall surface paneled	0.45 b	
50% wall surface paneled	0.31 c	69
25% wall surface paneled	0.19 d	38
100% floor surface with particleboard	0.26 b	
50% floor surface with particleboard	0.16 c	62
25% floor surface with particleboard	0.11 d	35

[a]Means identified by dissimilar letters are significantly different at a probability level of 0.01.

Use of MDF is limited to cabinet and furniture manufacture. Substitutes include cabinetry and furniture made out of solid wood, metal, or wood products completely overlain with formaldehyde barrier materials such as vinyl or products generically similar to Formica. Any breaks in the barrier, such as drill holes and edges exposed by saw cuts, will cause significant decreases in barrier effectiveness.[3] Alternative wood products should not have been finished with U-F resin–based coating materials.

Selective Avoidance

Product avoidance can also be selective. For example, significantly higher formaldehyde levels will occur when surface-to-volume (product surface to indoor air volume) ratios of formaldehyde-releasing wood products are high.[4] By minimizing this ratio, described commonly as the loading factor, potentially health-affecting indoor formaldehyde levels can be avoided without absolutely excluding the use of a product with desirable characteristics. As an example, the effect of a single piece of furniture made of hardwood plywood, medium-density fiberboard, or particleboard on indoor formaldehyde levels may be negligible and therefore its avoidance or exclusion may not be warranted. On the other hand, a high loading of hardwood plywood, MDF, or particleboard furniture in a closed bedroom may result in unacceptably high formaldehyde concentrations and exposures. Formaldehyde levels of 0.10 ppm and 0.70 ppm have been measured in bedrooms with high load factors of new hardwood plywood and MDF furniture, respectively.[5]

Other applications of the loading factor principle include the use of particleboard as subflooring and hardwood plywood as decorative wall covering. The use of particleboard on most floor surfaces may produce unacceptably high formaldehyde concentrations, whereas use in a single room may result in levels which are acceptable. The same holds true for hardwood plywood. The effect of different loading factors on formaldehyde levels associated with particleboard and hardwood plywood paneling applied singly in a conventional house can be seen in Table 3.1.[6] Doubling the amount of formaldehyde-

Table 3.2 Equilibrium Formaldehyde Concentrations of Three Foam Insulation Samples[7]

Sample	Equilibrium Concentration (mg/mL)
A	5.10
B	4.68
C	2.85

releasing material was observed to slightly less than double formaldehyde levels. Because of the low availability of UFFI (which resulted from attempts by the Consumer Product Safety Commission to ban it and from the associated adverse publicity) it represents a relatively unlikely product choice for residential installation to consumers in the United States and Canada (where it has been banned). Therefore, avoidance of this product in North America is highly likely. Nevertheless, in Europe, particularly in Great Britain, it is still widely available and widely used.

In the United States, UFFI is experiencing a small resurgence as an insulating medium for concrete block and panels used in the construction of industrial and commercial buildings. In addition, a UFFI variant, comprised of a phenol-urea-formaldehyde tripolymer, is presently being marketed for retrofit residential insulation installation. Distributors claim that no formaldehyde is used in the manufacture of this product. As can be seen in Table 3.2, this product (sample C), compared to two conventional UFFI products, emitted lower quantities of free formaldehyde. Emission rates are nevertheless sufficient to produce indoor levels of 0.10–0.18 ppm immediately, and at least up to a year, after installation.[7] The characterization of this product as not being made from formaldehyde is misleading, since consumers may (and often do) purchase and install it in their homes believing it is free of formaldehyde. In such cases, the attempt to avoid a potent formaldehyde-releasing product is controverted by misleading distributor claims.

The use of fiberglass batts for sidewall insulation in new house construction, and cellulose for retrofit installations, is preferred. However, the use of cellulose for sidewall insulation suffers from the fact that blown-in cellulose has a tendency to settle, producing voids in the upper portion of wall cavities.

Urea-formaldehyde–based finishes are widely used in North America on wood cabinetry and furniture. To a lesser extent, urea-formaldehyde/melamine-formaldehyde finishes are used on hardwood floors. Numerous product substitutes are available, including a variety of varnish products. They do not, however, have the scratch and chip resistance that makes U-F resin-based finishes so attractive to manufacturers of wood cabinetry and furniture.

One of the major problems with the use of avoidance as a formaldehyde control technique is that it reduces consumer choices to products which may be more expensive and/or less aesthetically appealing. Solid hardwood furniture, for example, is too often prohibitively expensive. Veneer furniture with a wood, particleboard, or MDF core provides a means of providing an aesthetically appealing product at a relatively attractive price. Hardwood veneer cabinets have high aesthetic and social appeal. All-wood cabinets made of pine or other relatively inexpensive wood materials, or metal cabinetry, may have neither aesthetic nor social appeal.

Product Improvement

Significant reductions in potential formaldehyde exposures can be achieved by the development and use of low-emitting or non-emitting wood products. Low-emission U-F–based materials have been the choice of wood product manufacturers in North America, Europe, and Asia. The underlying premise is that production of low-emission products will reduce formaldehyde levels sufficiently in buildings to warrant continued use of U-F–based adhesives; odors will be obviated and health complaints reduced.

There are a variety of approaches to reducing formaldehyde emissions from U-F–based wood products such as hardwood plywood, particleboard, and MDF. These include changes in resin formulation and production variables, addition of scavenging compounds, attention to quality control, and a variety of post-production steps.

Resin Formulation

Significant reductions in formaldehyde emissions from U-F–based wood products have been achieved in North America and Europe since the late 1970s by changing the molar ratio of formaldehyde to urea (F:U ratio) in the resin formulation.[3,8,9] A decade ago, most U-F resins marketed as wood adhesives had F:U ratios in excess of 1.5:1. Such high ratios were used to ensure that sufficient formaldehyde was present to achieve crosslinkage of all primary, and most secondary, amino groups. The presence of unreacted and primary amino groups made U-F resins hygroscopic, an undesirable resin characteristic. Low-emission resins produced today are manufactured in three or more steps, with large excesses of formaldehyde used initially. Successive additions of urea bring the F:U ratio down sufficiently to retain unreacted amino groups, which act as scavengers of unreacted formaldehyde and formaldehyde produced by hydrolysis.[10] By 1984, molar ratios in particleboard products were reduced to a range of 1.15–1.3:1; hardwood plywood, 1.2–1.5:1; and MDF, 1.4–1.65:1.[8]

The use of low molar ratios of formaldehyde and urea has been the primary reason for the decline in formaldehyde emissions and levels associated with products shortly after they have been manufactured. The effect of changes in molar ratios on formaldehyde emission rates from newly manufactured particleboard has been reviewed by Sundin[3] and Myers.[11] A lowering of the molar ratio from 1.4:1 to 1.2:1 was observed to reduce emissions by 69%.

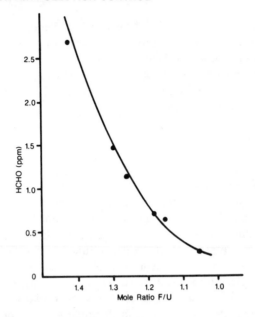

Figure 3.2. Effect of formaldehyde:urea molar ratios on formaldehyde emissions from urea-formaldehyde–bonded particleboard.[3]

Lowering the molar ratio further, from 1.2:1 to 1.05:1, reduced emissions by an additional 68%, with a total reduction in emission rates of 90% from a beginning molar ratio of 1.4:1 to a final ratio of 1.05:1.[3] The relationship between molar ratio and formaldehyde emissions can be seen in Figure 3.2.

Manufacturers and users of U-F resin systems as wood adhesives are reluctant to push the use of lower F:U ratios to reduce formaldehyde emissions too far lest they cause deterioration in strength properties of the adhesives and board products made from them. Additional reductions are therefore achieved by other means. These include more careful control of curing/production conditions, addition of a variety of additives (scavengers) to the resin formulation before and during board production, and application of a variety of postprocess treatments.

Curing/Production Conditions

Urea-formaldehyde adhesives are thermosetting. When subjected to heat (a process called curing), they undergo crosslinking and other changes. Free formaldehyde present in the resin occurs as a consequence of incomplete crosslinking during resin cure. This free formaldehyde is the most significant source of formaldehyde immediately after product manufacture.

Significant reductions in emissions of this free formaldehyde can be achieved by attention to curing and production conditions, including wood moisture and acidity, resin concentration, and press temperature and duration.[12] Desirable particleboard curing/production parameters include:

1. lowest possible moisture content in unblended particles
2. lowest possible moisture content in resin-blended particles
3. use of particle types designed to give open surfaces in boards
4. highest possible pressing temperature
5. longest possible pressing time

Formaldehyde is also released from the hydrolysis of formaldehyde-bearing species in U-F resins like methylene ureas, urea methylene ethers, and cellulose crosslinked species. The rate of hydrolysis of U-F bonded wood products can be reduced several fold during the curing/production process by neutralizing acidic conditions (e.g., from pH 3.0 to 6.5) and increasing cure time (from 5 minutes to 15 minutes at 150°C).[13] Low F:U ratio resins have low free formaldehyde contents and are as a consequence more susceptible to hydrolysis than are high F:U ration resins.[8]

Additives/Scavengers

A variety of reactive chemicals can be used in wood product manufacturing to bind unreacted formaldehyde, or that which is released during curing. Chemical additives scavenge free formaldehyde, forming stable complexes. They typically include urea, ammonium compounds, and sulfites. They are added to the resin or wood furnish or are components of the wax sizing.[3] The effect of the addition of reactive chemicals to furnish/veneer on formaldehyde emissions has been extensively reviewed by Myers.[13] Much of his review is based on the patent literature.

The use of additives suffers in part from the fact that they may interfere with curing and thus produce products with undesirable physical properties. To prevent this, furnish or veneer substrates may be treated with formaldehyde-reactive additives prior to or after coating the substrates with adhesive. In this way, availability of the formaldehyde-scavenging ingredient is reduced during curing while maintaining its longer-term availability for scavenging formaldehyde.[14,15]

Much attention has been focused on the use of urea as a scavenger. Desirable properties include its availability, low cost, water solubility, relative safety, and effectiveness as a scavenger. Its effectiveness can be seen in Figure 3.3, where a 0.4% urea addition to boards of two different emission rates is shown to reduce formaldehyde emissions by 45–60%.[15]

The use of urea as a scavenger has its limitations, however. It involves an additional increase in moisture, and use at higher levels may affect product physical properties. Meyer et al.[16] indicate that addition of excess urea makes the resin more susceptible to formaldehyde release by hydrolysis, since it increases the concentration of methylol end groups in the resin.

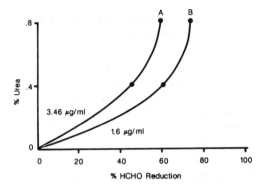

Figure 3.3.   Effect of urea additions on percentage formaldehyde reductions in particleboard.[15]

Postcure Treatments/Barriers

The use of reactive chemicals as formaldehyde scavengers can also be applied to products after they have been cured. These are best described as secondary treatments or aftertreatments. The effectiveness of a variety of chemical aftertreatments is summarized in Table 3.3.[17]

Chemical aftertreatments may include exposure to a reactive gas such as ammonia or painting or spraying wood panels. A variety of postcuring ammonia treatment processes have been developed and patented. These include the Verkor FD-EX, RYAB, Swedspan, BASF, and Weyerhaeuser processes. Application of the Swedspan and RYAB methods is reported to reduce formal-

Table 3.3   Effectiveness of a Variety of Chemical Aftertreatments Applied to Hardwood Plywood Veneer[17]

Treatment	Formaldehyde Emissions (ppm) (Japanese dessicator method)
Control	13.20
Polyurethane	0.50
Urea	0.07
$(NH_4)_2SO_3$	0.15
Na_2SO_3	0.00
$NaHSO_3$	0.25
$Na_2S_2O_4$	0.34
$Na_2S_2O_5$	0.26
NH_4HCO_3	0.37
$(NH_4)_2S_2O_3$	0.64
H_2NCONH_2	0.06

dehyde emissions by a factor of 4–10 in equilibrium tests and 2–3 in perforator tests made 1–2 months after treatment.[8]

Postcuring ammonia treatment is described here using the Verkor FD-EX system as an example. Cured, cooled, and trimmed particleboard panels are passed through two chambers that are sealed from each other. The first chamber is the entrance chamber; the second is the absorption chamber. In the latter, particleboard panels are exposed on all sides to a homogenous atmosphere of ammonia at controlled concentrations. Reaction of ammonia with free formaldehyde produces hexamethylenetetramine. The ammonia reacts with all free acids, thus stabilizing bonds and achieving a long-term controlled neutralization. Controlled ventilation removes most of the free ammonia from boards' surface layers. Finished boards leave the chamber through an exit seal with their mechanical properties unchanged.[18]

Surface-applied aftertreatments are also commonly used. Commercial treatments are typically applied by roller coaters. Treatment effectiveness depends on the treatment formulation, application rate, and nature of the product being treated. Most treatment formulations are believed to contain ammonia or urea compounds. They are applied at rates that achieve a 30–85% reduction in formaldehyde emissions.[9]

Most industrial-grade particleboard is laminated or coated on one or both sides when it is to be used for furniture or cabinets. Prime-and base-coated boards show a 20–30% reduction in formaldehyde emissions; conventional and UV-cured, filled boards, a 50% reduction; conventional and UV-cured, filled boards with base coat, a 75% reduction; and boards with UV fill and acrylic top coat, a 40% reduction.[15]

Decorative surface finishes on hardwood plywood can also have an effect on formaldehyde emissions. Groah et al.[19] evaluated the effect of a 2-mil vinyl film overlay on emissions from high-emitting hardwood plywood panels. Emissions were reduced by 90%.

Alternative Resin Systems

An attractive approach to reducing or eliminating formaldehyde emissions from wood products is the use of low-emitting or non-emitting resin adhesives. Alternative resins include phenol-formaldehyde, polyvinylacetate, lignosulfonates, methyl diisocyanates (MDI), and combinations of MDI and U-F resins.

Phenol-formaldehyde is used to bond softwood plywood, waferboard, and oriented-strand board products made to an indoor/outdoor grade. It is more stable than U-F and emits negligible quantities of free formaldehyde (Figure 3.1). It can be (and has been) used to manufacture particleboard. In 1976, P-F particleboard made up 10% of the particleboard produced in the common market countries of Europe. Its disadvantages relative to U-F particleboard include higher cost, a 5–10% higher density, and a high rate of thickness swelling.[20]

Phenol-formaldehyde resins are not suitable for hardwood plywood manufacture. Because face veneers are thin (typically 0.05–0.01 inches, 1.3–0.25 mm), a colorless glue line is required to prevent discoloration of the face. Phenol formaldehyde is reddish and has a tendency of bleeding through the face veneer and discoloring it. It is for this reason that U-F–based adhesives are so attractive.[9]

Low-emission or formaldehyde-free particleboards can be manufactured using isocyanate (4,4 diphenyl-methonediisocyanate) resins or combinations of isocyanate and U-F resins. Advantages of isocyanate resins, in addition to no formaldehyde emission (other than that associated with the wood chips themselves), include board densities which are 10–15% less than P-F boards of comparable strength, reduced press time, and high weather resistance. Isocyanate resins are considerably more expensive than U-F resins. However, this economic disadvantage is offset by a 20–30% reduction in resin needs, board density, and press time. The primary problem of isocyanate use in the manufacture of particleboard and other wood products is that it is too good an adhesive. As a consequence, resinated wood chips have a tendency to stick to the metal parts of presses.[20]

The problem of board/press sticking can be overcome by the use of U-F/isocyanate resin combinations and coatings that facilitate the release of boards from metal press parts. In the former, the core binder is an isocyanate resin with face layers containing conventional U-F or P-F resins. This combination obviates the problem of isocyanate sticking during pressing. Self-releasing isocyanate binders designed specifically for replacing U-F resins have been developed in both Europe and the United States.[21]

A major concern associated with the use of isocyanate resins is their potential for posing a health hazard by releasing unreacted methyl diisocyanate (a sensitizing agent more potent than formaldehyde) and hydrogen cyanide during a structural fire. Galbraith et al.[21] have reported that no isocyanate vapors were detectable when measured over a period of 14 days to one year in chamber studies with high loading rates. In combustion tests, boards manufactured with isocyanates were observed to release less hydrogen cyanide than U-F–bonded particleboards.

Despite the results reported by Galbraith et al., considerable uncertainty exists about the potential toxic hazards that may be associated with isocyanate binders used for interior-grade wood products. Based on its formaldehyde experience, the wood products industry is understandably concerned about the possibility of substituting a new toxic hazard for an old one, particularly when they believe that the hazards of formaldehyde have been exaggerated.

Particleboard made with isocyanate binders is available commercially in both Europe and the United States. In the latter, production is limited to one mill in Oregon.[22] The product is commonly referred to as iso-board.

Table 3.4   Formaldehyde Levels in Residences Before and After UFFI Removal[25]

Residence No.	Preremoval Concentration (ppm)	Postremoval Concentration (ppm)	% Change
1	0.12	0.09	− 25
2	0.13	0.15	+ 15
3	0.11	0.10	− 5
4	0.14	0.16	+ 17

Retrofit Control Measures

Up to this point, source control measures have been described in the context of avoiding or reducing exposures to formaldehyde before potential formaldehyde-releasing products are brought into indoor spaces. Because of past and continuing use of formaldehyde-emitting products of high contamination potential and associated health problems, a need exists for effective control methods which can be applied on a retrofit basis. Applicable source control methods include source removal, product treatment, and climate control.

Source Removal

In theory, the most effective method of reducing indoor formaldehyde levels is to remove the source or sources from the building environment. Studies on the effectiveness of source removal on formaldehyde levels associated with particleboard or hardwood plywood wall paneling present singly were reported by Godish and Rouch.[6] As can be seen in Table 3.1, a decrease in product loading is equivalent to an increase in product removal. These data suggest that a 50% reduction in source materials will provide an approximate 35–40% reduction in formaldehyde levels.

Removal of urea-formaldehyde foam insulation has been widely practiced in Canada and, to a lesser degree, in the United States. In most instances, removal practices are based on detailed recommendations of the National Research Council of Canada in their Building Practice Note No. 23.[23] In these recommendations, wall cavities are treated with a 3% solution of sodium bisulfite after UFFI removal. This treatment is necessary to prevent formaldehyde release from the U-F resin–impregnated wall studs, brick, gypsum board, etc.

Little information is available on the effectiveness of UFFI removal in reducing formaldehyde levels. Broder et al.[24] reported that significant decreases in reported health complaints were associated with UFFI removal in Canadian houses. However, they observed no significant differences in pre- and postremoval formaldehyde levels. In four houses in which formaldehyde measurements were made before and after UFFI removal, little or no reduction in formaldehyde concentrations was observed by Godish[25] (Table 3.4). In

Table 3.5   Model Prediction of Formaldehyde Levels (ppm) Associated with Stepwise Removal/Treatment of Formaldehyde Sources in a Conventional Residence Containing a Variety of Major Formaldehyde Emitters[31]

Air Change Rate (ACH)	All Emitters[a]	UFFI Removed	Tile Flooring Installed	Replacement with Low Emission Paneling[b]	MDF Removed
0.25	0.38	0.32	0.28	0.20	0.13
0.50	0.28	0.21	0.17	0.12	0.07
1.00	0.18	0.13	0.10	0.07	0.04

[a]1. Urea-formaldehyde foam insulation: 1.5 walls, 14 m2.
2. Carpet-covered particleboard underlayment: 16 m2.
3. Decorative hardwood plywood paneling: 1 wall, 10 m2 print overlay.
4. Cupboards, industrial particleboard: 2 m2.
[b]Replacement paneling with domestic veneer overlay.

two of the four houses, formaldehyde levels were actually observed to have increased slightly after foam removal. The latter phenomenon is not uncommonly reported on an anecdotal basis. It is most likely associated with disturbance of the foam material and the increase in surface area of resin-impregnated wall cavity materials available for formaldehyde release.

Formaldehyde sources in a multisource environment interact, resulting in a variety of complex outcomes. Interaction effects observed in laboratory chamber and whole-house environments have included suppression of formaldehyde emissions from one source by another, slight augmentations of formaldehyde levels associated with source combinations, and complete additivity.[26-30] Complete additivity for source combinations has only been reported for source combination studies of particleboard subflooring and hardwood plywood paneling conducted in an uninsulated experimental house.[30] Such additivity was not observed with the same source materials evaluated under laboratory chamber conditions, and is at variance with other reported laboratory chamber studies.[26-29] Apparently, complete additivity may occur under some circumstances; however, how common it is under real-world conditions is not known. A model to predict the effect of step-by-step removal or covering of formaldehyde sources in multisource environments has been developed by Matthews et al.[31] The model is based on principles of molecular diffusion and the effect that formaldehyde levels have on emission rates from U-F resin–based sources. Model predictions of formaldehyde reductions associated with stepwise source removal/covering are summarized in Table 3.5.

Because of interaction effects, the apparent simplicity of source removal as a control measure is unfortunately deceptive. Removal of a simple major source in a multisource environment will, in many cases, reduce formaldehyde levels only marginally. Therefore, source removal must include identification and removal of all major sources to be effective. Major sources include particleboard subflooring, hardwood plywood wall paneling, UFFI, cabinetry, and, in some cases, furniture. The reported apparent ineffectiveness of UFFI removal may in part be due to interaction effects. It may also be partially caused by the continued presence of U-F resin impregnated in wall materials and by the close proximity in time after UFFI removal that formaldehyde measurements are made.

Source Treatment

In general, source treatment represents a potentially lower-cost approach to retrofit formaldehyde control than source removal. In theory, it should also be less demanding, both timewise and physically. There are two approaches to source treatment. The first involves surface treatments, that is, application of coating materials or interposition of physical barriers between the source and surrounding airspace. Such chemical and physical barriers prevent the release or penetration of formaldehyde through the barrier material. In some instances, treatment coating materials prevent the penetration of moisture into the source material, thus reducing formaldehyde emissions due to hydrolysis. The second approach is to treat the source material with ammonia gas, applying the principle that some wood product manufacturers use to reduce formaldehyde emissions after board curing. (See "Postcure Treatments/Barriers," above.) This approach is called ammonia fumigation.

Surface treatment/surface barriers. Source treatment by the application of coatings and a variety of physical barriers is most appropriate when the major sources of formaldehyde are unfinished wood products or wood products with one or more unfinished surfaces. Applications may include particleboard subflooring/decking, hardwood plywood panel backs, particleboard shelving, cabinet joints and edges, particleboard countertop undersurfaces, and unfinished furniture components.

The efficacy of surface treatments has been evaluated in tests conducted in small laboratory chambers, large climate-controlled chambers, and under whole-house conditions.

Haneto[32] has reported on the relative effectiveness of a variety of surface treatments and overlays applied to particleboard in chamber studies. Results for surface finishes and overlays commonly used in the building trade are summarized in Table 3.6; results obtained with overlays used by the furniture and joinery industries are presented in Table 3.7. Alkyd resin paint and vinyl wall paper were observed to produce significant reductions in formaldehyde emissions and chamber formaldehyde concentrations (>85%). Needle felt carpeting, cushion floor covering, and carpeting with foam backing surprisingly were reported to reduce formaldehyde levels to a high degree (90–95%) The two most effective furniture/cabinet material overlays were a paper/plastic laminate and short-cycle melamine-faced paper, with reductions of 67% and 51%, respectively.

Seymour[33] evaluated the effectiveness of water-based paints with formaldehyde scavengers added on emissions from both new (currently produced)

Table 3.6   Effect of Surface Treatments/Barriers on Formaldehyde Levels Associated with Particleboard Panels Under Controlled Chamber Conditions[32]

Finish/Barrier Type	Steady State Concentration (ppm)	
	Calculated	Measured
Reference board	1.39	1.39
Alkyd paint	0.08	0.09
Latex paint	0.80	1.12
Wallpaper	0.76	1.37
Wallpaper (vinyl)	0.09	0.22
Carpeting (needle felt)	0.05	—
Floor cushion	0.04	—
Carpeting with foam backing	0.07	—

and one-year-old hardwood plywood samples. Sealing with formaldehyde-scavenging coatings was observed to reduce formaldehyde emissions from current production hardwood plywood and particleboard by 88–91% and 93–97%, respectively, and 97–98% for the one-year-old hardwood plywood samples. Seymour concluded that formaldehyde-scavenging paints were more effective on older materials, suggesting that such sealants reduced formaldehyde emissions from U-F resin hydrolysis rather than by a diffusion barrier effect. In one formulation the scavenger was known to be 5% urea added to a 42–45% solids acrylic emulsion.

Sundin[2] reported on the effectiveness of a variety of surface treatments and covers on formaldehyde release from both low-emission and high-emission particleboards. Significant reductions in formaldehyde emissions were observed for vinyl carpet, vinyl wallpaper, formaldehyde-absorbent paint, short-cycle melamine-formaldehyde paper, and polyethylene foil. Effectiveness was observed to decrease with time over an eight-day evaluation period (Figure 3.4).

Matthews et al.[34] investigated the effectiveness of a variety of potential formaldehyde permeation barriers over particleboard underlayment under controlled whole-house conditions (Table 3.8). These were vinyl linoleum, 6-mil polyethylene, and two carpet/cushion combinations. Vinyl linoleum and 6-mil polyethylene were very effective, resulting in emission rate and formaldehyde level reductions of 90% and 80%, respectively, in one set of evaluations.

Table 3.7   Effect of Overlay Materials on Formaldehyde Levels Associated with Particleboard Panels Under Controlled Chamber Conditions[32]

Overlay Type	Calculated Steady State Concentration (ppm)
Reference board	0.49
Melamine short cycle paper	0.21
Paper/plastic laminate	0.16
Finish foil, 100 g/m2	0.40
Finish foil, 50 g/m2	0.57

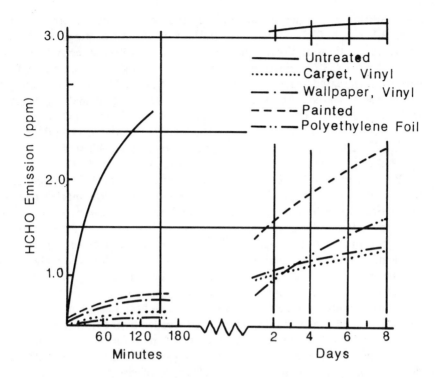

Figure 3.4. Effectiveness of surface treatments and covers in reducing formaldehyde emissions from particleboard.2

Unlike the studies of Haneto,[32] carpet and cushion backing were observed to be totally ineffective as formaldehyde barriers. Reductions observed by Haneto may have been due to the fact that measurements were made immediately after placing carpets and cushion pads over source materials. Formaldehyde penetration of carpeting materials may have been time-dependent.

Godish et al.[35] evaluated the effectiveness of a variety of varnishes and specially formulated formaldehyde-scavenging coatings applied to particle-

Table 3.8   Formaldehyde Emission Rates and Airspace Levels Associated with Different Underlayment Coverings34

Barrier/Covering	Formaldehyde Emission Rate (mg/m2/h)	Formaldehyde Concentration (ppm)
Bare underlayment	0.059	0.124
Vinyl linoleum	0.006	0.014
6-mil polyethylene	0.012	0.025
Carpet and cushion I	0.060	0.127
Carpet and cushion II	0.057	0.120

board subflooring under whole-house conditions. Results are presented in Tables 3.9 and 3.10. Treatment effectiveness of varnish coatings was highest for two applications of nitrocellulose-based varnish (70% reduction), followed by alkyd resin (53% reduction) and polyurethane (43% reduction). Of the specially formulated sealants, a single application of Valspar formaldehyde sealant was observed to be highly effective in reducing formaldehyde levels (78–87% reduction). Hyde-Chek was somewhat less effective (57–67%). One coat of Tri-Con AP-10 was shown to be ineffective, but the product did reduce formaldehyde levels by 65% when applied in two coats. All surface coatings appeared to provide long-term reductions in formaldehyde levels.

Both Valspar and Hyde-Chek formaldehyde sealants are no longer commercially available. Suspended sales have resulted from low market demand and manufacturer liability concerns. Though nitrocellulose-based varnishes are effective treatment coatings, their use, as well as the use of other varnish coatings, is limited to relatively small surface areas at a given time because of significant emissions of volatile vehicle materials. Formaldehyde sealants, on the other hand, are water-based and have low volatilities.

A formaldehyde-scavenging paint has been used in West Germany to treat the rooms of housing structures in which walls and ceilings are made of particleboard. One of the most popular of these is called by the trade name Falima F. It is a dispersion paint containing volatile liquid and solid formaldehyde reactive chemicals. Roffael[1] reported that particleboard treated with such paints exhibits very low levels of formaldehyde release and that such paints were more effective than treatments with gaseous ammonia.

Ammonia fumigation. Jewell,[36,37] of Weyerhaeuser Corporation, has conducted studies of the effectiveness of the application of gaseous ammonia treatment as a retrofit control measure for mobile homes. In this method, the interior of a closed mobile home is exposed to ammonia gas released from strong solutions of ammonium hydroxide (28–30% ammonia) for a period of 12+ hours at an indoor temperature of 80°F (26°C).

Because of the large volume of formaldehyde-emitting wood products used in mobile home manufacture, this method has the potential of achieving significant reductions in formaldehyde levels and occupant exposures. As a consequence, it has been widely used by mobile home manufacturers and others in an attempt to mitigate odor and health complaints.

In early studies, Jewell[36] reported that the treatment effectiveness of ammonia fumigation (measured over a 32-week period) on mobile homes was in the range of 70–75%. In follow-up measurements made 3.5 years after ammonia fumigation, some of the original effectiveness appeared to be lost, but effectiveness still averaged 60%.[37]

Due to concerns that gaseous ammonia will cause stress corrosion and cracking of brass connections on critical appliances such as gas stoves, water heaters, and furnaces as well as of electrical connections, the use of ammonia fumigation as a formaldehyde mitigation measure has been questioned by staff

Table 3.9   Treatment Effectiveness of Varnish Coatings Applied to Particleboard Underlayment in a Conventional House[35]

Coating	Immediate (1–2 wks)		Retest		
	% HCHO reduction to baseline	% HCHO reduction total	Time after application (months)	% HCHO reduction to baseline	% HCHO reduction total
Polyurethane (1 coat—#1)	29	17	—	—	—
Polyurethane (2 coats)	43	34	15	68	53
Nitrocellulose (1 coat—#1)	64	48	16.5	62	46
Nitrocellulose (1 coat—#2)	46	42	—	—	—
Nitrocellulose (2 coats)	70	64	11.5	84	76
Alkyd resin (1 coat)	40	31	—	—	—
Alkyd resin (2 coats)	53	40	—	—	—

Table 3.10   Treatment Effectiveness of Specially-Formulated Formaldehyde Sealants Applied to Particleboard in a Conventional House[35]

Coating	Immediate (1–2 wks)		Retest		
	% HCHO reduction to baseline	% HCHO reduction total	Time after application (months)	% HCHO reduction to baseline	% HCHO reduction total
Hyde-Chek (1 coat—#1)	67	57	15	93	79
Hyde-Chek (1 coat—#2)	57	44	—	—	—
Valspar (1 coat—#1)	87	67	7.5	79	60
Valspar (1 coat—#2)	78	56	9	68	49
Tri-Con AP-10 (1 coat)	17	12	—	—	—
Tri-Con AP-10 (2 coats)	65	46	8	65	46

Table 3.11   Effect of Temperature and Relative Humidity on Formaldehyde Levels in a
Mobile Home under Controlled Conditions of Indoor Climate[39]

Temperature °C ± 1°	Relative Humidity % ± 5%	Formaldehyde Concentration ppm	% of Maximum Value
30	70	0.36	100
25	70	0.29	81
30	50	0.28	78
30	30	0.23	64
25	50	0.17	47
25	30	0.14	39
20	70	0.12	33
20	50	0.09	25
20	30	0.07	19

members of the Consumer Product Safety Commission. Responding to these concerns, Jewell[37] conducted a series of tests and found that cracking only occurred when brass was in contact with free water and exposed at high concentrations of ammonia. He concluded that no stress cracking of brass connections on critical appliances or electrical connections should occur if liquid water is absent from such connectors.

The use of ammonia fumigation has also been reported by Muratzky.[38] Formaldehyde levels in rooms of a prefabricated house were observed to decrease from 0.20–0.30 ppm before treatment to 0.05–0.07 ppm measured over a period of 1–48 months after treatment. Muratzky reported that because of residual ammonia the house was not habitable for a period of 2–4 weeks after treatment and that irritating ammonia odors appeared up to four months after treatment under high temperature and low air exchange rates conditions.

Climate Control

A novel approach to controlling levels of formaldehyde (and other potential contaminants) is to control environmental factors such as temperature and relative humidity (RH), as these have significant effects on source emissions. The effectiveness of climate control on reducing formaldehyde exposures has been reported by Godish and Rouch.[39] As can be seen from Table 3.11, formaldehyde levels at a temperature and humidity combination of 20°C and 30% RH are only 20% of those at 30°C and 70% RH. Reducing temperature from 30 to 20°C alone resulted in a reduction of formaldehyde levels of 70%; a 40% reduction was observed when reducing relative humidity from 70% to 30%. Analysis of energy consumption and associated costs indicated that temperature reduction from 25 to 20°C during the cooling season would increase energy usage by about 20%. Temperature reduction during the heating season would reduce both formaldehyde levels and energy costs. Although effective, humidity reduction to 30% during summertime conditions would be prohibitively costly.

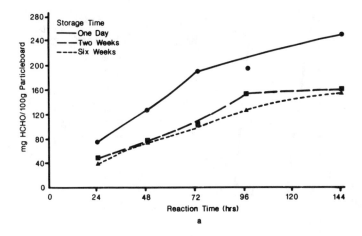

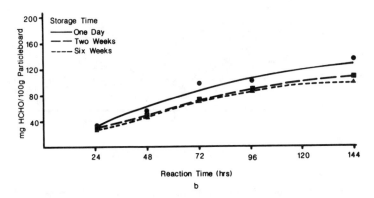

Figure 3.5.   Effect of storage time on formaldehyde release from particleboard panels bonded with two different F:U ratios.[1]

Product Aging

Significant decreases in formaldehyde emissions and associated airspace concentrations occur with time. This phenomenon has been reported by a number of investigators in both laboratory tests of wood product samples and measurements made in indoor environments. The effect of storage time on formaldehyde emissions from particleboards manufactured from U-F resins formulated with F:U ratios of 1.55:1 and 1.27:1 can be seen in Figure 3.5. A significant decrease in formaldehyde emissions in the first two weeks can be seen for U-F–bonded particleboard panels with molar F:U ratios of 1.55:1.

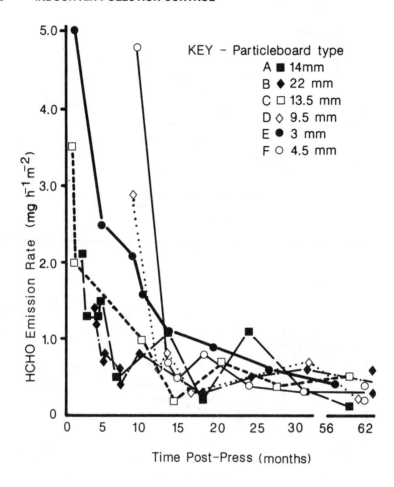

Figure 3.6. Time-dependent changes in formaldehyde levels associated with six grades of particleboard.[40]

However, the decrease with time for boards produced with resin F:U ratios of 1.27:1 are comparatively small.[1] Storage beyond 2–6 weeks was associated with a smaller reduction in emissions in both board types as compared to the first two weeks.

Kazakevics[40] evaluated time-dependent changes in formaldehyde emission rates of six grades of particleboard over a period of five years postmanufacture. As can be seen in Figure 3.6, there is a rapid decrease of formaldehyde emissions with time, particularly during the first year. The rapid first-year decrease in formaldehyde was suggested to be due to the release of residual formaldehyde present in the resin, while longer-term release may be due to resin hydrolysis.

Exponential decay in formaldehyde levels has been observed with a urea-formaldehyde/melamine coating applied to wood stock and in a UFFI house,[41] and in a population of UFFI houses.[42] In the decay model of Cohn et al.,[42] the first predicted half-life was 40 weeks, and the second, 4.5 years. Exponential decreases in formaldehyde levels with time were also reported by Hanrahan et al.;[43] high correlations ($R^2 = 0.80$) were observed between formaldehyde levels and mobile home age when the log formaldehyde concentration was plotted against the log of mobile home age. Using this plot and a starting point mobile home age of 3 months, half-lives for 167 randomly selected mobile homes were calculated to be 5, 8, 13, and 83 months, corresponding to the first, second, third, and fourth half-lives.

The above results are consistent with those reported by Haneto[32] for chamber studies in which formaldehyde levels associated with particleboard decreased by 50% in 8–9 months; the data of Groah et al.,[44] in which formaldehyde levels in relatively new mobile homes decreased by approximately 40–50% in a period of 8 weeks; and the dynamic chamber studies of Matthews et al.[45] In the latter case, observed decay rates of boards evaluated over 6–12 months after manufacture were as follows: hardwood plywood paneling, 10.7 ± 3.0 months and medium-density fiberboard, 18.8 ± 4.4 months. Formaldehyde decay rates generally correlated with the order of increasing product emission strength and thickness.

The rapidity of formaldehyde level decline in a given structure depends on a variety of variables. These include product emission strength,[45] presence of multiple sources,[30] and factors such as local climate and occupant behavior. Decay rates may vary substantially from one structure to another, as emission rates are affected exponentially by indoor temperature and linearly by relative humidity and formaldehyde concentration.[39] The latter would be affected by the extent of ventilation determined by building tightness, outdoor meteorological variables, and occupant behavior. In structures where average temperature, relative humidity, and ventilation rates are high, rapid formaldehyde decay rates can be expected. This is likely to be true in Sunbelt states where high temperatures and relative humidities are common and where relatively high temperatures (78–82°F, 25–27°C) are maintained during air conditioning. Decay rates would therefore be expected to reflect each structure's individual history.

This aging phenomenon can be used both actively and passively as a means of reducing formaldehyde levels and human exposure. An active approach would be the storage of U-F–based materials for several weeks or more by board manufacturers, distributors, industrial board users, and, ultimately, consumers. The longer the interval between board manufacture and its placement and use in indoor spaces, the lower initial levels and exposures will be. This, of course, may increase costs because of the need for additional storage facilities.

On a de facto basis, the passive approach is the one most widely used. The homeowner or building occupant who takes no remedial action will realize decreased exposures, often with the assumption that formaldehyde contamination will go away. The latter view is a mistaken one, as formaldehyde emissions will continue to occur as long as U-F-based products are present. In addition, the time period required to achieve an additional 50% reduction in emission rates and levels increases with time.

VOLATILE ORGANIC COMPOUNDS

The application of source control techniques (or any technique, for that matter) is made difficult, from a practical standpoint, by the fact that so little information is available relative to health problems specific VOCs or VOC combinations may be causing or have the potential to cause. This, in great measure, reflects the fact that VOCs vary widely in number, sources, and individual and total concentrations. They are time-consuming and expensive to identify and quantify. It is difficult to control something if one does not know what to control. Even if such knowledge existed, source identification would in many cases pose a difficult problem. This would be particularly the case when sources are building materials and furnishings, as a wide variety of organic solvents are used by manufacturing industries in the industrial processes employed in a product's making.

In the following discussion we will assume that total VOC concentration elevated above some undefined norm is undesirable and that minimal levels of exposure are desired for VOCs that have been identified as carcinogens or potential carcinogens.

Avoidance

In theory, the best way of controlling exposure to total VOC levels or specific compounds is to avoid products or activities producing high concentrations of VOCs indoors. In the case of products, avoidance would require knowledge that alternatives have lower emissions and do not themselves pose any unique health hazards. In other cases avoidance may be a relatively simple task. For example, cigarette smoking is known to be a significant source of VOCs, particularly benzene, indoors.[46] Limitations on smoking would limit VOC exposures from this source. Emissions from paints, solvents, gasoline, and newspapers/magazines can be reduced by storage of these materials in outbuildings or well-ventilated spaces.[47] Since higher VOC levels in residential spaces are associated with attached garages,[48] detached garages would, in theory, be preferable. As higher propane, benzene, and toluene levels are associated with fossil fuel heat,[49] all-electric homes would be similarly preferable. The caveat "in theory" is appropriate to the gas heat and attached garage situations since application of such avoidance principles would, in many instances, be considered extreme, particularly in the context of retrofit application, cost, and the relatively low anticipated health risk associated with such VOC exposure.

Table 3.12 Total VOC Concentrations and Emission Rates Associated with Various Floor/Wall Coverings and Coatings[50]

Material Type	Concentration (mg/m³)	Emission Rate (mg/m²/hr)
Wallpaper		
Vinyl and paper	0.95	0.04
Vinyl and glass fibers	7.18	0.30
Printed paper	0.74	0.03
Wall Covering		
Hessian	0.09	0.005
PVC	2.43	0.10
Textile	39.60	1.60
Textile	1.98	0.08
Floor Covering		
Linoleum	5.19	0.22
Synthetic fibers	1.62	0.12
Rubber	28.40	1.40
Soft plastic	3.84	0.59
Homogeneous PVC	54.80	2.30
Coatings		
Acrylic latex	2.00	0.43
Varnish, clear epoxy	5.45	1.30
Varnish, polyurethane, 2-component	28.90	4.70
Varnish, acid-hardened	3.50	0.83

Product Selection—Total VOCs

Avoidance of elevated total VOC levels and, in theory, of specific VOC compounds, can be realized by judicious product selection. In practice this is difficult, since emission characteristics of specific product types are, in general, unknown on a generic basis and may vary considerably from one product manufacturer to another.

However, sufficient research has been conducted to at least anticipate which products, or circumstances of product use, lead to undesirably high indoor VOC levels. Molhave[50] tested chamber air concentrations and emission rates from 42 commonly used building materials. Adhesive and filler products such as ethylene vinyl acetate and polyvinyl acetate were particularly notable in terms of total VOC emissions. Total VOC concentrations and emission rates associated with various floor/wall coverings and coatings are summarized in Table 3.12. It is apparent that significant product differences in VOC emissions exist and that products could be selected on the basis of their potential for VOC emissions. Under floor covering, it can be seen that rubber and homogeneous polyvinylchloride (PVC) products have the highest total VOC emissions, with emissions from linoleum, synthetic fiber, and soft plastic floor covering being considerably less. Volatile emissions from a two-component polyurethane varnish were considerably higher than two other varnish products and an acrylic latex paint.

Latex-based paints typically use water as a vehicle and should therefore be expected to release much lower quantities of VOCs than varnishes and other oil-based paints. However, this appears to be product-dependent. Sheldon et al.,[51] for example, reported high emissions of aliphatic, oxygenated, aromatic, and halogenated hydrocarbons from Glidden latex paint; Bruning latex paint had no detectable emissions of aliphatic, oxygenated, and halogenated hydrocarbons, and one-sixteenth the aromatic hydrocarbon emissions of the Glidden product. They also reported low or nondetectable VOC emissions of target compounds in fiberglass insulation, cement block, interior mineral board, ceiling tile, red clay brick, and a plastic laminate. Presumably such products would be considered acceptable if a prospective new home builder wished to achieve a low VOC residential environment. However, it must be pointed out that Sheldon et al. only measured 24 target VOCs and that other VOCs may have been present and were neither identified nor quantified.

The intuitive characterization of water-based products as low VOC emitters compared to those that are solvent-based may in some cases be erroneous. Girman et al.[52] investigated VOC emissions from a variety of solvent- and water-based adhesives. VOC emissions from water-based rubber adhesives were observed to be significantly higher than those of solvent-based adhesives.

Studies of Molhave[50] indicate that vinyl, plastic, and rubber-type floor coverings have relatively high VOC emission rates. These emissions are even more significant in the context of the relatively high loading factor (surface-to-volume ratio) for such products in actual use. VOC levels can therefore be significantly minimized by limitations on loading factor or product use. The same principle would apply to the use of moderate-to-high VOC-emitting surface coatings and wall and floor coverings.

Product Selection—Specific VOCs

Vinyl floor and wall coverings may liberate VOCs of specific concern. These may include residual monomers of vinyl chloride; plasticizers such as diisobutyl phthalate, butyl benzyl phthalate and di(ethylhexyl) phthalate;[53] and plasticizer contaminants such as benzyl and benzal chloride. Benzyl and benzal chloride are mucous membrane irritants and carcinogenic as well. They have been suggested as a cause of temporarily sick buildings.[54] In some buildings phthalate plasticizers have been observed to cause a whitening of green plants and for this reason have been suggested to be a potential health risk as well. Assuming phthalate plasticizers and their contaminants pose the health risks suggested, avoiding them, particularly in the context of a large loading factor, would be desirable in new construction.

Products that release polychlorinated biphenyls (PCBs) may be of particular concern. Defective fluorescent light ballasts emit PCBs and are a significant source of indoor PCB contamination.[55] In buildings with PCB transformers, PCB levels are double the levels in buildings without such transformers.[56]

Avoidance is not an issue for new buildings, since the EPA banned the manufacture, processing, distribution, and use of PCBs in 1977.[57]

Aging

Although most indoor spaces are contaminated with VOCs at any particular time, elevated VOC levels (both total and specific) are typically short-lived. Both emission rates and air concentrations decrease with time. VOC decay rates appear to be significantly more rapid than formaldehyde decay associated with U-F resin–based products. In addition, the reservoir-type emission phenomenon observed with U-F resins does not apparently occur with other VOCs and products which contain them. Berglund et al.[58] studied VOC levels in a newly built preschool. Fifteen selected VOCs were observed to decline from an average of 15-20 ppb to 2-3 ppb within six months. The trend over time differed for specific compounds. The strongest declines were shown for toluene and n-butanol, which were reduced by a factor of 5-10. Qualitative measurements indicated that the number of VOCs detected changed from 160 to 53 after a 6-month period. Hartwell et al.[59] measured VOC levels in a newly constructed office building. Samples were taken prior to occupation and two additional times over a 6-month period. Significant declines were observed for o-, m-and p-xylene, n-undecane, n-dodecane, 1,1,1-trichloroethane, ethylbenzene, and n-decane.

The effect of aging on average concentrations of aliphatic, aromatic, and chlorinated hydrocarbon species in two new office buildings and a nursing home was investigated by Wallace et al.[60] Significant declines in aliphatic and aromatic hydrocarbons were observed over the period from 7 to 23 weeks after building completion (Table 3.13). Decreases in VOCs with time on the order of 3- to 10–fold were apparently due to decreased emissions from building materials and higher rates of building ventilations when samples were collected. However, in several cases chlorinated hydrocarbon species increased in concentration with time. The increase in chlorinated VOCs with time was likely due to solvent-using building activities such as photocopying and cleaning. Tichenor and Mason[61] studied time-dependent changes in emission rates from selected materials. Changes in VOC emission rates with time for a floor adhesive and a vinyl caulk can be seen in Figure 3.7.

Observed rapid decreases in total and specific VOC emissions and levels have two practical source control applications. In the first instance, VOC-related comfort or health problems (if they occur) can be expected to be associated only with the first few months of occupancy. They can justifiably be expected to diminish or go away completely in a relatively short period of time (6 months or less). Knowledge of this phenomenon can be used to schedule building occupancy. The sooner a building is occupied after completion, the more likely it is that occupants will experience VOC-related health and comfort effects. Some degree of building aging under normal ventilation conditions prior to occupancy would be desirable. The aging process could, in

Table 3.13 Three-Day Average VOC Concentrations ($\mu g/m^3$) in Three New Buildings[60]

	Recently Constructed			Following Occupancy				Maximum Outdoor Concentration
	Office 1	Office 2	Nursing Home	Office 1	Office 2	Office 1	Nursing Home	
Time Since Completion (weeks)	1	2	6	7	15	22	23	—
Aliphatics								
Decane	380	440	68	38	15	4	4	4
Undecane	170	210	69	48	34	13	4	2
Dodecane	47	150	31	19	24	5	ND[a]	1
Aromatics								
Xylenes	214	59	33	27	19	12	7	6
Ethylbenzene	84	51	8	6	5	5	2	2
Benzene	5	3	2	7	5	7	2	5
Styrene	8	3	3	7	3	4	1	4
Chlorinated								
1,1,1-Trichloroethane	380	13	4	100	39	49	2	10
Trichloroethylene	1	ND	3	38	8	27	1	1
Tetrachloroethylene	7	ND	1	2	2	3	1	1

[a]ND = none detected.

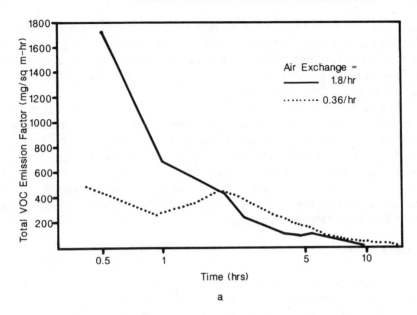

a

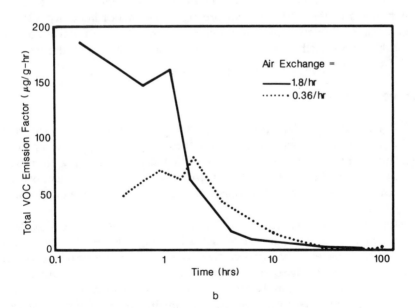

b

Figure 3.7.   Time-dependent changes in VOC emissions associated with (a) floor
adhesive and (b) vinyl caulk.[61]

theory, be accelerated by employing high ventilation rates (in a mechanically ventilated building) in the first 3–6 months, followed on a sustained basis by rates sufficient to maintain CO_2 levels below 1000 ppm. This is in keeping with the Scandinavian decision to require 100% outdoor makeup air for the first 6 months of a building's existence.[60]

Knowledge of the VOC aging phenomenon can also be applied to mitigate other indoor air quality problems, most specifically those associated with carpeting in residential and non-residential environments. Health complaints associated with carpeting are most common when factory-ordered materials are installed immediately upon receipt. They have had little time to age. In residential situations, installation of carpeting is most appropriate during the spring and summer months when natural ventilation by opening windows can be used to accelerate the aging process.

Climate Control

Emission of VOCs from building products and furnishings can be expected to increase with increased building temperatures. This is likely the reason why VOCs such as toluene may be higher during summer months.[52,62] Studies by Berglund et al.[63] of VOCs in a library indicate a strong negative correlation between relative humidity and the concentrations of 2-butanone, 1-butanol, 2-ethylcyclobutanol, and n-butylacetate. This implies that emissions from building materials, books, etc. are high when indoor humidity is low. In theory, the manipulation of either building temperature or relative humidity (or both) can be used as a VOC source control measure to reduce exposure levels in the short term or to accelerate aging.

Source Removal

Some VOC-related indoor air quality problems could be controlled or minimized by identifying and removing them. If we were, for example, to assume that PCB exposures associated with the use of existing fluorescent lighting ballasts and PCB-cooled transformers were excessive, removal and replacement of the ballasts and transformer coolant would be an appropriate control measure.

Two instances of source removal to control total or specific VOC levels have been reported. In a Connecticut office building Hijazi et al.[64] were able to identify plastic panels used as office dividers as a significant source of a potent lachrymator, dimethyl acetamide. Removal of the panels was accompanied by a reduction, but not a complete elimination, of occupant complaints.

The author has received numerous complaints over the past eight years from homeowners who associate illness symptoms with the installation of new wall-to-wall carpeting. It is not uncommon in such instances for homeowners to respond by removing the carpeting. Unfortunately, there have been no systematic studies associated with VOC emissions from carpeting, their effect on human health, and the effectiveness of carpet removal in resolving this alleged problem.

Bake-out

An interesting and novel approach to reducing VOC levels in a new building is the use of what is called a bake-out procedure. In a "bake-out," an unoccupied new building is maintained at elevated temperatures and normal ventilation conditions for a period of time. It is based on the theory that by elevating temperatures the vapor pressure of residual solvents will be increased, and, if maintained long enough, cause the depletion of solvents, with a corresponding reduction in VOC emissions. Such a procedure has been recommended by the California State Architect for all new state office buildings. Effectiveness of the bake-out procedure on VOC levels in San Francisco area office buildings has been reported by Girman et al.[65] In one case, total VOC levels were reduced by 71% after a bake-out at 31°C for 72 hours at four air changes per hour (ACH). A bake-out was conducted in a second office building at temperatures in the range of 29–32°C for a 24-hour period and an air exchange rate of 1.59 ACH. Total VOC concentrations were 29% lower after the bake-out. The decrease, in one-third the time, was approximately one-third of that reported for the first bake-out building. Large increases in VOC concentrations observed during the bake-out period suggested that residual solvents were being depleted at a rapid rate. The greatest decreases in specific VOC concentrations occurred for those compounds showing the largest increases in concentration during the bake-out period.

PESTICIDES

The introduction and subsequent contamination of indoor air by pesticides is in most cases under the control of a building owner and is therefore intentional. This is in contrast to most VOCs, whose introduction into building spaces is secondary to the intended use of some product or furnishing. Pesticide users accept the fact of indoor contamination as a means to an end and assume that any health risks are small.

Avoidance

With few exceptions, pesticide use indoors and applications which cause indoor contamination are discretionary. Pesticide use, particularly for insect control, is directed toward nuisance pests such as cockroaches, flies, mosquitoes, and wood-boring insects like termites and carpenter ants. None of these pest problems represents a serious public health concern. Wood-boring insects can cause significant structural damage and therefore represent a problem much more serious than other insects.

Table 3.14. Air Concentrations (μg/m³) of Insecticides in Dormitory Rooms Following Application[68]

Insecticide	Vapor Pressure (mm Hg)	Days After Treatment	
		0	3
Propoxur (1.1%)	3×10^{-6} (20°C)	15.4	0.7
Bendiocarb (0.5%)	5×19^{-6} (25°C)	7.7	NDa
Fenitrothin (1.0%)	6×10^{-6} (20°C)	3.3	0.5
Diazinon (1.0%)	1.4×10^{-4} (20°C)	1.6	0.4
Acephate (1.0%)	1.7×10^{-4} (24°C)	1.3	0.3
Chlorpyrifos (0.5%)	1.9×10^{-5} (25°C)	1.1	0.3

aND = None detected.

Because pesticide use indoors is discretionary, most indoor pesticide contamination can be easily avoided if a building owner or manager chooses to do so.

Selective Avoidance

The degree or extent of pesticide contamination and potential health risks can be reduced by one's choice of pesticide formulation, the method and rate of application, and adherence to good application principles and practices.

Pesticides differ considerably in volatility. Highest short-term concentrations will be associated with pesticides that have the highest vapor pressures. For example, application of a termiticide containing chlordane and heptachlor at a ratio of 2:1 was observed to result in indoor concentrations of heptachlor eight times greater than those for chlordane. The vapor pressure of heptachlor is 3×10^{-4} mm Hg, for chlordane 1×10^{-5} mm Hg. Differences in vapor pressure were suggested to be responsible for the higher heptachlor levels, despite the fact that heptachlor was a relatively minor component in the pesticide formulation.[66] Similar results were reported by Louis and Kisselbach.[67]

Wright et al.[68] monitored pesticide levels in dormitory rooms after treatment with six insecticides applied as emulsions. Highest concentrations measured immediately after application were observed for those pesticides having the highest vapor pressures (Table 3.14). Acephate, diazinon, and chlorpyrifos were found in similar concentrations, corresponding to their similar vapor pressures. By the third day after application, however, concentrations of the six pesticides did not differ significantly from each other.

Differences in safe reentry time for two moderately toxic insecticides used in home foggers were studied by Goh et al.[69] A safe home reentry time of two hours was calculated for a propoxur-based home fogger, compared to a 7- to 10-hour safe reentry time for dichlorvos (DDVP). Propoxur was considerably less volatile than DDVP.

From a safety standpoint, use of chlorpyrifos for termite control would, in theory, be more desirable than formulations with chlordane and heptachlor, as the latter have been shown to be carcinogenic in laboratory animal tests.[70]

Application Practices

The rate and method of pesticide application can affect air concentrations. Wright and Leidy[71] observed that less chlorpyrifos was recovered from room air treated with a 0.5% chlorpyrifos formulation as compared to a 1% formulation. They also observed that, when measured on the day of treatment, air concentrations were four times higher when pesticides were applied with aerosol sprayers than when they were applied with compressed-air sprayers. However, airborne chlorpyrifos levels decreased more rapidly over a period of three days in rooms treated with an aerosol sprayer. Chlorpyrifos levels decreased by factors of 10 and 2 between days 0 and 1 in the aerosol- and compressed air–treated rooms, respectively.

The problem of significant air contamination associated with spray application of pesticides can, in theory, be reduced or minimized by use of controlled-release formulations. These include pest control strips containing propoxur, diazonin, or chlorpyrifos and paint-on formulations containing chlorpyrifos.[72,73] Both are intended for cockroach control for up to six months. They kill crawling insects by contact. The effect of pest control strips on room pesticide concentrations was investigated by Jackson and Lewis.[72] Contrary to manufacturers' beliefs, pest control strips did release a significant amount of pesticide vapor. Highest room concentrations of propoxur, diazinon, and chlorpyrifos were 0.8, 1.4 and 0.25 $\mu g/m^3$, respectively. Ware and Cahill[73] compared the air concentrations of a 2% slow-release paint-on formulation to a 0.5% emulsion spray of chlorpyrifos applied in a closed room. The spray resulted in an approximate two-fold increase of chlorpyrifos over the paint-on formulation.

Many reported instances of significant contamination of indoor air by pesticides, particularly by termiticides, have resulted from application practices that in retrospect can be described as faulty (if not outright dumb!). The most notable of these is the injection of termiticides such as chlordane into or below slabs containing heating/cooling ducts. To minimize the likelihood of significant indoor contamination by chlordane and other termiticides, the State of New Jersey promulgated regulations describing the permitted use of termiticides.[74] They required that a certified applicator be onsite for termiticide treatments and prohibited the use of organochlorine termiticides (such as chlordane) on interior portions of slabs containing heating ducts.

New Jersey Department of Environmental Protection staff evaluated the effectiveness of their termiticide application requirements in 12 homes.[67] In 100 posttreatment samples of living spaces, 57% had no detectable levels of chlordane and 32% had levels up to 0.5 $\mu g/m^3$ or one-tenth of the National Academy of Science (NAS) guideline of 5 $\mu g/m^3$.[75] Samples above 0.5 $\mu g/m^3$ were associated with a house previously treated with chlordane. Based on these studies, Louis and Kisselbach[67] concluded that termiticides can be applied with current technology and result in living space contamination levels substantially below the 5-$\mu g/m^3$ NAS guideline. In a larger study, Leidy et al.[76] measured chlordane and heptachlor levels in 120 rooms in 60 houses treated with chlordane. Eight rooms in seven houses (6.7%) had chlordane levels (5.3–9.9 μg/m^3) exceeding the NAS guideline; the heptachlor guideline was not exceeded.

A special application practice is recommended for plenum houses. Plenum houses were developed in the 1960s. They employ a downflow furnace and air conditioning system that distributes conditioned air to a plenum beneath a wood floor from which air is circulated to various parts of the house through vents in the baseboard or floor. Termiticides are applied to the footings and soil adjacent to the foundation in such structures. The termiticide is applied to both the inside and outside of the plenum space. Unlike crawl space houses, underfloor air in plenum houses is constantly recirculated throughout living spaces when the heating/cooling system is being operated. Maye and Malagodi[77] investigated the use of vapor barriers in preventing the entry of chlordane, heptachlor, and chlorpyrifos into plenum houses. Saranex-15 (4 mil), a combination of polyvinyldene chloride and polyethylene, was applied to the entry plenum area during the construction of two plenum houses. The application of this vapor barrier was observed to effectively inhibit the entry of chlordane and chlorpyrifos after the application of termiticidal treatments.

Retrofit Measures

Termiticides

Excessive levels of chlordane and other termiticides have been typically associated with sub-slab injections in which heating ducts were penetrated or where pesticides migrated into heating ducts. In such cases, the most common retrofit control measure is to reroute heating ducts through an attic space and seal off slab or sub-slab ducts with concrete. This control method is based on the observation by Lillie and Barnes[78] that 83% of houses having heating/cooling ducts in the attic had no detectable level of chlordane; no such house exceeded 2 μg/m^3. It is highly effective. In 36 houses in which new ducts were installed in the attic, heaters were replaced, and intraslab ducts were sealed, chlordane levels were reduced by 93–97%.[79] In crawl space houses, chlordane can enter heating ducts (particularly the cold air return) through gaps in the sheet metal assembly, between the sheet metal and wooden floor joists, or between wooden joists and the subfloor. Barnes[80] reported that either eliminating cold air return ducts or sealing the soil under the crawl space with concrete was effective in reducing indoor chlordane levels.

Jurinski[66] applied a silicone-based waterproofing compound on 50% of the previously untreated concrete block basement walls in a building in which heptachlor levels were excessive. This partial treatment was observed to reduce heptachlor levels by 60%. In a second building where considerable termiticide contamination had occurred, interior walls, floors, and ceilings were washed twice with an aqueous trisodium phosphate cleaner and the floor was sealed with a urethane floor sealer. Though pesticide surface contamination was reduced by approximately 90%, airborne concentrations were not proportionally reduced.[66]

Pesticide contamination of residences is not limited to air contamination. Significant surface contamination can occur from particulate-phase materials following application/condensation and volatilization/recondensation of low-vapor pressure pesticides such as chlordane. Reduction in airborne levels in heavily contaminated indoor environments may require surface decontamination as well as measures such as rerouting of heating/cooling ducts. Jurinski[66] observed (in a single house) that the use of an aqueous trisodium phosphate cleaner was highly effective in reducing surface contamination by chlordane. This is consistent with studies of pesticide residue removal from clothing and other fabrics.[81,82]

Wood Preservatives

Few studies have been conducted to evaluate the effectiveness of retrofit control measures on reducing elevated indoor levels of pesticides applied as wood preservatives. The Federal Health Office of the Federal Republic of Germany recommends (in addition to continuous ventilation and moving to noncontaminated spaces) source control measures including removal of treated wood; vacuum cleaning of house dust; surface cleaning with alkaline solutions; and washing carpets, curtains, bedding, and clothing.[83] Wood treated with preservatives such as pentachlorophenol (PCP) can, in theory, be sealed to reduce emissions to indoor spaces. In preliminary studies, Levin and Hahn[84] observed that application of one coat of polyurethane would reduce PCP emissions by 80%; two coats, 95%. They then applied polyurethane varnish to a large proportion of PCP-treated timbers in a large office building. Such treatment resulted in a reduction of airborne concentrations from $27\mu g/m^3$ to $5.9\mu g/m^3$, a reduction of 78%.

REFERENCES

1. Roffael, E. 1978. "Progress in the Elimination of Formaldehyde Liberation from Particleboards." 233–249. In: T.M Maloney (Ed.). *Proceedings of the WSU Twelfth Annual Particleboard Symposium.* Washington State University. Pullman, WA.
2. Sundin, B. 1978. "Formaldehyde Emission from Particleboard and Other Building Materials: A Study from the Scandinavian Countries." 251–273. In: T.M. Maloney (Ed.). *Proceedings of the WSU Twelfth Annual Particleboard Symposium.* Washington State University. Pullman, WA.
3. Sundin, B. 1985. "The Formaldehyde Situation in Europe." 255–275. In: T.M. Maloney (Ed.). *Proceedings of the WSU Nineteenth International Particleboard/ Composite Materials Symposium.* Washington State University. Pullman, WA.
4. Myers, G.E. and M. Nagooka. 1981. "Emission of Formaldehyde by Particle-

board: Effect of Ventilation Rate and Loading on Air Contamination Levels." *For. Prod. J.* 31:39–44.

5. Godish, T. 1988a. Unpublished data.

6. Godish, T. and J. Rouch. 1984. "Efficacy of Residential Formaldehyde Control. Source Removal." 127–132. In: *Proceedings of the Third International Conference on Indoor Air Quality and Climate.* Swedish Council for Building Research. Stockholm. Vol. 2.

7. Godish, T. 1988b. Unpublished data.

8. Versar, Inc. 1986. "Formaldehyde Exposure in Residential Settings: Sources, Levels and Effectiveness of Control Options." EPA Contract No. 68-02-3968.

9. Groah, W.J. 1986. "Formaldehyde Emissions: Hardwood Plywood and Certain Wood-Based Panel Products." 17–25. In: B. Meyer et al. (Eds.). *Formaldehyde Release from Wood Products.* ACS Symposium Series 316.

10. Meyer, B. and K. Hermanns. 1986. "Formaldehyde Release from Wood Products: An Overview." 1–15. In: B. Meyer et al. (Eds.). *Formaldehyde Release from Wood Products.* American Chemical Society Symposium Series 316.

11. Myers, G.E. 1984. "How Mole Ratio of UF Resin Affects Formaldehyde Emission and Other Properties: A Literature Critique." *For. Prod. J.* 34:35–41.

12. Sundin, B. 1982. "Present Status of Formaldehyde Problems and Regulations." 3–19. In: T.M. Maloney (Ed.). *Proceedings of the WSU Sixteenth Annual Particleboard Symposium.* Washington State University. Pullman, WA.

13. Myers, G.E. 1985. "Effect of Separate Additions to Furnish or Veneer on Formaldehyde Emission and Other Properties: A Literature Review (1960–1984)." *For. Prod. J.* 35:57–62.

14. Myers, G.E. 1982. "Hydrolytic Stability of Cured Urea-Formaldehyde Resins." *Wood Sci.* 15:127–138.

15. McVey, D.T. 1982. "Great Strides Forward—Formaldehyde Emission from the Production Standpoint." 21–33. In: T.M. Maloney (Ed.). *Proceedings of the WSU Fourteenth Annual Particleboard Symposium.* Washington State University. Pullman, WA.

16. Meyer, B. et al. 1980. "Formaldehyde Release from Sulfur-Modified Urea-Formaldehyde Resin Systems." *For. Prod. J.* 30:2431.

17. Imura, S. et al. 1977. "Prevention of Formaldehyde Liberation from Plywood and Particleboard by Secondary Treatment." *Renson Shikenyor Gippor* 1:305–333.

18. Simon, S.I. 1980. "Using the Verkor FD-EX Chamber to Prevent Formaldehyde Emission from Boards Manufactured with Urea-Formaldehyde Glues." 163–169. In: T.M. Maloney (Ed.). *Proceedings of the WSU Fourteenth Annual Particleboard Symposium.* Washington State University. Pullman, WA.

19. Groah, W.J. et al. 1984. "Effect of a Decorative Overlay on Formaldehyde Emissions." *For. Prod. J.* 34:27–29.

20. Deppe, H.J. 1977. "Technical Progress in Using Isocyanate as an Adhesive in Particleboard Manufacture." 13–31. In: T.M. Maloney (Ed.). *Proceedings of the WSU Eleventh Annual Particleboard Symposium.* Washington State University. Pullman, WA.

21. Galbraith, C.J. Jr. et al. 1983. 263–282. "Self-Releasing Emulsifiable MDI Isocyanate: An Easy Approach for an All-Isocyanate Bonded Board." In: T.M. Maloney (Ed.). *Proceedings of the WSU Seventeenth Annual Particleboard Symposium.* Washington State University. Pullman, WA.

22. Sundin, B. 1987. Personal communication.

23. Bowen, R.P. et al. 1981. "Urea-Formaldehyde Foam Insulation: Problem Identification and Remedial Measures for Wood Frame Construction." Building Practice Note No. 23. National Research Council, Ottawa, Canada.

24. Broder, I. et al. 1987. "Comparison of Health of Occupants of Control Homes and Homes Insulated with Urea-Formaldehyde Foam Before and After Corrective Work." 605–609. In: *Proceedings of the Fourth International Conference on Indoor Air Quality and Climate.* Institute for Water, Soil and Air Hygiene. West Berlin. Vol. 2.

25. Godish, T. 1988c. Unpublished data.

26. Singh, J. et al. 1982. "Evaluation of the Relationship Between Formaldehyde Emissions from Particleboard Mobile Home Decking and Hardwood Plywood Panelling as Determined by Product Test Methods and Formaldehyde Levels in Experimental Mobile Homes." U.S. Dept. of Housing and Urban Development. Contract No. HC-5222.

27. Newton, L.R. 1982. "Formaldehyde Emission from Wood Products: Correlating Environmental Chamber Levels to Secondary Laboratory Tests." In: T.M. Maloney (Ed.). *Proceedings of the WSU Sixteenth Annual Particleboard Symposium.* Washington State University. Pullman, WA.

28. Godish, T. and B. Kayner. 1985. "Formaldehyde Source Interaction Studies." *For. Prod. J.* 35:13–17.

29. Matthews, T.G. et al. 1983. "Formaldehyde Release from Pressed Wood Products." 179–202. In: T. Maloney (Ed.). *Proceedings of the WSU Seventeenth Annual Particleboard Symposium.* Washington State University. Pullman, WA.

30. Godish, T. and J. Rouch. 1987. "Formaldehyde Source Interaction Studies Under Whole-House Conditions." *Environ. Poll.* 48:1–12.

31. Matthews, T.G. et al. 1983. "Formaldehyde Emissions from Combustion Sources and Solid Resin-Containing Products: Potential Impact on Indoor Formaldehyde Concentrations and Possible Corrective Measures." In: Proceedings of the ASHRAE Symposium—Management of Atmospheres in Tightly Enclosed Spaces. Santa Barbara, CA.

32. Haneto, P. 1986. "Effect of Diffusion Barriers on Formaldehyde Emissions from Particleboard." 202–208. In: B. Meyer et al. (Eds.). *Formaldehyde Release from Wood Products.* American Chemical Society Symposium Series 316.

33. Seymour, J.W. 1979. "Formaldehyde Problems in Manufactured Housing." In: *Proceedings of the 33rd Annual Meeting: Forest Products Research Society.* San Francisco. Forest Products Research Society. Madison, WI.

34. Matthews, T.G. et al. 1986. "Preliminary Evaluation of Formaldehyde Mitigation Studies in Unoccupied Research Homes." In: D.S. Walkinshaw (Ed.). *Transactions: Indoor Air Quality in Cold Climates.* Air Pollution Control Association. Pittsburgh.

35. Godish, T. et al. 1987. "Control of Residential Formaldehyde Levels by Source Treatment." 221–225. In: *Proceedings of the Fourth International Conference on Indoor Air Quality and Climate.* Institute for Water, Soil and Air Hygiene. West Berlin. Vol. 3.

36. Jewell, R.A. 1980. "Reduction of Formaldehyde Levels in Mobile Homes." 102–108. In: *Wood Adhesives—Research, Application and Needs.* Forest Products Research Society. Madison, WI.

37. Jewell, R.A. 1984. "Reducing Formaldehyde Levels in Mobile Homes Using 29%

Aqueous Ammonia Treatment or Heat Exchangers." Weyerhaeuser Corp., Tacoma, WA.

38. Muratzky, R. 1987. "Formaldehyde Injuries in Prefabricated Houses: Causes, Prevention, and Reduction." 690–694. In: *Proceedings of the Fourth International Conference on Indoor Air Quality and Climate.* Institute for Water, Soil and Air Hygiene. West Berlin. Vol. 2.

39. Godish, T. and J. Rouch. 1986. "Mitigation of Residential Formaldehyde Contamination by Indoor Climate Control." *Amer. Indust. Hyg. Assoc. J.* 47:792–797.

40. Kazakevics, A.A.R. 1984. "Studies on the Reduction of Formaldehyde Emission from Particleboard by Polymers." Ph.D. Dissertation. University of New Zealand, Auckland.

41. Godish, T. and C. Guindon. 1988. "Time-Dependent Changes in Formaldehyde Concentration." Unpublished data.

42. Cohn, M.S. et al. 1984. "Sources Contributing to Formaldehyde Indoor Levels." 133–138. In: *Proceedings of the Third International Conference on Indoor Air Quality and Climate.* Swedish Council for Building Research. Stockholm. Vol. 3.

43. Hanrahan, L.P. et al. 1985. "Formaldehyde Concentrations in Wisconsin Mobile Homes." *JAPCA* 36:698–704.

44. Groah, W.J. et al. 1985. "Factors that Influence Formaldehyde Air Levels in Mobile Homes." *For. Prod. J.* 35:11–18.

45. Matthews, T.G. et al. 1985. "Modeling and Testing of Formaldehyde Emission Characteristics of Pressed Wood Products." Technical Report XVIII. U.S. Consumer Product Safety Commission.

46. Wallace, L.A. and E.D. Pellizzari. 1986. "Personal Air Exposures and Breath Concentrations of Benzene and Other Volatile Hydrocarbons for Smokers and Nonsmokers." *Toxicology Letters* 35:113–116.

47. Seifert, B. and H.J. Abraham. 1982. "Indoor Air Concentrations of Benzene and Some Other Aromatic Hydrocarbons." *Ecotox. Environ. Safety.* 6:190–192.

48. Gammage, R.B. et al. 1984. "Residential Measurements of High Volatility Organics and Their Sources." 157–162. In: *Proceedings of the Third International Conference on Indoor Air Quality and Climate.* Swedish Council for Building Research. Stockholm. Vol. 4.

49. Pleil, J.D. et al. 1985. "Volatile Organic Compounds in Indoor Air: A Survey of Various Structures." 237–249. In: D.S. Walkinshaw (Ed.). *Transactions: Indoor Air Quality in Cold Climates.* Air Pollution Control Association, Pittsburgh.

50. Molhave, L. 1982. "Indoor Air Pollution Due to Organic Gases and Vapors of Solvents in Building Materials." *Environ. Int.* 8:117–127.

51. Sheldon, L.S. et al. 1986. "Volatile Organic Emissions from Building Materials — Results of Preliminary Headspace Experiments." In: Proceedings of the 80th Annual Meeting of the Air Pollution Control Association. Minneapolis.

52. Girman, J.R. et al. 1984. "Volatile Organic Emissions from Adhesives with Indoor Applications." 271–276. In: *Proceedings of the Third International Conference on Indoor Air Quality and Climate.* Swedish Council for Building Research. Stockholm. Vol. 4.

53. Rittfeldt, L. et al. 1984. "Indoor Air Pollutants Due to Vinyl Floor Tiles." 297–302. In: *Proceedings of the Third International Conference on Indoor Air Quality and Climate.* Swedish Council for Building Research. Stockholm. Vol. 3.

54. Vedel, A. and P.A. Nielsen. 1984. "Phthalate Esters in the Indoor Environment."

309-314. In: *Proceedings of the Third International Conference on Indoor Air Quality and Climate.* Swedish Council for Building Research. Stockholm. Vol. 3.

55. Oatman, L. and R. Roy. 1986. "Surface and Indoor Air Levels of Polychlorinated Biphenyls in Public Buildings." *Bull. Environ. Contam. Toxicol.* 37:461-466.

56. MacLeod, K.E. 1981. "Polychlorinated Biphenyls in Indoor Air." *Environ. Sci. Technol.* 15:926-928.

57. *Code of Federal Regulations.* 1977. 40 CFR 761.

58. Berglund, B. et al. 1982. "A Longitudinal Study of Air Contaminants in a Newly Built Preschool." *Environ. Int.* 8:11-115.

59. Hartwell, T.D. et al. 1985. "Levels of Volatile Organics in Indoor Air." In: Proceedings of the 79th Annual Meeting of the Air Pollution Control Association. Detroit.

60. Wallace, L. et al. 1987. "Volatile Organic Chemicals in Ten Public Access Buildings." 188-192. In: *Proceedings of the Fourth International Conference on Indoor Air Quality and Climate.* Institute for Water, Soil and Air Hygiene. West Berlin. Vol. 1.

61. Tichenor, B.A. and M.A. Mason. 1986. "Characterization of Organic Emissions from Selected Materials in Indoor Use." In: Proceedings of the 80th Annual Meeting of the Air Pollution Control Association. Minneapolis.

62. Jungers, R.H. and L.S. Sheldon. 1987. "Characterization of Volatile Organic Chemicals in Public Access Buildings." 144-148. In: *Proceedings of the Fourth International Conference on Indoor Air Quality and Climate.* Institute for Water, Soil and Air Hygiene. West Berlin. Vol. 1.

63. Berglund, B. et al. "Air Quality in a Sick Library Over a Period of Sixteen Weeks." 537-541. In: *Proceedings of the Fourth International Conference on Indoor Air Quality and Climate.* Institute for Water, Soil and Air Hygiene. West Berlin. Vol. 2.

64. Hijazi, N.R. et al. 1983. "Indoor Organic Contaminants in Energy-Efficient Buildings." 471-477. In: *Proceedings: Measurement and Monitoring of Noncriteria (Toxic) Contaminants in Air*: A Specialty Conference. Air Pollution Control Association. Pittsburgh.

65. Girman, J. et al. 1987. "Bake-Out of an Office Building." 22-26. In: *Proceedings of the Fourth International Conference on Indoor Air Quality and Climate.* Institute for Water, Soil and Air Hygiene. West Berlin. Vol. 1.

66. Jurinski, N.B. 1984. "The Evaluation of Chlordane and Heptachlor Vapor Concentrations Within Buildings Treated for Insect Pest Control." 51-56. In: *Proceedings of the Third International Conference on Indoor Air Quality and Climate.* Swedish Council for Building Research. Stockholm. Vol. 4.

67. Louis, J.B. and K.C. Kisselbach. 1987. "Indoor Air Levels of Chlordane and Heptachlor Following Termiticide Applications." *Bull. Environ. Contam. Toxicol.* 39:911-918.

68. Wright, C.G. et al. 1981. "Insecticides in the Ambient Air of Rooms Following Their Application for Control of Pests." *Bull. Environ. Contam. Toxicol.* 26:548-533.

69. Goh, K.S. et al. 1987. "Dissipation of DDVP and Propoxur Following the Use of a Home Fogger: Implication for Safe Reentry." *Bull. Environ. Contam. Toxicol.* 39:762-768.

70. 1982. "IARC Monographs on the Evaluation of Carcinogenic Risk of Chemicals to Humans." Suppl. 4. 80. International Agency for Research on Cancer. Lyons.

71. Wright, C.G. and R.B. Leidy. 1980. "Chlorpyrifos Residues in Air After Application to Crevices in Rooms." *Bull. Environ. Contam. Toxicol.* 19: 340–343.
72. Jackson, M.D. and R.G. Lewis. 1981. "Insecticide Concentrations in Air After Application of Pest Control Strips." *Bull. Environ. Contam. Toxicol.* 27:122–125.
73. Ware, G.W. and W.P. Cahill. 1978. "Air Concentrations of Chlorpyrifos (Dursban) from a 2% Slow-Release Paint-On Formulation vs. a Standard 0.5% Emulsion Spray." *Bull. Environ. Contam. Toxicol.* 20:413–417.
74. New Jersey Department of Environmental Protection. 1985. New Jersey Pest Control Code. Title 7. Chapter 30. Subchapter 10.
75. National Research Council. 1982. *An Assessment of the Health Risk of Seven Pesticides Used for Termite Control.* National Academy Press, Washington, D.C.
76. Leidy, R.B. et al. 1985. "Subterranean Termite Control: Chlordane Residues in Soil Surrounding and Air Within Houses." 265. In: R.C. Honeycutt et al. (Eds.). *Dermal Exposure Related to Pesticide Use.* American Chemical Society Symposium Series 273. Washington, D.C.
77. Maye, H.A. and M.H. Malagodi. 1987. "Levels of Chlordane and Chlorpyrifos in Two Plenum Houses: Saranex-15 as a Vapor Barrier." *Bull. Environ. Contam. Toxicol.* 39:533–540.
78. Lillie, T.H. and E.S. Barnes. 1987. "Airborne Termiticide Levels in Houses on United States Air Force Installations." 200–204. In: *Proceedings of the Fourth International Conference on Indoor Air Quality and Climate.* Institute for Water, Soil and Air Hygiene. West Berlin. Vol. 1.
79. Lillie, T.H. 1982. "Chlordane Contamination of Air Force Family Housing." 181–191. In: *Proceedings of the 14th Bioenvironmental Engineering Symposium.* USAF School of Aerospace Medicine. Brooks AFB, Texas.
80. Barnes, E.S. 1984. "An Evaluation of Engineering Modifications Designed to Remove Airborne Chlordane from Crawl Space Houses at McConnell AFB, KS." USAF Occupational and Environmental Health Laboratory Tech. Rep. 84-085EH118APB. Brooks AFB, Texas.
81. Kim, C.J. et al. 1982. "Removal of Pesticide Residues by Laundering Variables." *Bull. Environ. Contam. Toxicol.* 29:95–100.
82. Lillie, T.H. et al. 1982. "Effectiveness of Detergent and Detergent Plus Bleach for Decontaminating Pesticide Application Clothing." *Bull. Environ. Contam. Toxicol.* 29:89–94.
83. Krause, C. et al. 1987. "Pentachlorophenol-Containing Wood Preservatives— Analysis and Evaluation." 220–224. In: *Proceedings of the Fourth International Conference on Indoor Air Quality and Climate.* Institute for Water, Soil and Air Hygiene. West Berlin. Vol. 1.
84. Levin, H. and J. Hahn. 1984. "Pentachlorophenol in Indoor Air: The Effectiveness of Sealing Exposed Pressure-Treated Wood Beams and Improving Ventilation in Office Buildings to Address Public Health Concerns and Reduce Occupant Complaints." 123–129. In: *Proceedings of the Third International Conference on Indoor Air Quality and Climate.* Swedish Council for Building Research. Stockholm. Vol. 5.

4	**SOURCE CONTROL—**
	BIOGENIC PARTICLES

Control of viable and nonviable particles of biological origin (biogenic) and the health problems they cause can be achieved on both preventative and retrofit bases by the application of a variety of source control methodologies. Control of biogenic particles is discussed here in the context of specific health problems. These include allergies/asthma, hypersensitivity pneumonitis/ humidifier fever, and Legionnaires' disease. Allergies/asthma are notable in that disease syndromes are primarily associated with exposure to aeroallergens in residential environments. Cases of hypersensitivity pneumonitis and Legionnaires' disease are, on the other hand, typically associated with air contamination of large nonresidential buildings.

ALLERGIES/ASTHMA

Animal Danders/Dust Mites

Avoidance

A variety of allergies and allergic-type asthmatic conditions can be controlled, in whole or in part, by avoiding exposure to airborne particles that initiate allergic reactions in sensitized individuals (or those who by family history have a high probability of becoming sensitized). Avoidance, in the case of allergies/asthma caused by exposure to animal danders or urine, would require exclusion of allergen-producing pets from indoor residential environments. Such avoidance is commonly recommended by allergists for patients who react strongly to animal-related allergens in standard allergy tests.

A variety of avoidance measures are recommended to dust mite-sensitive allergy and asthma patients by their treating physicians. Recommendations focus on dust control; this is designed to reduce the concentration of dust-borne allergens in the patient's living environment by controlling both allergen production and the dust which would serve to transport it. Sarsfield et al.[1] studied the effectiveness of a variety of dust control measures on symptom expression in mite-sensitive asthmatic children. Avoidance measures were primarily directed at the bedroom environment. These included replacement of feather and down pillows with those having synthetic fillings; enclosing the mattress top and sides with a plastic cover; thoroughly vacuuming mattress, pillows, and the base of the bed; daily damp dusting of the plastic mattress cover; weekly changing and washing of pillow cases, sheets, and under-blankets; and vacuuming of the bed base and around the covered mattress. Other recommended measures included the replacement of woolen blankets by nylon or cotton cellulose ones, frequent blanket and curtain washing, carpet removal and replacement with vinyl floor covering, and the use of a vacuum cleaner with disposable bags. The recommended measures should, in theory, reduce dust mite levels, allergen production, allergen exposure, and allergen-transporting dust levels. Covering mattresses with plastic materials is of particular importance.[1] The impervious plastic prevents allergen-containing fine dust particles from entering the mattress substrate where they are usually deeply imbedded and minimally removed by bed vacuuming. The plastic cover also facilitates vacuuming.

Application of dust control measures resulted in a significant reduction in dust mite populations. Average dust mite populations prior to implementation of control measures were 80/g dust, as compared to 2 afterwards. Significant improvements in the children's asthmatic status were universally observed, as evidenced by a drop in symptom scores. Symptoms of associated allergic rhinitis, however, were unaffected, as were specific IgE antibody levels. A reduction in IgE specific to dust mite allergen would have been expected after a reduction in allergen exposure. Application of similar dust control measures were evaluated in a controlled study of 20 dust mite–sensitive asthmatic children by Murray and Ferguson.[2] Unlike the studies of Sarsfield et al.,[1] which used each patient as his/her own control, Murray and Ferguson separated patients into (1) a control group and (2) a group whose hospital bedroom environment was modified to include dust control. Asthmatic children with dust-controlled bedrooms had fewer days when (1) wheezing was observed, (2) medication was given, and (3) abnormally low peak expiratory flow rates were recorded. Bronchial tolerance to aerosolized histamine was also significantly reduced.

Burr et al.[3,4] investigated the effectiveness of a variety of mite and dust control measures on symptom expression in asthmatic adults and children. In contrast to the studies previously described, they observed no significant improvement in asthma symptoms as evidenced by daily peak flow readings and drug usage. The contradictory results of Burr et al. were suggested to have been due to inadequate reduction in dust exposure achieved in their studies.[2] It was not known to what extent mites or their products were removed from the bedding by the procedure used.

Korsgaard[5] evaluated the effectiveness of a variety of dust control measures on dust mite populations in the homes of 23 patients allergic to dust mites, comparing them to the homes of 75 randomly selected control subjects. Though cleaned more frequently, significantly higher dust mite populations were found in samples collected from patients' bedroom carpets as compared to those of controls. A similar trend was observed for mattress surfaces and living room carpets. No differences were observed in populations of dust mites relative to the amount of dust collected, mattress type, or cleaning frequency. Despite careful instruction, patients did not strictly adhere to the program of mite preventative measures. Compliance typically occurred with procedures that were easy (e.g., replacement of quilts and pillows), but not with difficult or uncomfortable measures (e.g., frequent vacuuming).

Similar results have been reported by Arlian et al.[6] In a study of 26 homes, they observed no correlation between mite abundance and frequency of house cleaning, sheet changing, and mattress pad laundering. Their studies indicated that vacuuming of carpeted floors, fabric-covered furniture, and mattresses resulted in little or no reduction in mite densities, as both dust and dust mites are difficult to remove by vacuuming.

Arlian et al.[6] observed that carpeted floors contain significantly more dust mites than vinyl or wood floors. Mulla et al.[7] also reported that wood and vinyl floors have low mite populations. Arlian et al. noted that long-pile carpets contain significantly more dust mites than short-pile ones; the conclusion was drawn that long-pile carpet provides an excellent microhabitat for mite survival and breeding. They recommended that carpeting be removed from homes of dust-sensitive patients or, if this were not desirable, short-pile carpets be used instead.

Encasing mattresses with plastic is known to significantly reduce mite abundance on beds,[6-8] and this measure is widely recommended by allergists to allergy and asthma patients in the United States. Arlian et al.,[6] however, observed that the mattress is not the principal source of mites in the American midwest. Significantly higher mite populations were observed on the floor beside the bed, the family room couch, and the floor around the couch. The fact that the mattress was not observed to be a major source of mites may have been due to the fact that mattresses were covered with fitted sheets, which apparently prevent the accumulation of human skin scales on the surface. Fitted sheets could prove to be a suitable alternative to the practice of encasing mattresses with plastic while affording more patient comfort. The use of plastic encasing materials would apparently provide little benefit to patients who already have fitted sheets in place.

Based on studies described above, application of frequent vacuuming as a dust (and therefore mite) control measure does not appear to be effective. Indeed, it is more likely to aggravate allergic asthmatic conditions because conventional vacuums are very inefficient. Lehti[9] has reported that vacuuming results in a significant increase in airborne dust concentrations. Ironically, dust collection by conventional vacuums is least efficient for those particle sizes

Table 4.1   Mite Populations per g of House Dust as a Function of Absolute Humidity[10]

Absolute Humidity (g/kg)	Mattress		Bedroom Floor	
	Nov.–Dec.	Feb.–March	Nov.–Dec.	Feb.–March
<7	0	0	0	0
7–8	15	10	5	10
>8	40	100	35	130
	June–July	Sept.–Oct.	June–July	Sept.–Oct.
<9	10	10	10	10
9–10	10	10	0	10
>10	60	110	30	30

(< 2 μm) which are likely to have the greatest allergenic or asthmatic significance. Vacuuming would be best accomplished by the application of HEPA-type cleaners and those that entrain dust in a liquid medium. This would, in theory, prevent or reduce the suspension and dissemination of allergenic dust particles by the act of vacuuming itself.

Climate Control

Korsgaard[10] has proposed that dust control measures are unlikely to be effective in controlling dust mites and allergen production because they ignore the major environmental factor that affects dust mite population survival and growth. He has observed a strong correlation between absolute humidity and the concentration of dust mites (Table 4.1). Few mites were found at absolute humidities below 7 g/kg, i.e., 45% RH, at 20°C. When patient and control homes were compared, both humidity levels and dust mite numbers recovered from bedroom carpets were significantly higher in the former. A tendency was also observed towards greater numbers of mites from mattress surfaces and living room carpets in patient homes. A correspondingly higher indoor air humidity level was recorded in patient homes.

Despite a similar outdoor climate throughout Denmark, the level of indoor humidity (particularly in the colder seasons) was a well-defined characteristic of each individual apartment Korsgaard[10] studied, and these differences were largely responsible for the variation in dust mite populations observed from one apartment to another.

The significant correlation between absolute humidity and dust mite populations was observed only for the Danish cold season period from November to March. Interestingly, homes with few or no mites in winter continued to have low mite counts in summer and autumn irrespective of a rise in humidity.[10] It appears, therefore, that the effect of indoor humidity conditions is to permit the survival of mite populations in a few critical dry winter months.

Stressing the fact that avoidance measures have been clinically disappointing, Korsgaard suggested that it would be more reasonable to focus on decreasing indoor humidity (particularly during the winter period) as a means of controlling dust mite populations.[10]

Korsgaard[11] suggested that the two main determinants of indoor humidity levels are magnitude of the household moisture load and failures in construction that cause water vapor from the underlying soil to be absorbed by building materials. In newer homes, the cause of elevated moisture levels is suggested to be, in part, lower air infiltration, and therefore lower ventilation, rates.

Humidity control, i.e., reduction in household humidity levels to support minimal dust mite populations, can in theory be achieved by the application of a variety of moisture control measures. These include forsaking humidifier use during winter periods, use of dehumidifiers during high-humidity periods, use of central air conditioning,[12] and a variety of techniques that reduce moisture entry and facilitate moisture egress from the building environment. An expanded discussion of moisture control measures is presented at the end of this chapter. Effective control of mites would require the maintenance of relative humidities below 50%.

The effectiveness of air conditioning on controlling dust mite populations has been reported by Carpenter et al.[12] in a study conducted in Hawaii. They observed a significant correlation between percentage of time that the air conditioner was on and absolute humidity. Homes with air conditioning on constantly had significantly lower mite counts than non–air-conditioned homes. When air conditioners were disconnected, dust mite populations rose significantly.

Acaricides

Dust mite populations can be controlled by the application of aqueous acaricides to mite-infested materials such as bedding, carpeting, and upholstery. Combinations of benzoic acid esters and polymers or solid adsorbents (ACROSAN products, Werner and Mertz GmBH, Mainz, Germany) have been shown to be very effective acaricidal agents under real-world test conditions.[13] Bischoff[13] evaluated the effectiveness of acaricides on mite populations and excrement content of house dust from 10 houses. Within a few days after treatment, dust mite populations were shown to be reduced to 0–10% of their pretreatment numbers, depending on thickness and texture of the textile article treated. A single treatment was sufficient in some cases, while one or more after-treatments were necessary in others. Typically, the final mite destruction rate ranged between 95% and 100%.

Effects of acaricidal treatment on allergen content of dust over a period of 28 months can be seen in Table 4.2. For upholstery, allergen contamination is seen to decrease in every case. Results for mattresses, however, were mixed. In some instances the allergen contamination decreased; in other cases it did not. The presence of allergen contamination in mattresses in the absence of

Table 4.2   Effect of Acaricidal Treatments on Dust Mite Allergen Indicator Concentrations of Guanine (ACAREX score) in Upholstery and Mattress Dust in 10 Houses[13]

	House No.									
	1	2	3	4	5	6	7	8	9	10
Upholstery										
Before Treatment	3	3	0–1	3	2–3	2	1–2	1	3	2
Weeks After Treatment										
3	3	2–3	1	2		0–1		1		
7	2	2	1	2–3	1–2	0–1	1–2	1		
Months After Treatment										
4	1	2			1–2				1	
9	1	2	1	2	1	1.2			1	
12	0–1	1	1–2	1					1	
16	1	1		1					1	0
Mattresses										
Before Treatment	3	3	2–3	3	2–3	2	3	2–3	2	2
Weeks After Treatment										
3	3	3	2–3	3		2		3	2	1–2
7	2–3	3	3	2	1–2	2	3	2–3	1	1
Months After Treatment										
4	2	2	3		1–2					1
9	1–2	3	3	3	1	1–2		2–3	0	1
12		3	3	2					0	1
16			3	2–3					0	0

dust mites was suggested to be due to the persistence of pretreatment contamination. Mattress replacement was advised.[13]

Kniest[14] applied an acaricidal cleaning agent to floor coverings, mattresses, and furniture in three Dutch homes. Mite and allergen contents of dust were reduced by approximately 80%. In addition to benzoic acid esters (e.g., benzyl benzoate), a variety of other substances have been shown to significantly reduce dust mite populations when evaluated under laboratory or field conditions. These include pesticides that are toxic to mites, fungicides that inhibit the microbial decomposition of human skin scales, and liquid nitrogen. Acaricides that have been shown to have significant toxicity to dust mites include pirimiphos-methyl, benzene hydrochloride, lindane, synthetic pyrethrins, dibutyl phthalate, and diethyl *m*-toluamide.[15-17] Dust mites can also be controlled indirectly by fungicidal treatments to inhibit the mold conversion of human skin scales into a mite-edible product.[18-20] Significant control of dust mites in mattresses has been reported for Paragerm AK, a bactericidal and fungicidal spray used commonly in French hospitals.[20] Paragerm AK is a complex solution of balsamic essences associated with benzoic acid, salol, thymol, terpineol, citrus fruit natural essence, natural essences from *Syringa* and *Nardus*, two halogenated phenyl alcohols, and light liquid paraffin. It is anhydrous, very volatile, and claimed to be nontoxic. Natamycin, a fungicide applied to mattress dust, has also been reported to significantly reduce dust mite populations.[19,20] The effectiveness of Natamycin in controlling dust mites is apparently due to inhibition of the growth of *Aspergillus penicilloides,* which is important to mite growth and survival.[18]

While a variety of acaricides appear to be effective in reducing dust mite populations (some on sustained bases after a few application treatments),[13] a major question remains as to the advisability of using pesticides whose safety is unknown under the conditions of exposure in a patient's living environment, particularly where the applied chemicals are so immediate to a patient's breathing zone (e.g., in bedding materials). Chlorinated hydrocarbons such as lindane and benzene hydrochloride would pose health risks somewhat similar to chlordane and pentachlorophenol, as they would tend to accumulate in fat tissue even at low levels of exposure. Chlorinated hydrocarbons as a group also tend to be carcinogenic. Evaluation of liquid nitrogen as a safe acaricidal treatment alternative to pesticides has been conducted by Colloff.[21] While effective, liquid nitrogen treatment requires significant care in application, has a potential for damaging materials to which it has been applied, and appears to provide only short-term control. The latter may be due to reinfestation from nontreated materials or incomplete eradication.

Benzoic acid esters such as benzyl benzoate have been shown to be very effective acaricides in both laboratory and field evaluations. Health risks, based on what is presently known, appear to be slight, as benzoates are rapidly metabolized in the body to hippuric acid, which is excreted in urine.[16]

All reported acaricidal studies for dust mite control have been conducted in Western European countries. Before any of the pesticidal treatments can be promoted for use in the United States, they will have to be approved by the EPA. Without such approval, the use of pesticides to control dust mites, and thereby mite allergen–induced allergy and asthma, will not be an option in the United States. The potential immediate benefit to patients seems to be considerable.

Mold Contamination

Avoidance

The principle of avoidance is applicable to initial prevention, remediation, and future prevention of mold contamination problems. Excessive mold contamination of indoor spaces, causing allergies and asthma, can be avoided by consideration of factors that contribute to its growth and dispersion. Water is in almost all instances a key factor. Mold growth is favored by the presence of liquid water and high-humidity environments. Many mold species require a minimum relative humidity of 70% for spore germination.[22] Avoidance of mold allergies and of asthmatic conditions caused by mold contamination in residential environments can be best achieved by maintaining a relatively dry home environment and preventing condensation on indoor surfaces.

Remedial Measures

Application of remedial measures to control mold infestation of residential environments requires that materials and systems that are the source of mold growth and/or dispersal be identified. Short-term control measures would necessarily focus on removing and/or disinfecting mold-damaged materials and mold-contaminated systems.

From the homeowner's perspective, disinfection is the most desirable approach to controlling a mold contamination problem, since it may require little effort and expense and infested materials can, in theory, be salvaged. While a variety of disinfectants is commercially available, there is very little evidence that their effectiveness goes beyond those organisms that are on the surface of contaminated materials or are easy to reach. Where materials are amenable to disinfection, the use of household bleach or sodium hypochlorite at a dilution of one cup to a gallon is recommended.[23] Disinfection may arrest mold growth, but will not reduce the allergenicity of spores, because allergenicity is independent of viability.[24]

In many cases, mold infestation or damage is so extensive that removal of damaged materials is necessary. In theory, removal is the best choice. How effective it is in reducing airborne levels of mold on a sustained basis, however, depends on accurate identification of contaminated materials and systems and implementation of measures to control the problem that caused mold infestation in the first place.

In residential environments, mold growth may occur on walls, carpeting, ceiling tiles, shower curtains and tiles, books and magazines, shoes, furniture, clothing, storage boxes, etc. It can also occur in humidification and air conditioning systems. In the latter case, and in the case of structural materials such as walls, floors, etc., removal is unlikely to be an acceptable approach to homeowners (unless significant structural damage has occurred.) Kozak et al.[24] have investigated mold levels and mold damage to materials in the homes of seven asthma patients. High mold levels were observed to be associated with jute-backed carpeting repeatedly wetted by bathroom spills, frequent carpet shampooing, pet urination, etc. In one house, a dramatic reduction in a patient's asthmatic symptoms occurred after damaged carpets and pads were removed. In a second house, a significant reduction in mold counts occurred after removal and replacement of mold-damaged carpeting and a wicker basket.

In another study, Kozak et al.[25] observed that significantly higher house mold levels were associated with high shade levels and high levels of organic debris nearby because of poor landscaping and maintenance. The effect of shade on indoor mold counts can be seen in Table 4.3. By implication, reduction in shade level and removal of organic debris from around a home would

Table 4.3   Effect of Shade Levels on Indoor Mold Spore Counts—CFU[a]/m3 25

| | Shade Level | | |
	Minimal	Moderate	Heavy
Mean	408	421	2,228
Minimum	36	36	1,109
Maximum	3,828	2,004	5,984
Homes/Category	40	22	6

[a]CFU = colony-forming units.

be expected to result in lower indoor mold concentrations. Lower mold levels were also reported when allergic patients complied with good dust control practices.

Remedial control of residential mold contamination often requires the implementation of practices to change the environmental conditions that initiate and sustain mold growth. These practices, which for the most part focus on moisture control, include the prevention of condensation on interior surfaces and wall covers, and maintenance of indoor relative humidities at desired levels.

Elevated levels of mold and other microbial organisms indoors can result from the use of domestic humidifiers and cold mist vaporizers.[26,27] Solomon[26] observed that the efflux from newly purchased cold mist vaporizers, or those that had been cleansed, dried and stored for considerable periods, was relatively free of viable mold spores. Vigorous cleaning, or simply draining the unit and refilling with tap water, greatly reduced yeast levels, though filamentous mold forms were relatively unaffected. The reservoir fluid apparently was the source of yeasts, while filamentous forms grew on components that could not be easily cleaned.

Burge et al.[27] studied, among other things, the effectiveness of antifoulants added to humidifier fluid reservoirs on microbial populations. No significant differences were observed in microbial recoveries in humidifier units (furnace and console) that contained antifoulants as compared to those that did not. The most common antifoulant used was sodium hypochlorite. The apparent ineffectiveness of antifoulants, and the potential adverse health effects of aerosolizing antifoulants currently available, makes such use highly undesirable.

The best approach to controlling mold and other microbial contamination from humidifiers is to terminate their use. Use recommendations by physicians to alleviate allergy-type symptoms are based on "seat-of-the pants" medicine and have no scientific data base to support them.

HYPERSENSITIVITY PNEUMONITIS

Avoidance

Hypersensitivity pneumonitis (HP) outbreaks in offices and other public access buildings can be avoided by consideration of those factors that contribute to microbial growth, most notably water. Microbial growth responsible for HP outbreaks is favored by the presence of liquid water in cooling coil drip pans, evaporative condensers, humidifiers, and cooling towers. Avoidance of HP outbreaks in office spaces and other large buildings requires selection of systems that decrease the likelihood of microbial growth and subsequent contamination of building air handling systems, as well as fastidious adherence to the prevention of microbial growth.

Recommendations for the prevention and mitigation of HP outbreaks in large mechanically-ventilated buildings have been made by Morey et al.[22,28] From a systems standpoint, they recommend that humidification units based on recirculated water should not be used, as these almost always become rapidly contaminated by organic dusts and organisms. Steam as a moisture source in HVAC systems is preferred. If cool water humidifiers are to be used, water should originate from a potable source and run to a drain instead of being recirculated. Cool water humidifiers should be regularly inspected, cleaned, and disinfected. Because they are known to readily contaminate indoor air with microorganisms, cool mist vaporizers are not recommended for building humidification.

Any water-using or -generating system associated with HVAC systems should be inspected periodically so that preventive maintenance practices can be implemented. Stagnant water should not be permitted to accumulate in cooling coil condensate drip pans. Drip pans should be properly inclined to provide continuous drainage. If contaminated with microbial slime, the slime should be mechanically removed and the pan cleaned with detergents or by steam lancing. Slimicides containing chlorine and proprietary biocides can be used for disinfection as long as they are removed prior to reactivation of the air handling system.[29]

Building relative humidity levels should not exceed 70%. Relative humidity can be lowered by either reducing air moisture content or raising air temperature. Moisture levels can be reduced by running cooling coils at a low enough temperature to cause conditioned air to be further dehumidified. Humidity levels may also be reduced by decreasing the amount of recirculated air and increasing the amount of outside air;[22,28] this is dependent on outdoor humidity being lower than that indoors. Another avoidance practice would be to maintain plumbing and other fluid systems so that no water incursions occur in buildings. When such incursions occur, they should be rapidly cleaned up and water-damaged materials discarded. Microbially contaminated materials (such as ceiling tiles and carpeting) resulting from a water incursion should be discarded and replaced.

Remedial Measures

A variety of remedial actions have been proposed or applied by investigators of HP outbreaks. Morey et al.[28] conducted detailed investigations of five office buildings. In one of these, microbial contamination found in the HVAC system was considered to be responsible. For this building they recommended that all nondisposable building contents, including books, desks, carpets, drapes, HVAC system ductwork, and water spray-direct expansion surfaces, be thoroughly cleaned with a HEPA-type vacuum cleaner and that those not amenable to cleaning be discarded and replaced. In a second building where an outbreak of HP appeared to be associated with a flood from a cafeteria, recommended mitigation measures included (1) structural changes in plumbing; (2) discarding of damaged carpeting and ceiling tiles; (3) cleaning of all upholstered furniture, wall partitions, and office materials that needed to be reused with a HEPA-type vacuum cleaner; and (4) disinfection of the floor with chlorine bleach. Despite the implementation of these remedial measures, a second outbreak occurred as a consequence of the liberation of large amounts of dust when office partitions were subsequently handled. In a third building, remedial measures recommended included (1) providing adequate drainage of cooling coil condensate; (2) installing deep, sealed, water-filled traps; and (3) cleaning and disinfecting cooling coils and drain pans of fan coil units. In a fourth building, installation of adequately-sized access doors to the cooling deck portion of air handling units was recommended to facilitate a preventive maintenance program of removing slime and stagnant water. A fifth office building had a history of floods from roof leaks, with relative humidities exceeding 70% during the summer air conditioning season. Remediation recommendations included preventing moisture incursions into occupied spaces from drain pan overflows, replacing filters from fan coil units and air handling units routinely and frequently, and operating the main air handling units during all occupied times.

In other HP outbreaks, Weiss and Soleymani[30] removed a water spray unit from the HVAC system and replaced it with an air-cooled system and the symptoms ceased. Banazak et al.[31] removed water systems and steam cleaned the HVAC system. Aranow et al.[32] replaced the HVAC system and replaced all furnishings in the occupied space. Scully et al.[33] cleaned ductwork. Bernstein et al.[34] cleaned components of the HVAC system with ammonia and chlorine and provided regular maintenance; despite significant reductions in mold levels, HP symptoms reoccurred. Ganier et al.[35] tried unsuccessfully to decontaminate the humidifier with a fungicide and subsequently removed it from the HVAC system.

Cockcraft et al.[36] observed an outbreak of humidifier fever among staff in a hospital operating theatre. Draining the static spray-type humidifier twice a week and cleaning it every two weeks were not effective in preventing the buildup of antigenic materials. The process of running out as much water as was run in was observed to be effective but impractical because of the loss of

cooling effect when ambient temperature reached 21°C. Draining the humidifier tank daily and replenishing it with fresh water was effective but very labor-
intensive. Cockcraft et al. also evaluated three biocidal compounds: Halophan
(a bismethylene chlorophenol compound), Resiguard (a mixture of picloxy-
dine, octyl phenoxy polyethanol, and benzalkonium chloride) and Metroni-
dazole (an antiamoebic drug). All were observed to be ineffective in controlling the amoebae responsible for producing humidifier fever antigens.

Application of remedial measures in response to HP and humidifier fever-
type outbreaks in buildings is predicated on identifying sources of microbial
contamination (in some cases confirmed by serological testing of those
affected). Remedial measures are applied to prevent problems from reoccur-
ring. If outbreaks do not subsequently occur, remedial measures are judged to
be effective; conversely, if subsequent outbreaks do occur, measures are
judged to be ineffective or partially effective. Remedial measures recommended are based on intuitive determination and do not represent systematic
evaluations of efficacy.

LEGIONNAIRES' DISEASE

Controls applied for both remediation and prevention of outbreaks of
Legionnaires' disease focus on treating cooling tower and evaporative condenser waters with biocidal chemicals. A variety of proprietary compounds
specially formulated for use in large evaporative cooling systems is available
and many companies (75 national companies in the United States) specialize in
treating waters of cooling towers.[37] Because of cost, such services are usually
used only on larger systems.[38] A variety of biocidal compounds and formulations have been shown to be effective in controlling *Legionella pneumophila*
growth under laboratory conditions.[39-42] Biocides commonly used to treat
cooling tower waters include quaternary ammonium compounds, 1-bromo-3-
chloro-5,5-dimethyhydantoin, bis(tri-*n*-butyltin) oxide, *n*-alkyl-1,3-
propanediamine, methylene-bis(thiocyanate), dithiocarbamates, and
chlorine.

The effectiveness of biocidal treatments under field conditions appears to
fall considerably short of expectations based on laboratory results. Witherell
et al.[37] sampled 130 operating cooling towers in Vermont. Of these, 52%
practiced some type of biocidal water treatment. No statistically significant
differences in recoveries of *L. pneumophila* were observed between treated
and untreated cooling tower waters. A slightly increased prevalence of *L.
pneumophila*-positives were observed among units treated with quaternary
ammonium compounds on a once-per-week schedule. Witherell et al. concluded that use of biocidal treatments for control of *L. pneumophila* may be
of no benefit, and may in fact facilitate its growth by eliminating competing
organisms. Braun[43] tested a variety of commercially available biocides (including dithiocarbamates, bis(tri-*n*-butyltin) oxide, and *n*-alkyl dimethylbenzol

ammonium chlorides) on waters of cooling towers and evaporative condensers. All were ineffective. The apparent discrepancy in efficacy results observed for laboratory and field tests appears to be due in part to the difficulty in maintaining a residual concentration of biocides in continually operating systems.[44] For example, an initial concentration of quaternary ammonium salts of 20 ppm was observed to drop to less than 1 ppm in three hours.[43]

Barbaree et al.[45] have observed that *L. pneumophila* is ingested by amoebae and a cilate, *Tetrahymena* spp., in which it undergoes intracellular multiplication. This suggests a mechanism by which these disease-producing bacteria can persist through adverse conditions and then propagate rapidly under favorable conditions. It may also explain, in part, the observed ineffectiveness of biocides under real-world operating conditions.

The use of chlorine or chlorine-releasing compounds has been suggested for controlling *L. pneumophila* in cooling towers.[46] Fliermans and Harvey[44] claim this to be the most effective biocidal treatment under field use. A free chlorine residual of 2 ppm is apparently effective in controlling *L. pneumophila* under cooling tower operating conditions.[47] Chlorine is often applied as a shock treatment (i.e., application of high chlorine concentrations).[48] A rapid reduction of chlorine concentration can be expected as it reacts with and oxidizes the living and nonliving organic substances typically found in cooling tower waters. Chlorine will also cause corrosion and attack wooden packing materials in the cooling tower structure.[49]

In addition to biocides, cooling and evaporative condenser design factors may affect the growth of *L. pneumophila* and aid in its dispersion. In new and well-maintained systems, for example, mist eliminators limit drift loss to 0.05–0.2% of the circulating water flow rate. Older towers operated under poorer conditions, on the other hand, have drift losses which may be 5–6 times this amount.[38]

Mallison[50] reported that cooling towers having wooden components and/or that use open decks to distribute flow of cooling water favor the growth of *L. pneumophila*. However, in the Vermont study[37] no statistically significant association between cooling tower design features and recoveries of *L. pneumophila* were observed. Units with wooden structural components were no more likely to be associated with *L. pneumophila* than those with metal or other construction materials.

MOISTURE CONTROL

Moisture, present as a liquid or vapor, is the common denominator underlying contamination of indoor air by biogenic particles of high allergenicity, particularly mold and dust mites. Sustained control of such particles and of the illness syndromes associated with them can, in theory, be best accomplished by application of measures to prevent their growth. Control efforts directed to moisture control are therefore of considerable relevance. Moisture problems in residences typically result from occupant behavior and practices, and from defects in structural design. Occupant-induced moisture problems include (1) repeated accidental or intentional wetting of materials such as carpeting, flooring, etc.; (2) inattention to necessary maintenance associated with plumbing, roofing, etc.; (3) failure to turn on bathroom ventilation fans; and (4) excessive use of humidification. Defects in structural design include (1) inadequate provision for site and substructure drainage; (2) structural features that cause condensation on indoor building cavity and substructural surfaces; and (3) inadequate crawl space, attic, and occupied building space ventilation.

Occupant-Induced Moisture Problems

Moisture problems resulting from the activities of building occupants are unfortunately all too common. Their prevention or remediation basically requires a good application of common sense. Occupants must not respond with indifference to the need for immediate repair of leaking or blocked plumbing fixtures and leaks in roofs and walls. Supervision of children in preventing overflows in sinks and tubs is essential, as are quick cleanup responses when spills occur. Use of ventilation fans (when they are already installed) in bathing areas is essential and should not be viewed as discretionary. Humidification during the winter may result in condensation on cold windows and contribute to both survivability and growth of mold and dust mites. There are apparently few documented benefits of routine humidification that are worth the potential problems associated with its use.

Structural Factors

Common moisture problems due to structural defects include leaking roofs; water standing in basements and crawl spaces and under slabs; and condensation in attic areas, on floor joints, and on internal surfaces. Rain penetration of roofing is a common problem. Its minimization in new construction requires care in installing flashing materials.[51] Defects in site preparation, building design, and construction often result in relative humidities which are significantly above those associated with local climate.

Soil Moisture

Groundwater seepage into basements and ponding under crawl spaces or slabs represent inattention to site preparation, which is essential to the prevention of such problems. Basements invariably require a complete system of drain tile (with sufficient depth of gravel) underneath the slab and along the foundation perimeter. Drainage should be to a sump for rapid, pumped removal, or to a gravity surface drain.

Ponding of water beneath a crawl space or slab can be prevented by providing for drainage away from the building site (and, of course, not to someone else's site!). If, after grading, the site is still at risk (of being water-logged), field tile should be installed to intercept and divert surface water and groundwater.[52] The provision of adequate surface drainage can be easily accomplished in subdivisions with storm drains or where the homesites are large. In some subdivisions the provision of adequate surface drainage may be nearly impossible because of distance from stormwater and sewer systems, poor suitability of homesites, and the small size of building lots. In such instances, drainage problems cannot be remedied because of either physical or cost constraints.

Remediation of a drainage problem around an existing structure is, as with so many other things, much more difficult and costly than designing and constructing it properly in the first place. In some instances a drainage problem is not evident when construction is occurring on a particular site. This happens in the case of a seasonal high water table; high surface water levels may only occur during a limited time of the year. The problem of water seepage into basements and standing under substructures represents only one soil moisture problem. Soil moisture in the absence of standing water can be the dominant source of moisture in residences.[53] For example, soil moisture flux from the ground in a crawl space may be as high as 50–80 L/day in a 130-m² house.[54,55] Daily moisture generation by a 3- to 4-member household is considerably less, on the order of about 14 L/day. This latter moisture is generated from humans directly and indirectly by bathing, cooking, dishwashing, plants, and a variety of miscellaneous activities.[53,55]

One effect of soil moisture (whether it is standing or in the vapor phase) is to increase humidity levels below crawl spaces and within structures. This may lead to mold growth on basement materials, crawl space timbers, and soil surfaces. It may cause elevated indoor relative humidities, which promote mold growth and dust mite populations. It may also result in condensation of water vapor on crawl space building timbers and on cold indoor surfaces such as windows and walls.

The high water production by soils under crawl spaces can cause condensation to occur on cold surfaces of house floors. Of the approximately 40-L/day daily moisture generation rate measured for a New Zealand house, only 3% could be absorbed by subfloor timbers without them becoming excessively wet.[52]

Recommendations for control of elevated moisture levels under crawl spaces and of the structural damage they may cause focus on providing sufficient subfloor ventilation. Subfloor ventilation is provided passively by placement of grills in the foundation perimeter concrete walls. In New Zealand's building code, crawl space ventilation requirements include the provision of a 100- by 35-mm clear ventilation opening for every square meter of floor area. Vents are to be positioned no more than 1.8 m apart and 0.75 m from foundation corners. Vents or grills are to have a high ratio (greater than 50:50) of perforations to solid materials. Where it is difficult to ensure an adequate flow of air (e.g., under buildings enclosing large spaces), a moisture barrier (such as polyethylene) placed over the ground area is recommended.[52]

The University of Washington Energy Extension Service recommends 1 ft^2 (0.092 m^2) of net free ventilation for every 1500 ft^2 (138 m^2) of floor area or 1.5 ft^2 (0.138 m^2) of net free ventilation for every 25 linear feet (8.5 m^2) of foundation wall.[56] Net free ventilation is defined as the total area of the vent minus the area taken by louvers and screens.

Adequate subfloor ventilation requires that foundation plantings and other obstructions be kept away from vent openings to maintain airflow. Good ventilation practice would keep subfloor vents open at all times, including during the heating season. In the author's experience (in the midwest and northeast), crawl space vents are often obstructed by foundation plantings and homeowners typically close crawl space vents during the heating season. The latter poses few risks in houses on dry sites (with the possible exception of radon). Vents are often closed to prevent discomfort (from cold floors) and to conserve energy. Such concerns should be addressed by insulating the cavities between floor joists.

Heating systems can play a significant role in increasing the rate of moisture entry into indoor spaces and can serve as a means of mold entry from basements, slabs, and crawl spaces into living spaces. This is more likely to occur when cold air returns are both leaky and located in the substructure. The large negative pressures induced by furnace fans will cause moisture-laden (and possibly mold-laden) air to be drawn in and circulated throughout the house. In houses with heating ducts located within or below the slab, seasonal high water tables may result in moisture entering the ducts, with subsequent vaporization and excessively high humidity levels.

Cold air returns in moist and/or moldy basements or crawl spaces should be inspected for leaks and sealed with duct tape. A leaky cold air return may render a dust cleaner (installed to control mold levels) ineffective, since such cleaners are installed in cold air returns upstream of the large furnace fan, and contaminated air can be easily drawn in downstream of it. A furnace in a moist/moldy basement may also be a problem since it too can be an entry point for moisture-laden and mold-contaminated basement air. In such instances, it may be necessary to isolate the furnace and cold air return from the basement.

The need for a well-sealed cold air return duct applies to attics as well. Though attics are not usually a source of moisture, cellulose materials used as attic insulation provide an excellent medium for mold growth and yield high mold recoveries.[27]

Condensation

The condensation of water on internal surfaces, in interstitial spaces, and in attics can result in environmental conditions conducive to the growth of mold and other microorganisms and may result in structural damage. Indeed, in most instances, it is the latter which receives the attention of governmental housing authorities and building scientists.[57-63] Because condensation provides a good medium or environment for mold growth, its control is an important factor in controlling illness syndromes that have a mold etiology.

Water vapor condenses when interior surfaces are at a temperature below the dewpoint of indoor air. In winter (in northern climates), surface condensation is most common on windows and window frames, at wall corners, on floors, and over the gypsumboard face at or near the stud connection of walls.[58] Runoff of condensate from windows has been observed to cause damage and a mold-promoting environment on window frames, gypsumboard, and flooring. Condensate may also penetrate wall cavities where mold growth and structural damage occurs.[51]

Three factors are involved in the process of surface condensation: indoor temperature, indoor relative humidity, and surface temperature. Surface temperature is dependent on outdoor temperature and thermal resistance of structural elements. Surface condensation can also occur on basement floors when basements are ventilated under summertime conditions. Because of high moisture contents of outdoor air and high dewpoints during warm weather, uninsulated basement floors may be colder than the dewpoint.[58]

Control of surface condensation can be accomplished by either increasing the thermal resistance of structural elements subject to surface condensation or reducing the humidity level of house air. In the former case, thermal resistance of windows can be increased by using double or triple-paned windows and installing insulation where necessary. Surface condensation may also be reduced during winter by increasing the ventilation rate or reducing the number of moisture sources. Condensation on basement walls may be controlled by insulating the basement foundation walls so as to prevent warm, humid air from reaching the cool or cold concrete. The most practical approach to controlling condensation on basement floors is to dehumidify basement air.[58]

Condensation can occur in wall cavities where such condensation and attendant mold growth and structural damage may not be readily apparent to building occupants. Such interstitial condensation results when moist air moves from a building interior into a cavity and comes in contact with surfaces below the dewpoint of indoor air. This movement may occur by diffusion through wall or ceiling materials and by air leakage through holes or cracks.

Diffusion of water vapor into building cavities depends on the permeability of wall materials to water vapor. It may be controlled by placing a vapor barrier on the warm side of walls and ceilings. Typical vapor barrier materials include polyethylene plastic and aluminum foil. Low-permeability paints can also be used, particularly in retrofit applications. Though major attention is given to the installation of vapor barriers in new construction, air leakage through the building envelope is by far the major mechanism of moisture transfer into building cavities. This air leakage can be controlled by reducing the number and size of cracks and holes.[58,63] However, such tightening of the building envelope can be expected to cause increases in indoor humidity levels. Other recommendations to control condensation in building cavities include (a) ventilating the building to reduce interior vapor pressure and (b) ventilating cavities.[62]

Condensation often occurs in attic areas. Such condensation is due to positive air pressure at the ceiling, which causes infiltration of moisture-laden air through cracks and holes. Flat-roofed houses appear to be more susceptible to condensation problems than houses with sloped roofs. Because of the limited attic space in the former, air leaking through openings in the ceiling comes into immediate contact with the cold sheeting and condenses.[64] Mobile homes are a notable example of flat-roofed structures which have a high frequency of roof/ceiling condensation problems.

In studies conducted in Canada,[51] the frequency of attic condensation was high in two types of climatic regions. In northern regions, attic condensation was associated with a prolonged, cold winter, lots of sunshine throughout the year, and a strong drying potential during the spring. Under such climatic conditions, moisture escaping into attic spaces froze on sheeting and truss elements and when it thawed out rapidly in spring, melt water ran down the walls and ceiling to living spaces below. In maritime areas, attic condensation was associated with a relatively mild but prolonged heating season, high outdoor humidities, and a minimal amount of sunshine.

A variety of factors has been observed to contribute to attic condensation problems. These include (1) higher-than-average indoor humidity levels, (2) inadequate attic ventilation, (3) interior partitions/party walls venting into attic spaces, and (4) the stack effect, which promotes flow of humid air through cracks and openings into the cold roof space.

Control measures recommended for preventing attic condensation include pressurizing the roof space using a fan, sealing leaks, and increasing the airtightness of ceiling construction.[58,61] Prevention of attic condensation can be achieved by the provision of sufficient ventilation to flush out moisture that enters from the living space below. The effectiveness of ventilation depends on the quantity of moisture present in the attic, the capacity of outside air to absorb extra moisture, and the air flow rate.[58] Moisture can only be removed when outdoor air is drier than attic air. The rate of removal will depend on the difference in moisture content between the indoor and outdoor air. Installation of ventilators in the soffit on eaves on both sides of the roof is one of the most effective ways of improving roof ventilation.[59]

The University of Washington Energy Extension Service[56] recommends that the attic be provided with 1 ft² (0.092 m²) of net free ventilation for every 300 ft² (27.6 m²) of attic space if a vapor barrier is present, and 1 ft² (0.092 m²) for every 150 ft² (13.8 m²) of attic space when a vapor barrier is not present. Different types of attic vents are illustrated in Figure 4.1. Roof vents using motor-driven fans are also available.

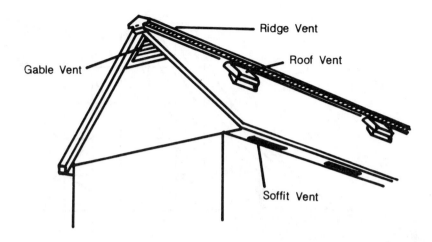

Figure 4.1. Attic ventilators.[55]

Humidity Control

The best approach to controlling a house moisture problem is, in theory, to identify the major moisture sources and then control moisture at the source. In many instances this may be difficult, impractical, or too costly. In such circumstances it may be appropriate to reduce vapor concentration of the air by other means. There are three basic approaches to humidity control: temperature control, ventilation, and dehumidification.

Germination of mold spores is favored by relative humidities of 70 + %. Mold spore germination can be minimized, therefore, by maintaining an indoor climate in which relative humidity does not exceed 70%. Since relative humidity is a function of temperature, it can be controlled to some degree by manipulating indoor temperature.[59] This is important in the context of night-time setback temperatures for occupied buildings, and temperature maintenance in buildings which are unoccupied for long periods of time (e.g., summer homes, unoccupied homes offered for sale, etc.). In the latter instances, wintertime heating is often only sufficient to protect plumbing.

Let us assume that we wish to maintain an indoor climate in which relative humidity does not exceed 70%. From a psychrometric chart (Figure 4.2), we can determine which setback temperatures can accommodate different air moisture contents without exceeding a relative humidity of 70%. For example, if we begin with an indoor temperature and relative humidity of 22°C and 50% RH, a setback temperature of 16°C would result in a relative humidity of 70%. If, however, the relative humidity were 60% at 22°C, the setback temperature should then be no lower than 19°C. This is of course considerably higher than setback temperatures (7–13°C) commonly used in unoccupied residences and summer homes.

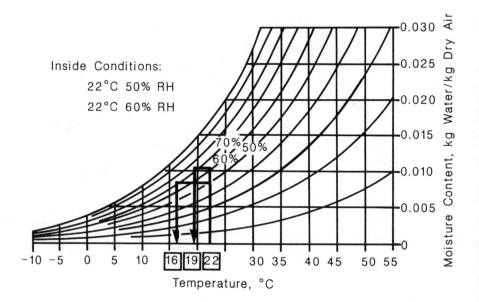

Figure 4.2. Psychrometric chart—relation between temperature and relative humidty.

As previously discussed in the context of crawl spaces and attics, ventilation by passive or active means is one of the most effective ways of reducing relative humidity. Ventilation cannot be described as a source control measure when the focus is moisture control alone. If, however, the focus is on controlling biogenic particles (which are produced in large numbers because of moisture levels favoring the development of organisms that produce them), then the application of ventilation can be seen as a source control measure. It is within that context that ventilation for moisture control is described here.

Ventilation for moisture control in living spaces of houses can be accomplished by a variety of approaches. One of the simplest of these would be to capture and exhaust moisture-laden air in high-production areas such as bathrooms, kitchens, and around automatic clothes dryers. This requires special exhaust fans, available in most modern houses and clothes dryers.

Humidity control during winter periods can be achieved by natural ventilation processes. Colder, drier outdoor air will both dilute indoor water vapor levels and carry water vapor to the outdoor environment by exfiltration. This process is facilitated by airflows up the chimneys of fuel-heated houses and is retarded in all-electric houses. Houses without chimneys, i.e., those heated electrically, tend to have higher indoor relative humidities and a greater prevalence of condensation problems.[64] The rate of air flow up the chimney may vary from a maximum of approximately 200 L/sec for an open fireplace, to 70 L/sec for an oil furnace with a barometric damper, to 25 L/sec for a conventional gas furnace, to 5 L/sec for an airtight stove. These would be equivalent to air change rates of 1.3, 0.5, 0.17, and 0.03 ACH, respectively, in a 252-m^3 house.[64]

In energy-efficient houses, special steps may have to be taken to provide sufficient outdoor air to control moisture levels in indoor spaces. In Great Britain, two types of passive ventilation systems have been developed to control water vapor levels for existing and new construction. They are referred to as stack and cross ventilation. Stack ventilation is suggested to be most suitable for two-story structures and for single-story dwellings with pitched roofs and a number of separate rooms. In this ventilation approach, a vent pipe is installed from the ceiling level to the ridge of the roof. Outdoor air is supplied by window head slot ventilators, with one to each room. Cross ventilation is recommended for houses with open plan room layouts, or with flat or severely limited roof access. Ventilators are incorporated into wall cavities and located diagonally opposite each other. They are designed to have a high vapor porosity and high resistance to air flow. Replacement air is provided by window head slot ventilators.[65]

Ventilation for vapor control in living spaces can also be achieved by using whole-house mechanical ventilation systems. The advantage of forced ventilation systems for water vapor control is that they can be relied upon to deliver sufficient outdoor air under most circumstances, unlike passive systems whose airflows vary with ambient weather conditions.[64] Such systems typically have heat recovery capabilities to conserve energy. (See Chapter 5.) Air-air heat exchangers should be of the sensible heat type for vapor control.

Relative humidity in closed indoor spaces can be controlled during warmer and more humid months by using dehumidification devices or an air conditioner. The advantage of dehumidification during the summer is that it will lower moisture content of interior materials, and consequently decrease stored moisture emissions during the fall and winter.[66]

Dehumidifiers remove moisture from air by either condensation or absorption. They are capable of reducing relative humidities to 40%. Condensation-type units will tend to ice up and have a very high energy usage when employed to reduce relative humidity to 30% under high humidity summertime conditions.[67]

Dehumidifiers are useful and effective under warm house conditions. In cool conditions (< 18°C) they tend to be ineffective and act instead as small heaters. In order to be effective, they must be able to extract at least 2 L/day of water.[66]

In studies conducted at Ball State University, when indoor temperature was kept constant, a single 40-pint dehumidifier coupled with central air conditioning was able to maintain relative humidity at a constant 50% under even the highest summertime ambient humidity loads.[67]

REFERENCES

1. Sarsfield, J.K. et al. 1974. "Mite-Sensitive Asthma of Childhood: Trial of Avoidance Measures." *Arch. of Disease in Childhood* 49:716–721.
2. Murray, A.B. and A.C. Ferguson. 1983. "Dust-Free Bedrooms in the Treatment of Asthmatic Children with House Dust or House Dust Mite Allergy: A Controlled Trial." *Pediatrics* 71:418–222.
3. Burr, M.L. et al. 1976. "Anti-Mite Measures in Mite-Sensitive Adult Asthma: A Controlled Trial." *Lancet* 333–355.
4. Burr, M.L. et al. 1980. "Effects of Anti-Mite Measures on Children with Mite-Sensitive Asthma: A Controlled Trial." *Thorax* 35:506–512.
5. Korsgaard, J. 1982. "Preventive Measures in House-Dust Allergy." *Amer. Rev. Respir. Dis.* 125:80–84.
6. Arlian, L.G. et al. 1982. "The Prevalence of House Dust Mites, *Dermatophagoides* spp. and Associated Environmental Conditions in Homes in Ohio." *J. Allergy and Clin. Immunol.* 69:527–532.
7. Mulla, M.S. et al. 1975. "Some House Dust Control Measures and Abundance of *Dermatophagoides* Mites in Southern California (*Acarina: Pyroglyphide*)." *J. Med. Entomol.* 12:5–9.
8. Lang, J.D. and M.S. Mulla. 1977. "Abundance of House Dust Mites, *Dermatophagoides* spp., Influenced by Environmental Conditions in Homes in Southern California." *Environ. Entomol.* 6:643–648.
9. Lehti, H. 1984. "Vacuum Cleaner—Friend or Foe." 107–110. *Proceedings of the Third International Conference on Indoor Air Quality and Climate.* Swedish Council for Building Research. Stockholm. Vol. 5.
10. Korsgaard, J. 1983. "House-Dust Mites and Absolute Indoor Humidity." *Allergy* 38:85–92.
11. Korsgaard, J. 1983. "Mite Asthma and Residency." *Amer. Rev. Respir. Dis.* 128:231–235.
12. Carpenter, M.D. et al. 1984. "Air Conditioning and the House Dust Mite." *J. Allergy and Clin. Immunol.* 75:121.
13. Bischoff, E. 1987. "Sources of Pollution of Indoor Air by Mite Allergen–Containing House Dust." 742–746. In: *Proceedings of the Fourth International Conference on Indoor Air Quality and Climate.* Institute for Water, Soil and Air Hygiene. West Berlin. Vol. 2.
14. Kniest, F.M. 1987. "Colorimetric Quantification of Inhalant Allergen Sources in House Dust." 732–737. In: *Proceedings of the Fourth International Conference on Indoor Air Quality and Climate.* Institute for Water, Soil and Air Hygiene. West Berlin. Vol. 2.
15. Platts-Mills, T.A.E. and M.D. Chapman. 1987. "Dust Mites: Immunology, Allergic Disease, and Environmental Control." *J. Allergy Clin. Immunol.* 80:755–775.
16. Heller-Haupt, A. and J.R. Busvine. 1974. "Tests of Acaricides Against House-Dust Mites." *J. Med. Entomol.* 11:551–558.
17. Report of an International Workshop. 1987. "Dust Mite Allergens and Asthma—A World Wide Problem." Bad Kreuznach, Federal Republic of Germany.
18. Saint-Georges-Gridelet, D. 1981. "Formulation of a Strategy for Controlling the House-Dust Mite (*Dermatophagoides pteronyssinus*)." *Acta Oecologia/Oecol. Applic.* 2:117–126.

19. Van de Maele, B. 1983. "A New Strategy in the Control of Dust Mite Allergy." *Pharmatherapeutica* 3:441–444.

20. Penaud, A. et al. 1977. "Results of a Controlled Trial of the Acaricide Paragerm on *Dermatophagoides* spp. in Dwelling Houses." *Clin. Allergy* 7:49–53.

21. Colloff, M.J. 1986. "Use of Liquid Nitrogen in the Control of House-Dust Mite Populations." *Clin. Allergy* 16:41–47.

22. Morey, P.A. et al. 1984. "Environmental Studies in Moldy Office Buildings: Biological Agents, Sources and Preventative Measures." *Ann. ACGIH* 10:21–36.

23. Burge, H.A. 1987. University of Michigan. Personal communication.

24. Kozak, P.P. et al. 1980. "Currently Available Methods for Home Mold Surveys. II. Examples of Problem Homes Surveyed." *Ann. Allergy* 45:167–176.

25. Kozak, P.P. et al. 1979. "Factors of Importance in Determining the Prevalence of Indoor Molds." *Ann. Allergy* 43:88–94.

26. Solomon, W.R. 1974. "Fungus Aerosols Arising from Cool Mist Vaporizers." *J. Allergy Clin. Immunol.* 54:223–230.

27. Burge, H.A. et al. 1980. "Prevalence of Microorganisms in Domestic Humidifiers." *App. Environ. Microbiol.* 39:840–844.

28. Morey, P. 1986. "Environmental Studies in Moldy Office Buildings." ASHRAE Trans. 92. Pt.1.

29. Brundrett, G.W. et al. 1981. "Humidifier Fever." *J. Chart. Inst. Build. Serv.* 3:35–36.

30. Weiss, N.S. and Y. Soleymani. 1971. "Hypersensitivity Lung Disease Caused by Contamination of an Air-Conditioning System." *Ann. Allergy* 29:154–156.

31. Banazak, E.F. et al. 1970. "Hypersensitivity Pneumonitis Due to Contamination of an Air Conditioner." *New England J. Med.* 283:271–276.

32. Aranow, P. et al. 1978. "Early Detection of Hypersensitivity Pneumonitis in Office Workers." *Amer. J. Med.* 64:236–242.

33. Scully, R.E. et al. 1979. "Case Records of the Massachusetts General Hospital — Case 47–1979." *New England J. Med.* 301:1168–1174.

34. Bernstein, R.S. et al. 1983. "Exposures to Respirable, Airborne *Penicillium* from a Contaminated Ventilation System: Clinical, Environmental and Epidemiological Aspects." *Amer. Indust. Hyg. Assoc. J.* 44:161–169.

35. Ganier, M. et al. 1980. "Humidifier Lung: An Outbreak in Office Workers." *Chest* 77:183–187.

36. Cockcraft, A. et al. 1981. "An Investigation of Operating Theatre Staff Exposed to Humidifier Fever Antigens." *Brit. J. Indust. Med.* 38:144–152.

37. Witherell, L.E. et al. 1986. "*Legionella* in Cooling Towers." *J. Environ. Health.* 49:134–139.

38. Miller, R.P. 1979. "Cooling Towers and Evaporative Condensers." *Ann. Int. Med.* 90:667–670.

39. Grace, R.D. et al. 1981. "Susceptibility of *L. pneumophila* to Three Cooling Tower Microbiocides." *App. Environ. Microbiol.* 4:233–236.

40. Skaily, P. et al. 1980. "Laboratory Studies of Disinfectants Against *L. pneumophila.*" *App. Environ. Microbiol.* 45:48–57.

41. Soracco, R.J. and D.H. Pope. 1983. "Bacteriostatic and Bactericidal Modes of Action of Bis(tributyltin) on *L. pneumophila.*" *Appl. Environ. Microbiol.* 45:48–57.

42. Soracco, R.J. et al. 1983. "Susceptibilities of Algae and *L. pneumophila* to Cooling Tower Biocides." *Appl. Environ. Microbiol.* 45:1254–1260.

43. Braun, E.B. 1982. Ph.D. Dissertation. Rensselaer Polytechnic Institute, Troy, NY.

44. Fliermans, C.B. and R.S. Harvey. 1984. "Effectiveness of 1-Bromo-3-chloro-5,5 dimethylhydontoin Against *L. pneumophila* in a Cooling Tower." *App. Env. Microbiol.* 47:1307–1310.

45. Barbaree, J.M. et al. 1986. "Isolation of Protozoa from Water Associated with a Legionellosis Outbreak and Demonstration of Intracellular Multiplication of *Legionella pneumophila.*" *App. Env. Microbiol.* 5:422–425.

46. DHSS. 1980. "Legionnaires' Disease and Hospital Water Systems." Health Notice HN(80) 39. Department of Health and Social Security, London.

47. Fliermans, C.B. et al. 1982. "Treatment of Cooling Systems Containing High Levels of *L. pneumophila.*" *Water Res.* 16:903–909.

48. American Society of Heating, Refrigerating and Air-Conditioning Engineers. 1984. "Systems." ASHRAE, Atlanta.

49. Ager, B.P. and J.A. Tickner. 1983. "The Control of Microbiological Hazards Associated with Air Conditioning and Ventilation Systems." *Ann. Occup. Hyg.* 27:341–358.

50. Mallison, G.F. 1980. "Legionellosis—Environment Aspects." *Ann. N.Y. Acad. Science* 353:67–70.

51. Building Research Association of New Zealand. 1985. "Subfloor Ventilation." Building Information Bulletin 245.

52. White, J.H. and P. Skvor. 1985. "The House as a Moisture Source." Presented at Air Pollution Control Association Specialty Conference. Indoor Air Quality in Cold Climates—Hazards and Abatements. Ottawa, Canada.

53. Reckitts, R. 1980. "How to Avoid Problems in the Crawl Space and Basement of a Home." Cooperative Extension Service. U.S. Department of Agriculture—University of Missouri, Columbia, MO.

54. Quirouette, R.L. 1984. "Moisture Sources in Houses." 15–28. In: Proc. Building Science Insight '83. No. 7. Division of Building Research—National Research Council of Canada, Ottawa.

55. Schaub, D. 1985. "Reducing Moisture Problems." Washington Energy Extension Service. EY3020.

56. Rousseau, J. 1984. "Rain Penetration and Moisture Damage in Residential Construction." 5–14. In: Proc. Building Science Insight '83. No. 7. Division of Building Research—National Research Council of Canada, Ottawa.

57. White, J.H. and R.L. Quirouette. 1985. "Condensation in Canadian Homes." Presented at Air Pollution Control Association Specialty Conference. Indoor Air Quality in Cold Climates—Hazards and Abatement Measures. Ottawa, Canada.

58. Rousseau, M.Z. 1984. "Control of Surface and Concealed Condensation." 29–40. In: Proc. Building Science Insight '83. No. 7. Division of Building Research—National Research Council of Canada, Ottawa.

59. Sanders, C.H. 1980. "Condensation and Its Treatment." *Building Technology and Management.* December, pp. 35–38.

60. Anonymous. 1984. "Condensation: Prevention Better than Cure." GLC Bulletin 142.

61. Tamura, G.T. et al. 1974. "Condensation Problems in Flat Wood-Frame Roofs." Research Paper 633. Division of Building Research—National Research Council of Canada, Ottawa.

62. Schaffer, E.L. 1980. "Moisture Interactions in Light-Frame Housing: A Review."

125–143. In: C.M. Hunt et al. (Eds.). *Building Air Change Rate and Infiltration Measurements.* ASTM STP 719. American Society for Testing and Materials, Philadelphia.

63. Johnson, K.A. 1986. "Controlling Condensation Risk." 178–182. RCI Directory.
64. Handegord, G.O. 1984. "Ventilation of Houses." 53–64. In: Proc. Building Science Insight '83. No. 7. Division of Building Research – National Research Council of Canada, Ottawa.
65. NCHA. 1986. "Passive Ventilation to Dwellings with Condensation." Building Technology and Energy Use in Housing Study Group. Northern Consortium of Housing Authorities.
66. Sanders, C.H. 1987. "Condensation and Mold in Housing." *BRE Digests.* January.
67. Godish, T. and J. Rouch. 1986. "Mitigation of Residential Formaldehyde Contamination by Climate Control." *Amer. Indust. Hyg. Assoc. J.* 47:792–797.

5   VENTILATION FOR CONTAMINANT CONTROL

Ventilation can be described as a process whereby air is supplied and removed from a space by natural or mechanical means. The desired effect of ventilation may be to remove heat or moisture or to reduce the concentration of one or more gaseous or particulate contaminants.

Ventilation may occur naturally, or it can be induced by mechanical means. Natural ventilation may occur as a consequence of infiltration, or it may be under the manual control of building occupants. Controlled natural ventilation occurs when windows or doors are opened. It is generally employed in residences and some public-access buildings during moderate to warm ambient weather conditions to reduce indoor temperatures or to cause air flow that may improve thermal comfort. To some degree, it is also used to reduce those "stuffy" feelings perceived by occupants during some closure conditions. The use of natural ventilation in the latter case is most probably an attempt to reduce levels of indoor contaminants.

INFILTRATION

Infiltration occurs as a consequence of the random flow of outdoor air through cracks and a variety of unintentional openings in the building envelope. It is caused by the dynamic pressure of the wind and buoyant forces resulting from indoor/outdoor temperature differences.[1]

Wind Effects

The wind is a major driving force for infiltration of air into a building. As wind approaches a building, it decelerates, producing a positive pressure on the windward side. As wind is deflected by the windward face of the building, the flow separates at building edges formed by corners and the

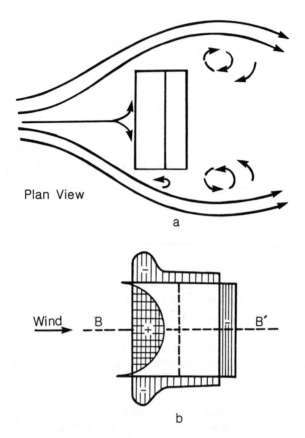

Plan View

a

Wind

b

Figure 5.1. Idealized air flow around (a) and pressure distribution on (b) a rectangular building oriented perpendicular to the wind—plan view.[2]

roof or building top. This flow results in internal building negative pressures which, while small in respect to atmospheric pressure, are sufficient to induce flows of large volumes of air. Negative pressures also occur on the lee side of the building in the wake area.[2] The effect of these pressure differences is to induce an inflow of air on the windward side where air pressure is positive and an outflow on the leeward side where air pressure is negative. A typical flow of wind around a rectangular building, with wind direction perpendicular to the normal face, is seen in two views in Figures 5.1a and 5.2a; pressure distributions in a similarly situated rectangular building are seen in Figures 5.1b and 5.2b.

Maximum positive pressure occurs at the center of the windward wall and decreases rapidly toward the building corners where flow separation takes place and highest negative pressures occur (Figure 5.1b). Upward deflection also causes negative pressure to occur at the beginning of the roof line. Tall

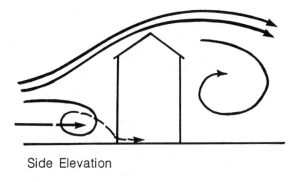

Side Elevation

a

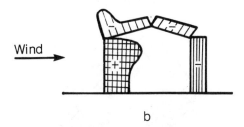

Wind

b

Figure 5.2. Idealized air flow around (a) and pressure distribution on (b) a rectangular building oriented perpendicular to the wind—side elevation.[2]

buildings have a particularly significant effect on upward deflection of air flow. As a consequence, roof areas in tall buildings are subject to relatively strong negative pressures.

The exact pressure distribution on a building will depend on a variety of factors. These include wind speed, building geometry and size, and incident angle of the wind. The effect of wind angle on air flow and pressure distribution can be seen in Figures 5.3a and 5.3b.

Wind-induced pressure differences, which affect the magnitude and distribution of infiltration air, may also be influenced by other factors. These include the type of building cladding and the presence of barriers to air movement such as other buildings, trees, and shrubbery.[2] The shielding effect of such barriers produces turbulence that both reduces wind speed and alters wind direction.

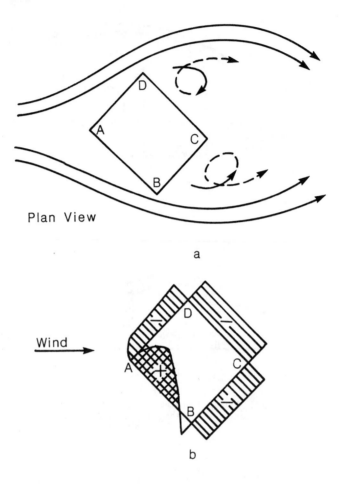

Plan View

a

Wind

b

Figure 5.3. Air flow around (a) and pressure distribution on (b) a rectangular building oriented at an oblique angle to the wind.[2]

Stack Effect

Pressure differences may also be caused by differences between indoor and outdoor temperatures. Under cool to cold outdoor temperature (heating season) conditions, warm air rises and flows out of the building near the top. This exfiltrated air is replaced by cooler/colder air that flows in by infiltration near the bottom of the building. During the cooling season, the direction of air flows due to infiltration/exfiltration are reversed, i.e., cool air flows out near the bottom and warm air flows in at the top.[1] The magnitude of air flow associated with the infiltration/exfiltration process is most significant during the heating season when indoor/outdoor temperature differences become very large.

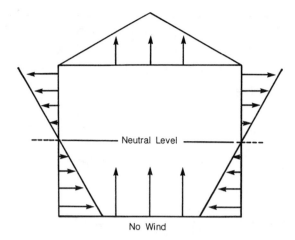

Figure 5.4.   Pressure differences on a building
associated with indoor/outdoor temperature
differences.[1]

The effect of indoor/outdoor temperature differences on pressure distribution on a building during the heating season is illustrated in Figure 5.4. The neutral pressure level (NPL) is the point at which interior pressure is equal to ambient pressure. Its location depends on the distribution of air leakage sites. The construction of Figure 5.4 indicates that leakage areas are uniformly distributed over the building envelope; therefore, the NPL is at mid-height. In single-family dwellings, NPL is usually above mid-height and, in the case of combustion heat source operation with a flue, NPL may be even higher, even above the ceiling.[3] In tall buildings, NPL may vary from 0.3 to 0.7 of the building height.[4]

It is generally agreed that stack effect increases with building height. The rule of thumb for estimating the magnitude of stack effect is that the pressure difference due to the stack effect is 10^{-3} inches H_2O (0.25 pascals)/story.[1] The stack effect in tall buildings may be particularly marked where there are vertical passages such as elevators and stairwells or other service shafts. Each story may act independently, i.e., have its own stack effect, if constructed in an airtight way.[1]

Combined Effects of Wind and Thermal Forces

So far, infiltration by wind and infiltration by thermal forces have been treated separately. Under normal conditions, both wind and thermal forces combine to produce infiltration air exchange. Wang and Sepsy[5] have attempted to relate both environmental parameters to the infiltration rate by a polynomial model (Equation 5.1).

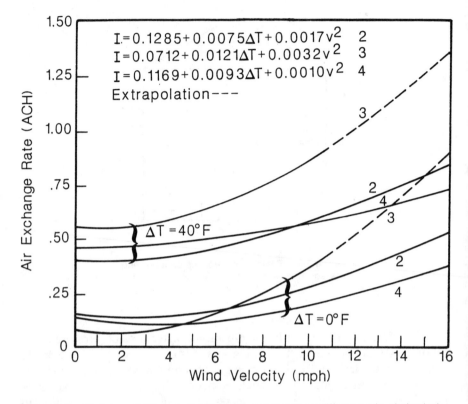

Figure 5.5. Model predictions of air infiltration rates based on differences in wind velocity and stack effect.[5]

$$I = A + B(\Delta T) + C(v^2) \tag{5.1}$$

where I = infiltration rate
A = intercept coefficient, ACH ($\Delta T = 0$, $v = 0$)
B = temperature coefficient
C = velocity coefficient
ΔT = indoor/outdoor temperature difference, °F
v = wind velocity, mph

In this model, the effect of temperature differences and the square of the wind velocity are seen to be additive. Curves and model equations for ΔT differences of 0°F (–18°C) and 40°F (4.5°C) are presented in Figure 5.5 for a range of wind speeds and for three different houses. Both the stack effect (represented by ΔT, the difference between indoor and outdoor temperatures) and the wind velocity are seen (Equation 5.1, Figure 5.5) to significantly affect the air exchange rate, expressed as air changes per hour.

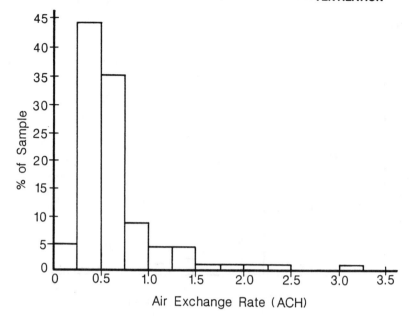

Figure 5.6.   Infiltration rates measured in 312 North American houses.6

Infiltration Values

Values for infiltration reported as air changes per hour vary considerably because of diurnal and seasonal ambient temperature and wind conditions. Infiltration-associated air change rates may vary from a low of 0.1–0.2 ACH in tight, energy-efficient houses to 3.0 ACH in leaky houses under high infiltration conditions.[1] A distribution of average seasonal infiltration rates for 312 houses drawn from many areas in North America is illustrated in Figure 5.6.[6] The median infiltration value is seen to be 0.50 ACH. Figure 5.7[6] represents a large sample of low-income housing with a median infiltration rate of 0.90 ACH. Differences apparently are due to energy efficiency and effective leakage area.

Although a building can be characterized by its average infiltration rate expressed as air changes per hour, it would be erroneous to assume that this air change rate is sufficient to provide for indoor air quality needs. As previously mentioned, air infiltration varies seasonally in response to outdoor temperature and wind conditions. Thus, a single building will have a range of air exchange rates depending on environmental conditions occurring at a particular time (Figure 5.8).[7] When winds are calm and indoor/outdoor temperature differences are small, energy-efficient houses may not differ substantially in air change rates from old, leaky houses. Note the low infiltration rates pre-

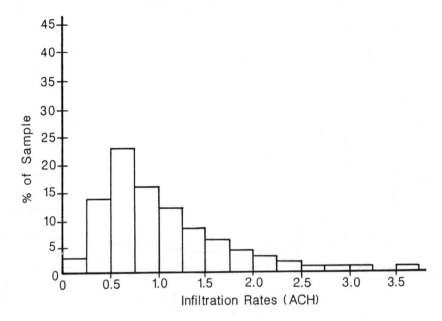

Figure 5.7. Infiltration rates associated with low-income housing.[6]

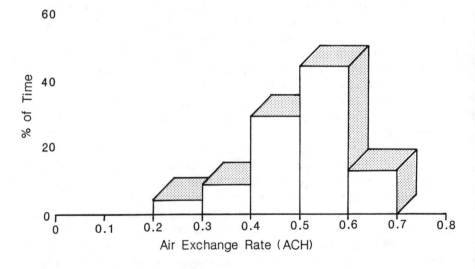

Figure 5.8. Variation of air exchange rates in a single residential building.[7]

dicted for $\Delta T = 0°F$ and low wind velocities in Figure 5.5. Under such conditions, even a leaky building would be subject to elevated contaminant levels.

Infiltration Measurements

Building infiltration rates resulting from normal pressures associated with wind and thermal forces can be measured directly using tracer gas techniques.[8] The tracer gas must be nonreactive, relatively nontoxic, and easily measurable at low concentrations. A variety of tracer gases have been used, the more common being sulfur hexafluoride, nitrous oxide, carbon dioxide, and perfluorocarbons. Both sulfur hexafluoride and perfluorocarbons are attractive as infiltration tracer gases because they can be detected in the parts per billion range.

The use of tracer gases is based on the mass balance assumption that the rate of change of the amount of tracer gas in the test space is equal to the amount of the tracer placed in the space minus the amount removed by ventilation or exfiltration.[9] It is also assumed that the only loss of tracer is due to ventilation or exfiltration.

Infiltration determination by the use of tracer gases can be conducted by any of three techniques: concentration decay, constant injection rate, and constant concentration.[8,9]

Concentration or tracer decay is the most widely used infiltration measurement method. It involves initial injection of a tracer gas into a space or building and assumes that the tracer gas is well mixed. Decay of the tracer gas is then measured over a period of time. From such measurements, and calculations from Equations 5.2, 5.3, and 5.4, the infiltration rate in ACH can be determined.

$$C_t = C_o e^{-\left(\frac{Q}{V}\right) t} \tag{5.2}$$

where C_t = tracer gas concentration at the end of the time interval, $\mu g/m^3$
C_o = initial tracer gas concentration at time (t) = zero, $\mu g/m^3$
V = volume of space, m^3
Q = ventilation rate, m^3/sec
e = natural log base
t = time, sec

The ratio Q/V, considered in the time context of one hour, yields the infiltration rate I in air changes per hour (Equation 5.3).

$$C_t = C_o e^{-lt} \tag{5.3}$$

The air change rate I is the slope of the line of the logarithm of concentration as a function of time plotted on semilog paper. The infiltration rate I can be calculated from Equation 5.4, derived from Equation 5.3.

$$I = \frac{\left(\ln \frac{C_o}{C_t} \right)}{t} \tag{5.4}$$

In the constant injection rate procedure, when steady-state conditions have been reached, the infiltration rate can be calculated from the ventilation rate (Q) which is the ratio of the rate of injection (F) to the concentration (C) (Equation 5.5).

$$Q = F/C \tag{5.5}$$

Since Q is expressed in either m^3/sec or cubic feet per minute (CFM), it must be multiplied by the appropriate time units and then divided by the volume of the air space to yield air changes per hour.

In the constant concentration method, injection rate is varied to maintain a constant concentration in the space. Ventilation rate Q is again calculated from Equation 5.5.

Leakage

Differences in air exchange rates resulting from wind and thermal forces are, to a significant degree, due to leakage characteristics of a building. The presence of cracks and unintentional openings, their sizes, and their distribution determine leakage characteristics of a building and potential for air infiltration. The different points of leakage in a single-family dwelling are illustrated in Figure 5.9. Major leakage sites include the sole plate where the building frame is attached to the foundation, cracks around windows and doors, exhaust vents, wall cavities, electrical outlets, exhaust fans and ducts, chimneys, etc. Leakage distribution in a building is a function of construction and architectural styles as well as regional and local climatic conditions. Leakage distribution is significant because it determines the magnitude of wind- and stack-driven infiltration and the nature of air flow patterns in buildings.[10]

The leakage potential of buildings can be assessed by fan pressurization techniques.[11] Unlike infiltration measurements using tracer gases, fan pressurization techniques characterize leakage and, indirectly, infiltration potential, independent of weather conditions. The unit of measurement or expression in both procedures is ACH. Though related, infiltration and leakage measurements are not equivalent. Conversion of values measured by one technique to those of the second cannot be made directly, but can only be accomplished by the use of models, discussed later in this section.

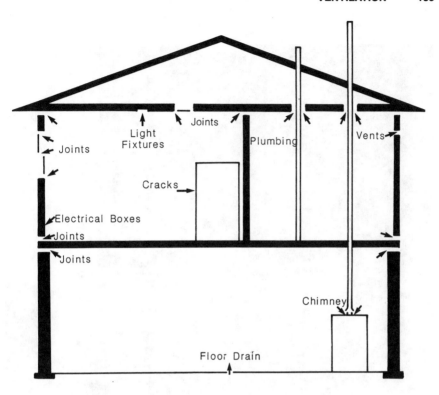

Figure 5.9. Air leakage sites in a residential dwelling.

In fan pressurization, a large blower fan is sealed into a window or door in order to move air into or out of a building at measured rates. Commonly, a fan is sealed into a door (blower door, see Figure 5.10) which can be easily attached and detached from each building under test. Indoor/outdoor pressure differences are measured as a function of flow rates.[1,11] Leakage tests are conducted at both positive and negative flows at pressure drop intervals of 10 pascals (0.04 inches H_2O) over the range of 10–60 pascals (0.04–0.24 inches H_2O). Results of such tests in both positive and negative pressure modes are illustrated in Figure 5.11.

Data from fan pressurization or evacuation measurements can be expressed in the form of the following equation:

$$Q = K\Delta P^n \qquad (5.6)$$

where Q = ventilation rate, m^3/hr
K = flow coefficient
ΔP = pressure difference, pascals
n = flow exponent

Figure 5.10. Blower door used to measure air leakage characteristics of a building. (Photo courtesy of Retrotec, Canada.)

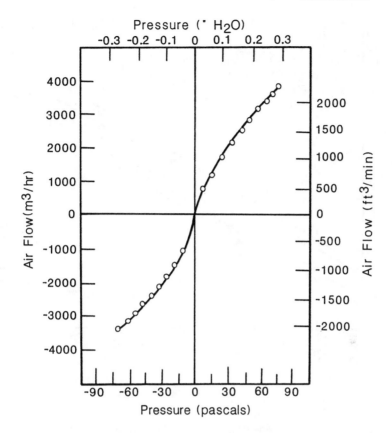

Figure 5.11.   Blower door pressure and air flow measurements.[8]

The ventilation or air flow rate Q is proportional to overall leakage area and the difference in static pressure ΔP. In a log-log plot, shown in Figure 5.12, the above equation forms a straight line with the flow exponent n, the slope. By dividing both sides of Equation 5.6 by the volume of the ventilated space, the flow rate is converted into ACH.[8]

Fan-induced and natural leakage rates differ considerably. The former are much larger than the latter. While fan-induced pressures are applied equally to all internal surfaces, wind and stack pressures are not.

Leakage measurements have the advantage of being relatively independent of weather conditions, so that leakage potentials of buildings can be easily compared. Comparisons are usually made at standard average pressure differentials such as 4 pascals (0.016 inches H_2O) or 50 pascals (0.20 inches H_2O). Such comparisons are particularly significant in the context of energy conservation and recommendations for weatherization measures.

Because of the significant influence of wind and stack effect on infiltra-

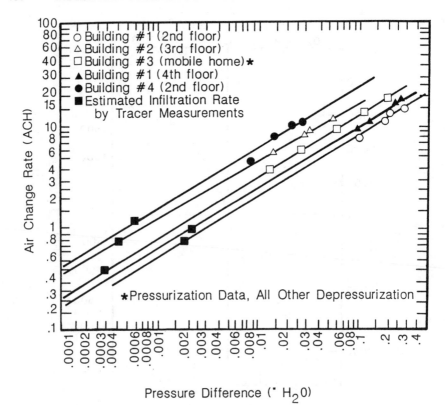

Figure 5.12. Air exchange rates determined from blower door leakage tests.[8]

tion measurements, comparison of infiltration rates among buildings can only be made if measurements are made under nearly equivalent atmospheric conditions (a task which may in many cases be difficult to accomplish), or if multiple seasonal measurements are made and averaged. It would be very useful, therefore, if fan pressurization data could be converted into infiltration rates, which reflect natural environmental conditions of wind and indoor/outdoor thermal differences. A variety of investigators have developed models to establish such a relationship. One of the most widely used models was developed at the Lawrence Berkeley Laboratory (LBL).[12-14]

The LBL model calculates infiltration rate I (ACH) by multiplying the effective leakage area (ELA) by the specific infiltration flow rate I_s (m³/hr-cm²).

$$I = ELA \times I_s \tag{5.7}$$

where

$$ELA = Q_o (2\Delta P/\rho)^{0.5} \tag{5.8}$$

where ELA = effective leakage area, m^2
       Q = ventilation flow at 4 pascals, m^3/hr
      ΔP = pressure difference during this flow, 4 pascals
      ρ = density of air, 1.2 kg/m^3

The LBL model assumes that air flow through leaks is similar to air flow through simple orifices whose leakage characteristics are quantifiable. The ELA is seen to be equivalent to an amount of orifice area of unit discharge coefficient that would allow air flow at a reference pressure of 4 pascals. The ELA of a particular crack would be equal to the area of a perfect nozzle that would pass the same amount of air at 4 pascals (0.016 inches H_2O). The air change rate Q_o can be derived by extrapolation from a fan pressurization curve (Figure 5.11).

The specific infiltration rate S is the ratio of infiltration rate to leakage area. It depends on leakage distribution, building height, ΔT (indoor/outdoor temperature difference), average wind speed, and effects of terrain and building shielding.[1,12-14] It can be derived from the following equation:

$$S = \sqrt{f_w^2 v^2 + f_s^2 (\Delta T)} \qquad (5.9)$$

The values of wind (f_w) and stack (f_s) parameters will depend on the distribution of leakage sites and characteristics of the building site. Sherman[14] reports values of f_w and f_s for a typical single-family house as

$$f_w = 0.047 \ m^2\text{-s}/cm^2\text{-hr}$$

$$f_s = 0.043 \ m^3/hr\text{-}cm^2\text{-}K^{0.5}$$

These values may vary as much as 50%, but generally do not.[14] Seasonal average values for the specific infiltration rate for major weather sites in North America have been found to lie between 0.16 and 0.38 $m^3/hr\text{-}cm^2$, with an average of 0.27 $m^3/hr\text{-}cm^2$.[14]

The LBL model described above attempts to relate leakage measurements to air infiltration under natural conditions, as conversions cannot be made directly.

OCCUPANT-CONTROLLED NATURAL VENTILATION

Before this modern era of year-round climate-controlled residential and nonresidential buildings, thermal comfort during seasons other than the heating season was achieved (or at least attempted) by allowing air to flow in and out of buildings by opening windows and doors.

The effectiveness of this form of natural ventilation varies widely; this variability, particularly in regard to thermal comfort and drafts, has been the major reason why year-round climate control has come into wide use even in residences.

In buildings in which air conditioning has not been installed, either because of cost considerations, climatic reasons, or personal preference, natural ventilation controlled by the occupants themselves is a significant factor in maintaining thermal comfort during periods of moderate to warm ambient weather conditions and eliminating the "stuffy, stale" air associated with some ambient conditions during building closure.

The inflow and outflow of air through building openings such as windows and doors is subject to both wind and thermal forces, as described previously. The quantity of inflowing and outflowing air will be affected by ambient wind speed and indoor/outdoor temperature differences. Though wind speed is important, the flow of air through an opening such as a window will also depend on the orientation of the open window to the microscale wind direction and any local obstructions, as well as on the size of the opening, the presence of outlets such as open windows or doors on walls in the building opposite to the wind direction, and wind-induced negative pressures associated with the roof and building edges. Open windows on opposite sides of a building provide excellent conditions for the natural flow of ventilation air.

Under conditions of low ambient wind speeds, the flow of air into a building is subject primarily to thermal forces due to indoor/outdoor temperature differences. The greater the temperature difference (ΔT), the greater the inflow of outdoor air. This inflow is also related to the size of openings and their relationship to each other. The greatest air flow occurs when inlet and outlet openings are of nearly equal area and the vertical distance between openings is significant.[1] Openings near the NPL are least effective in providing ventilation.

Natural ventilation through building openings varies widely in its effectiveness and acceptability. Under very warm weather conditions, it may serve to increase indoor temperatures in well-insulated buildings, at least for a portion of the daylight hours. Under breezy or windy conditions, the building may be subject to unacceptable drafts. Under conditions of calm winds and low ΔT, there may be little or no air flow through the building environment. Though the latter may be unimportant in terms of thermal comfort, it is likely to be very important in the context of contaminant levels and human exposures.

There have been no empirical studies quantifying the effectiveness of natural ventilation on reducing indoor contaminant levels. In most instances, effectiveness is inferred from occupant satisfaction, reduction in complaints of poor air quality, and a "common sense" understanding that fresh outdoor air must reduce contaminant levels. The questions to be answered are (1) "How effective is natural ventilation on specific contaminants?" and (2) "How does that effectiveness vary as a function of ambient weather conditions?" In complaint studies conducted by Godish et al.[15] associated with formaldehyde contamination of residences, occupants in 80 + % of households surveyed related that symptoms diminished in severity when their homes were ventilated by opening windows. Symptom severity, on the other hand, was observed to increase at the beginning of the heating season when closure conditions were reapplied. These studies suggest that, at least during a portion of the year, occupant-controlled natural ventilation is a reasonably effective mitigation measure in reducing the severity of a variety of acute symptoms associated with residential formaldehyde exposures. Because of energy costs, natural ventilation as controlled by occupants can only be considered to be an option when ambient weather conditions can be described as moderate or warm; energy costs would be prohibitive during most, but not all, of the heating season.

Natural ventilation can be a desirable means of reducing exposure to problem contaminants such as formaldehyde[16] or radon on an interim basis, prior to implementation of permanent abatement measures. Based on the author's experience, however, it is not unusual for homeowners to use natural ventilation for indoor contaminant control on a sustained basis, in fact having it become a long-term, if not permanent, control measure. Many individuals, in the absence of evidence to the contrary, conclude that this is the only indoor air pollution control measure they can afford.

MECHANICAL VENTILATION

In the past decade there have been numerous reports of illness symptoms in occupants of newly constructed office and other large nonresidential buildings. In over 203 health hazard evaluations of "sick building" complaints conducted through December 1983 (now in excess of 300), National Institute of Occupational Safety and Health (NIOSH) personnel concluded that the cause of illness or complaints in 50% of the cases was inadequate ventilation.[17] The classification of a building as having inadequate ventilation was based on evaluations of building ventilation design specifications, measurements of air flow, and comparison to ventilation guidelines.

Most large buildings constructed in the past decade have been designed to provide year-round climate control. Since windows are usually sealed, occupants cannot control ventilation. The availability of fresh or outdoor air is limited to that which is drawn in through a mechanical ventilation system and distributed throughout building spaces and that which makes its way in by infiltration.

The primary purpose of mechanical ventilation is to provide a healthy and comfortable indoor environment for building occupants. Other ventilation purposes include temperature and humidity control; enhancing air motion in warm spaces to improve thermal comfort; providing small pressure differences between zones or between the indoor and outdoor environments to control air exchange; and confining or exhausting smoke, heat, and toxic pollutants for fire safety.[18]

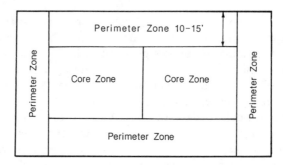

Figure 5.13. Temperature control zones in a large building.[18]

The provision of ventilation air to a mechanically ventilated building usually requires that outdoor air be conditioned by either heating or cooling before it enters occupied spaces. Conditioning outdoor air may account for 30–50% of the cost of heating or cooling a large building.[18] Ventilation air, therefore, is a significant cost factor in building operation. Increased energy needs are also associated with air leakage or infiltration; however, modern building design and construction practices have attempted to reduce air leakage in and out of such buildings as much as possible. Air exchange rates due to infiltration in several new government office buildings have been reported to be in the range of 0.2–0.7 ACH for maximum heating or cooling loads.[18]

Modern buildings are less well ventilated partly because they are designed and constructed to minimize air leakage, resulting in lower wind-induced and thermally induced air exchange rates. It is unfortunate, however, that the outbreaks of illness in such buildings should be described by the aphorism "Tight Building Syndrome," concluding that this is the cause of the problem. Indeed, there is little or no evidence that the tightening of the envelope of modern buildings is, in fact, responsible for the seeming explosion in air quality complaints.

HVAC Systems

In large buildings, the same mechanical system is designed to provide for heating, ventilation, and air conditioning needs. Such systems are referred to by the acronym HVAC. For purposes of heating or cooling, a building may be divided into zones, with the HVAC system designed to serve the core zones differently from those in the perimeter (Figure 5.13). During occupied hours, the core, because of heat produced by lights, people, etc., may require cooling; during unoccupied hours and just before occupancy, it may require heating. In the perimeter, which extends 4–7 m from exterior walls, the HVAC system must provide for additional heating or cooling in response to the energy load or demand that is placed on building faces by changes in weather conditions.

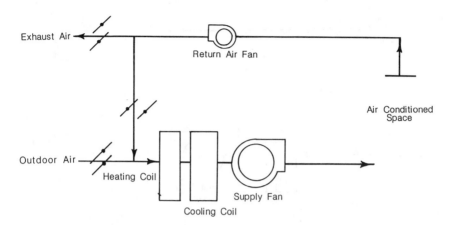

Figure 5.14. Constant-volume HVAC system design.[18]

The division into zones is designed to facilitate temperature control to maintain thermal comfort.

A variety of HVAC system designs is used in new and existing buildings. For purposes of illustration, let us look at two of the most commonly used: constant-air-volume (CAV) and variable-air-volume (VAV).

The single-zone CAV system is illustrated in Figure 5.14. This system has two fans: a larger one to distribute supply air and a smaller return fan used for recirculation and, as needed, to offset, by exhaust, outdoor air drawn in. Low-efficiency filters are usually placed before cooling coils and fans to protect them from fouling. Supply air is distributed through a series of ducts with inlet vents in the ceilings of spaces to be provided ventilation and conditioned air. Return air is typically drawn from the conditioned space through ceiling outlets into a large return air plenum, which is the open space above the ceiling. Dampers shown on the left side of Figure 5.14 can be operated to control the amount of intake and exhaust air and the percentage of air that is recirculated. These dampers may either be operated manually by building personnel or, as is the case in many newer buildings, automatically change positions or settings as ventilation needs change with weather conditions. The proper operation of dampers/damper systems is a critical element in supplying sufficient ventilation air.

Variable-air-volume HVAC systems, as illustrated in Figure 5.15, have become increasingly common in the past decade. They differ from CAV systems in that temperature control is achieved by altering the air volume to the space rather than varying the temperature of the air to the space. The use of such systems is attractive, since they have considerable energy-saving potential.

Each of the VAV boxes, seen on the right in Figure 5.15, has valve settings

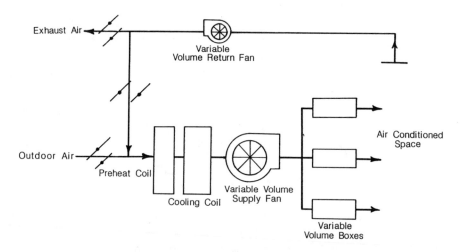

Figure 5.15. Variable-air-volume HVAC system design.[18]

that can be adjusted to provide the minimum ventilation requirement in each zone. Because outdoor air is only a fraction of the air supplied to a space at a given time, it is often difficult to ensure that the minimum outdoor air requirements are provided to each zone at all times.

Over the past decade or so, VAV systems have developed a reputation of being problem-plagued, particularly in the context of providing adequate ventilation and acceptable indoor air quality. This is due in part to the fact that these systems are more complicated than CAV systems to operate. For example, a not-uncommon problem associated with the operation of such systems is that VAV valves may be designed or set to close fully with no supply air when thermostat settings have been satisfied. Thus, when the desired temperature is attained, no ventilation is provided to the space. In response to complaints of poor air quality, VAV valves may be reset to provide a minimum ventilation rate, on the order of 20–40% of maximum flow to each space.[18,19]

The effect of VAV system operation on ventilation performance and indoor air quality can be seen in Figures 5.16a and 5.16b. Note the higher CO_2 levels in Figure 5.16a. In this case, numerous mechanical and operational problems were being experienced in a new classroom building. One of the major apparent operational problems was that VAV valves were completely closed when the thermostat setting was satisfied.[20] Figure 5.16b illustrates CO_2 levels in the same building after the VAV system was adjusted to maintain a minimum outdoor air ventilation rate in all occupied spaces.

Ventilation systems are designed to provide a range of ventilation rates. By adjusting dampers, the percentage of outdoor air introduced into the HVAC system may vary from 0 to 100%. Most HVAC systems are designed to operate on a minimum of outside air, typically 15–20% of the total air flow,

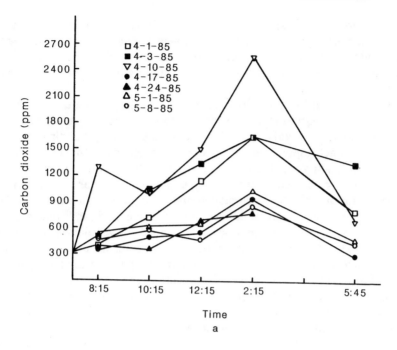

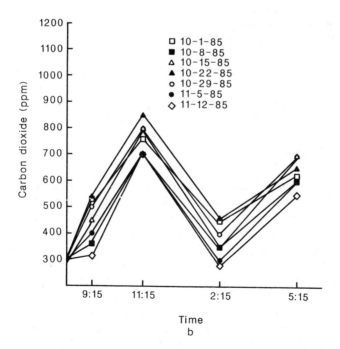

Figure 5.16. Ventilation performance in a VAV classroom building based on measurements of metabolic CO_2 levels.[20]

when outdoor temperature conditions are cold during the heating season, and very warm during the cooling season. When ventilation air is employed to reduce indoor temperatures during portions of the heating season, and during transitional periods between heating and cooling seasons, the percentage of outdoor air may reach 100%.

Ventilation problems experienced in mechanically ventilated buildings are often a result of poor system operation and maintenance failures. For example, in response to rapidly escalating energy costs, many building managers and/or personnel responsible for the operation and maintenance of building systems significantly modified the operation of their HVAC system, particularly in regard to the percentage of recirculated and outdoor air used. HVAC system operation on 100% recirculated air, although not the norm, has been common. In essence, many buildings have been operated (and some continue to be operated) without any outdoor air provided for ventilation. It is almost axiomatic that a building will have ventilation problems and uncomfortable and discontented occupants when outdoor air dampers are completely closed.

Another problem is the adequacy of ventilation air in all occupied spaces. In addition to the problem of ventilation efficiency (discussed below), inadequate ventilation of a space may occur as a result of system imbalance and building configurations that inhibit air movement into and out of a space. In regard to the latter case, many buildings are designed with movable walls. With such flexibility it is not uncommon, for example, to leave enclosed spaces with no cold air return outlets.

When the HVAC system is unbalanced, some spaces receive more ventilation air than is needed, while others receive less. System imbalance may result from a variety of factors, including poorly operating valves or dampers, dirty filters, improperly sized fans, inadequate provision for cold air return, and significant pressure differences between building spaces. The last factor may occur as a consequence of high exhaust rates in localized areas of the building.

Ventilation Efficiency

It is common in buildings served by mechanical ventilation systems to have supply air delivered to individual spaces through ceiling diffusers. Because of economics in building design and construction, it is also common to have exhaust or return air outlets in the ceiling as well. In small spaces such as offices, supply and return air outlets may be within a meter of each other. This proximity may result in "short-circuiting," wherein a portion of the supply air bypasses the occupied space below. The short-circuiting phenomenon can be observed in the smoke visualization demonstration in Figure 5.17. It is apparent that only a small portion of the smoke-laden supply air entered the occupied space. In this demonstration, air was supplied in the center of the space, and air exhausted in the wall just beneath the ceiling. The supply air temperature was elevated above that of the room.[21]

The fraction of supply air that short-circuits depends on a variety of

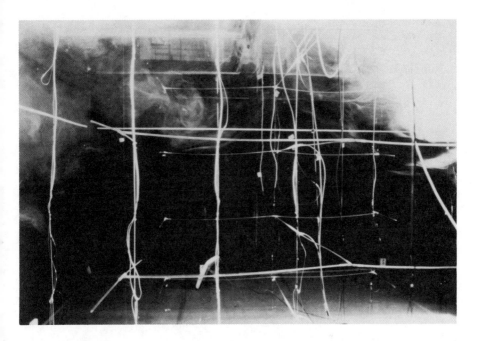

Figure 5.17. Smoke visualization demonstration showing short-circuiting.[21]

factors, including the location of supply and exhaust ports, supply air temperature, and the nominal ventilation rate. The limited data of Janssen et al.[22] suggest that 50% of the supplied air may short-circuit, and that VAV systems may be particularly prone to this phenomenon.

For the typical air distribution system illustrated in Figure 5.18, ventilation efficiency has been described[23,24] in the following form:

$$N_v = (Q_I - Q_E)/Q_I = (1 - s)/(1 - rs) \qquad (5.10)$$

where N_v = ventilation efficiency, %
           Q_I = outdoor air flow rate, L/sec
           Q_E = exhaust air flow rate, L/sec
           r = decimal fraction of recirculated air
           s = stratification factor or the decimal fraction of supply air that bypasses the occupied space to the return air system

If the stratification factor $s = 0.30$, and the decimal fraction of recirculated air $r = 0.50$, then solving the equation for the ventilation efficiency, N_v:

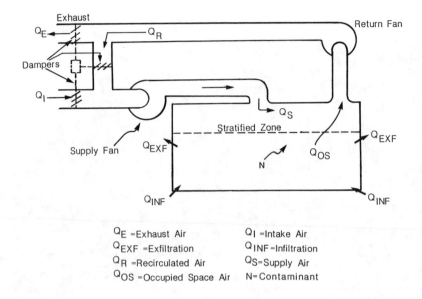

Q_E =Exhaust Air Q_I =Intake Air
Q_{EXF} =Exfiltration Q_{INF} =Infiltration
Q_R =Recirculated Air Q_S=Supply Air
Q_{OS} =Occupied Space Air   N=Contaminant

Figure 5.18. Diagrammatic representation of short-circuiting of building supply air.[24]

$$N_v = (1 -0.30)/[1 -(0.30)(0.50)] \qquad (5.11)$$
$$= 0.70/0.85$$
$$= 0.82 \text{ or } 82\%$$

Curves describing ventilation efficiencies as a function of different stratification factors and percentages of recirculated air are illustrated in Figure
5.19.[25] It is evident from these curves that stratification decreases ventilation
efficiency, but this effect is only significant when the percentage of outdoor air
used for ventilation is high and the percentage of recirculated air is low. At
high recirculation percentages (greater than 80%), ventilation efficiency
approaches 100% irrespective of the fraction of supply air that bypasses the
occupied space. The reasons for this apparent increase in ventilation efficiency
with increased recirculation percentages may not be obvious at first glance.
With the use of less outdoor air, contaminants that would have been exhausted
are recirculated, reducing the difference in contaminant levels in the supply air
and occupied space. As these differences decrease, the effect of stratification
becomes small. Although ventilation efficiencies appear high under conditions
of high recirculation rates, these belie the fact that, overall, ventilation may
not be adequate to control contaminants to acceptable levels.

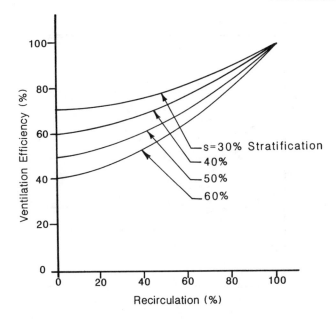

Figure 5.19. Effect of stratification on ventilation efficiency.[25]

Measurement of Ventilation Efficiency

Tracer gas techniques are commonly employed to quantitatively assess ventilation efficiency. The transient behavior of the concentration of a tracer gas is determined at different horizontal and vertical points in a space.

Relative ventilation efficiencies as a function of height, air exchange rate, and over temperature (excess of supply air temperature to average room air temperature, °C) for a mechanically ventilated room (41 m³) are shown in Figure 5.20. Supply air entered the room through a centrally mounted port; the exhaust port was mounted over the door 0.2 m beneath the ceiling. Note the significant effect of supply air temperature on ventilation efficiency.

Cross Contamination

A common indoor air pollution problem in multi-purpose buildings (e.g., hospitals, research laboratories, or industrial buildings with contiguous office space) is the movement of contaminants generated in a localized area to other parts of the building. This includes the entrainment of contaminants generated external to the building in the make-up air drawn into the building's HVAC system and the re-entry of exhausted contaminants through the make-up air inlet or by infiltration. These indoor pollution problems can be described or characterized as being ones of cross contamination or entrainment.[26]

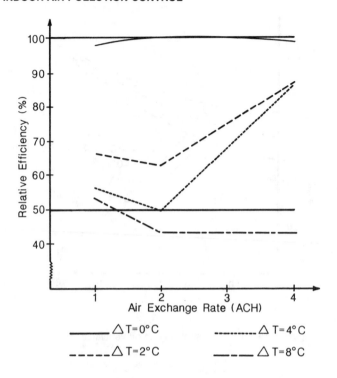

Figure 5.20. Ventilation efficiencies reported for different air exchange rates and over temperatures.[21]

In a number of reported cases of significant indoor air pollution (characterized by occupant complaints), the problem was caused by local exhaust ducts being vented upwind of outdoor air intakes, exhausts from idling motor vehicles being drawn into outdoor air intakes, motor vehicle exhausts from lower-level parking garages being swept by a strong stack effect through elevator shafts, etc. into upper level offices and residential spaces, movement of contaminants from one building space to another because of pressure differences, and/or failure of combustion gases to rise through a flue when the space was under negative pressure.[17,26]

Pressure imbalances will cause movement of contaminated air from one building space to another. Cross contamination will occur even when air handling units are separate or not interconnected. Flow of contaminants will be in the direction of negative pressure. The potential for such cross contamination may be determined from release of tracer gases,[27] release of odoriferous substances such as wintergreen, or simply watching the behavior of a cigarette smoke plume near an opening.

Microbial Contaminants

A special case of cross contamination or entrainment is that which occurs by a variety of microorganisms that cause outbreaks of illness in mechanically ventilated buildings. In many instances, sources of illness-producing agents include components of the HVAC system such as ductwork, filters, humidifiers, air washers, and fan coil units. A common source in building supply air is the microbial slime that develops on poorly drained condensation pans serving cooling coil units. As the slime dries out, it can become aerosolized and enter the building air supply stream;[28] organic deposits in ducts and on filters can serve as media for microbial growth and a source of heavy microbial contamination.

Re-Entry

Even though ventilation is typically effective in controlling indoor contaminant levels, ventilation systems themselves may cause significant indoor contamination to occur. This is particularly true if air intakes and exhausts are close together. Since designers are primarily concerned with the most efficient use of space and the appearance of the building, such occurrences are common in many large multi-purpose buildings. The greatest potential for serious indoor contamination from exhaust occurs where local ventilation is used to exhaust toxic gases or vapors from multi-purpose laboratory, hospital, and industrial buildings and where ventilation system imbalance occurs as a result of poorly planned reductions in general ventilation.

In laboratory environments, exhaust hoods are typically used to reduce the direct exposure of personnel to high concentrations of gases or volatile liquids. However, from the studies of Reible et al.,[29] it appears that the return of exhaust pollutants may cause higher exposures than those experienced directly. The effect of contaminant re-entry from fumehood exhaust is described in tracer gas experiments conducted in a laboratory building at the California Institute of Technology.[29] In Figure 5.21, concentrations of tracer gas (SF6) as a function of time are illustrated for locations near the fumehood and in the center of the room. Four to ten minutes after tracer release began, a steady increase in tracer gas concentration was observed near the fumehood and in the center of the room. Measurements in other parts of the building interior, as well as outside, indicated that the increase was a result of re-entry of fumehood exhaust. This re-entry resulted in a steady-state SF6 concentration of 6.6 ± 1.1 ppb, corresponding to a 10-fold higher concentration than direct exposure at the fumehood. In hundreds of experiments, exposures resulting from re-entry of the exhausted contaminant exceeded direct exposure when the fumehood was operated at normally recommended face velocities.

The re-entry problem has been described as being primarily a result of interaction between fumehood exhaust and the building wake. Since the building wake is only weakly coupled with free-flowing air, contaminants emitted

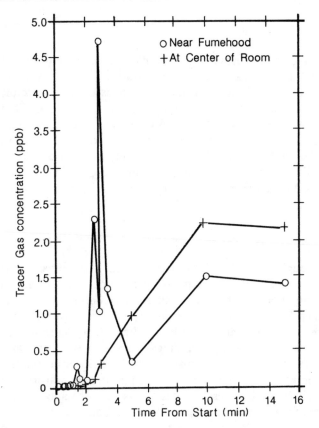

Figure 5.21.   Re-entry of a tracer gas associated with fumehood operation.[29]

into the wake are likely to be effectively mixed throughout the wake. This contaminated air may then be returned to the building interior through ventilation system intakes or by infiltration.

Significant factors that affect re-entry and subsequent building air contamination include the location of intakes relative to exhaust vents, the height of exhaust vents, stack gas velocity, and ventilation system imbalance.

Although exhaust gases will, in most instances, find a way to contaminate intake air for some specific wind directions, careful building design can minimize such contamination in mechanically ventilated buildings. Based on studies of Wilson,[30] a good design would place the intake on the lower one-third of the building and the exhausts on the upper two-thirds. In systematically evaluated flow patterns around high- and low-rise buildings, Wilson found little mixing around the surface flows of the upper two-thirds of the building with that of the lower one-third. This tendency is illustrated in Figure 5.22.

Such placement of air intakes may itself result in the creation of a differ-

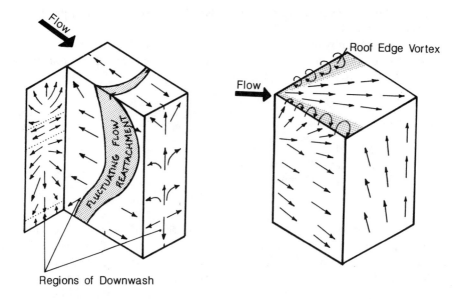

Figure 5.22. Air flow patterns around a high-rise building.[31]

ent type of indoor pollution problem—the entry of motor vehicle exhaust gases from loading docks and nearby streets. In heavily built-up areas, entrainment of motor vehicle pollutants in intake air may result in significant indoor contamination.

For both practical and aesthetic reasons, exhaust vents on large buildings are flush with the building surface or of very short height. Many of these vents are covered with rain shields or louvers which significantly reduce the upward momentum of exhaust gases. In such cases, penetration of the building wake by fumehood and other exhausts is unlikely to occur. Smith[32] has suggested that penetration of the building wake and concomitant reduction of exhaust gas re-entry can be achieved by the replacement of flush exhaust vents with small stacks on the order of 10% or less of the building height, and vent velocities of greater than twice the speed of the wind.

The jet momentum of an exhaust plume will cause it to rise and move away from the building surface until it is bent over by the wind. This plume rise and subsequent dispersion cannot be accurately estimated from plume rise and dispersion models normally applied to industrial smoke stacks. Because of the short exhaust-to-intake distances, the use of such models may result in errors on the order of 10- to 100-fold.

Wilson and Chui[33] and Wilson[30] have developed plume rise and dispersion models for exhaust vents on large buildings. Plume rise is modeled in the context of an initial dilution resulting from an apparent entrainment rate that occurs near the exhaust. This entrainment is a function of the ratio of the exhaust velocity (V_e) to the wind speed (v). At the exhaust vent it produces an initial dilution (D_o).

For a nonbuoyant momentum plume from a flush exhaust directed away from the building surface, the researchers proposed that the initial dilution could be described by the following equation:

$$D_o = 1 + 7.0 \, M^2 \, \sin^2\phi \qquad (5.12)$$

where $M = V_e/v$
D_o = initial dilution
V_e = exhaust velocity, m/sec
v = wind speed, m/sec

For uncapped, perpendicular (relative to the roof surface) vents, angle ϕ = 90°; for vents with rain caps or louvers, angle ϕ = 0. This initial dilution equation is useful only after the plume has bent over and travelled a downwind distance of 5–10 $A_e^{0.5}$, where A = the cross sectional area of the vent.

Dilution of the exhaust plume as it is carried downwind can be modeled theoretically as a turbulent diffusion from a point source.[34] Such modeling assumes plume speed increases linearly with distance. This distance dilution (D_d) can be described by the following equation:

$$D_d = \frac{B_1 \, vS^2}{Q_e} \qquad (5.13)$$

where   S = straight line exhaust-intake distance, m
Q_e = exhaust flow, L/sec
B_1 = constant

$$B_1 = \pi\alpha^2/2 \qquad (5.14)$$

where α is the ratio of the entrainment velocity (V_a) to wind speed (v). From wind tunnel data for a variety of exhaust-intake configurations, α has been determined to be 0.20.[31-33] Therefore,

$$D_d = 0.0625 \, \frac{vS^2}{Q_e} \qquad (5.15)$$

From an entrainment model, Wilson describes how the initial dilution from plume rise D_o and distance dilution D_d combine to produce the total minimum dilution:

$$D_{min} = (D_o^{0.5} + D_d^{0.5})^2 \qquad (5.16)$$

Taking into consideration the averaging time dependence of the entrainment constant B_1, Equation 5.14 becomes

$$B_1 = 0.0625 \left(\frac{T_a}{T_w} \right)^{0.33} \tag{5.17}$$

where T_a = actual averaging time, min
         T_w = reference averaging time, 10 min

for averaging time T_a in the range of one minute to several hours. By combining equations for B_1, D_o, and D_d and describing vS^2/Q_e as S^2/MA_e, the minimal dilution equation can be written as:

$$D_{min} = \left[(1 + 7.0\,M^2 \sin^2\phi)^{0.5} + \left(\frac{0.0625}{M} \left(\frac{T_a}{T_w} \right)^{0.33} \frac{S^2}{A_e} \right)^{0.5} \right]^2 \tag{5.18}$$

Note that the exit velocity term M occurs in both the initial dilution and distance dilution terms of Equation 5.18. As a consequence, a critical wind speed will exist at which an absolute minimum dilution will occur for a fixed exhaust-intake distance. The critical exhaust velocity ratio M lies in the range of 0.5–2.0, with a typical critical value of 1 for exhaust-intake distances $S/A_e^{0.5}$ of 5 to 50.[35]

From tracer gas[36] and wind tunnel studies,[33] it is apparent that there is no significant effect of building size on minimum dilution and that distance dilution decreases in direct proportion to the exhaust volume flow Q_e. The latter indicates the difficulty of achieving adequate dilution of high-volume exhausts.

Studies by Wilson[30,34] suggest that, with one major exception, minimum dilution is insensitive to where the exhaust/intake vents are located. Placing an intake around a corner on an adjacent building face has no apparent effect on minimum dilution. The exception apparently occurs when exhausts are located on the lower third of the upwind wall. In this situation, dilution may be only one-third of what might be expected. Wind direction changes, which alter building surface flow patterns, can significantly affect minimum dilution. When there is negligible exhaust momentum, for example, when $M\sin\phi < 0.1$, exhaust from roof vents in a 45° oblique wind will have a distance dilution which is three- to nine-fold less than wind normal to the upwind face of the building.[35] However, with increasing exhaust jet momentum, for example, $M\sin\phi > 1.0$, the effect of wind direction becomes insignificant. Such high-velocity exhausts not only avoid a three- to nine-fold decrease in distance dilution, but also gain a ten-fold increase in minimum dilution.

Wilson[34] conducted wind tunnel experiments simulating the roof of a wide low-rise building in order to determine the effect of exhaust-intake distance $S/A_e^{0.5}$ and exhaust velocity:wind speed ratio M on minimum dilution in practical situations of exhaust-intake distances. A vertically flush exhaust vent was

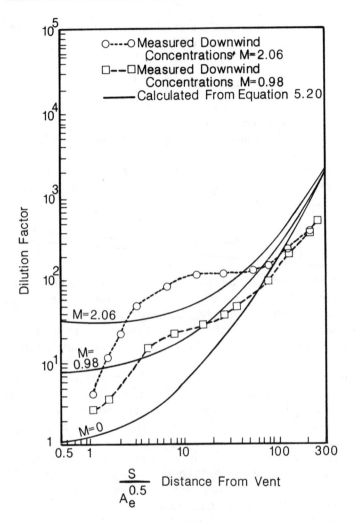

Figure 5.23. Dilution of exhaust from building vents as a function of the exhaust velocity.[34]

simulated at exhaust velocity:wind speed ratios of $M = 0.98$ and 2.06. Results are presented in Figure 5.23. Note the relative insensitivity of dilution for exhausts located at distances less than $10 \ S/A_e^{0.5}$. From Figure 5.23, it is evident that the exhaust velocity ratio M is the most significant factor affecting minimum dilution close to the exhaust vent.

The effect of capped and uncapped exhausts on minimum dilution is illustrated for a standard downwind distance in Figure 5.24. The effect of a rain cap or louver is to destroy vertical momentum. In this case, an uncapped vent at an exhaust velocity twice the wind speed will result in a minimum

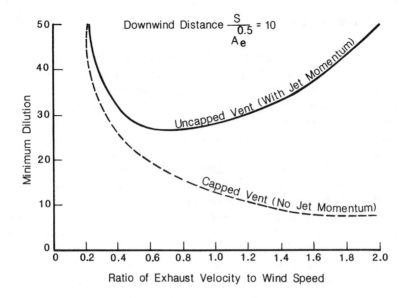

Figure 5.24. Maximum dilution associated with capped and uncapped building exhaust vents.[34]

dilution five times that of the capped exhaust. The entry of precipitation into uncapped exhaust vents can be prevented by operating exhausts continually or using elbows with rain traps and drains for intermittently operated vents.

Ventilation system imbalance. As described above, the location of intakes and exit velocity/wind speed ratio M are important factors affecting contaminant re-entry. The studies of Reible,[29] however, indicate that the most significant factor in the re-entry of exhaust pollutants is the imbalance between ventilation system makeup and exhaust air flow rates. Such imbalance can cause infiltration to occur through leakage sites over the entire building surface and subsequent intake of contaminated air from the building wake.

In a balanced ventilation system, approximately 1% of the exhaust gases typically return to the building. In buildings with extreme ventilation imbalance, and where building intakes and exhausts are close together, the re-entry of exhaust gases may be as high as 10–15%.

A major factor in causing ventilation imbalance is the use of hoods with high exhaust flow rates. The negative building pressures caused by such hoods produce rapid infiltration of exhaust gases through leakage sites. This phenomenon belies the conventional approach of reducing worker exposures to toxic gases and vapors by using relatively high hood face velocities. Careful attention must therefore be given to providing sufficient make-up or intake air to compensate for losses from hood and other exhausts.

VENTILATION STANDARDS

In designing a mechanically ventilated building, architects and architectural engineers must know the size of the HVAC system needed to provide sufficient ventilation air to maintain a healthy and comfortable indoor environment for those who will occupy the building. Because of energy concerns, designers need to know the minimum rate of ventilation necessary to provide for occupant comfort at reasonable energy costs.

The obvious approach to such needs is to have standards or guidelines. Typically in the United States, ventilation standards are developed (by a consensus process) by one or more voluntary professional organizations and/or standard-setting bodies. The lead organization for the development of ventilation standards has been the American Society of Heating, Refrigerating and Air-Conditioning Engineers, best known by its acronym, ASHRAE. Standards may be given further authoritative status by their acceptance and publication by the American National Standards Institute (ANSI).

The primary approach to standards is to specify a ventilation rate in the context of the quantity of outdoor air required in cubic feet per minute (CFM) or liters per second (L/sec) per person.[37] From 1936–1973, the ventilation rate standard of 10 CFM (5 L/sec) per person was based upon the amount of outdoor air needed to provide an odor-free environment (Figure 5.25); it was based on experimental studies of Yaglou.[38] Prior to that, minimum ventilation rates of 30 CFM (15 L/sec) per person were commonly required or recommended.[39] In response to energy concerns, ASHRAE, in its Standard 62-73,[40] recommended a minimum ventilation rate (of outdoor air) of 5 CFM (2.5 L/sec) per person in a nonsmoking general office environment. Higher ventilation rates were recommended for other environments, such as hotel rooms, auditoriums, bars, residential living areas, and industrial facilities. If smoking was permitted, a higher ventilation rate was specified (20 CFM, 10 L/sec). Similar ventilation rates were recommended in ASHRAE Standard 62-1981.[41]

Ventilation Rate Procedure

In both ASHRAE 62-73 and 62-1981, the minimum ventilation rate in a general office area was based on a maximum acceptable level of carbon dioxide (0.25%, 2500 ppm). This CO_2 guideline was based on the assumption that this level was significantly lower than those tolerated in nuclear submarines (1%) and bomb shelters (0.5%) and, therefore, conferred a margin of additional acceptability.

The ventilation guideline[24] was determined from the mass balance equation:

$$Q = \frac{G}{C_1 - C_a} \tag{5.19}$$

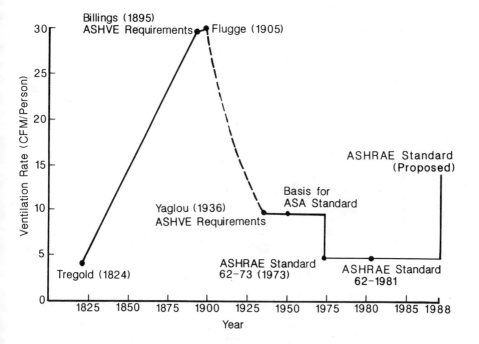

Figure 5.25. Historical development of ventilation guidelines in the United States.[39]

where Q = ventilation rate (L/sec)
C_i = indoor CO_2 concentration (0.25%)
C_a = ambient CO_2 concentration (0.03%)
G = CO_2 generation rate (0.005 L/sec)

Solving for this equation:

$$Q = \frac{0.005}{0.0025 - 0.003} \qquad (5.20)$$

$$= 2.27 \text{ L/sec} = 4.82 \text{ CFM}$$

Therefore, a minimum outdoor air flow rate of 5 CFM (2.5 L/sec) per person will yield a steady-state CO_2 concentration of 0.25%.

Experience has shown that this ventilation rate is insufficient to provide acceptable indoor air quality, particularly in the context of human bioeffluents. In ASHRAE standards revisions currently underway, the minimum recommended ventilation rate is proposed to be 15 CFM (7.5 L/sec) per person, which would yield a steady-state CO_2 level of 0.1% or 1000 ppm. A 1000-

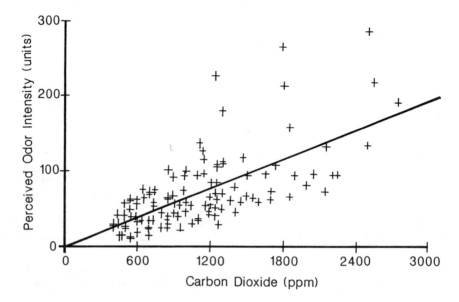

Figure 5.26. Relationship between building odor intensity and metabolic CO_2 levels.[43]

ppm CO_2 standard or guideline would be in conformity with that recommended by the World Health Organization.[42]

In the present context of assessment of ventilation needs, CO_2 is looked upon as an indicator of human bioeffluent levels rather than something that is itself harmful at low concentrations. What bioeffluents contribute to the stuffiness or discomfort associated with high occupant density/low ventilation environments is unknown. Because CO_2 is produced in the highest quantity and is easily measured, it appears to be a good indicator of bioeffluent levels and ventilation requirements. CO_2 levels also appear to be a relatively good indicator of human odor intensity. Comparisons of data for odor intensity and CO_2 levels (Figure 5.26) indicate good correlation between the two parameters.[43]

The effect of ventilation on metabolically enhanced CO_2 levels can be seen in Figures 5.27a and 5.27b. In Figure 5.27a, the ventilation system was being operated on 100% outside air; in Figure 5.27b, the ventilation system was on 85% recirculated and 15% outdoor air. Note also the effect of occupancy.[44]

Because CO_2 is easily measured with either portable real-time instruments or gas sampling tubes, measurement of CO_2 is recommended for evaluation of mechanical ventilation system performance.[20,27,44] Such measurements can provide a simple means of determining how well a ventilation system is doing in diluting and removing indoor contaminants. They are useful where there are no major unvented combustion sources present.

By continuously monitoring CO_2 in a building, and by coupling such

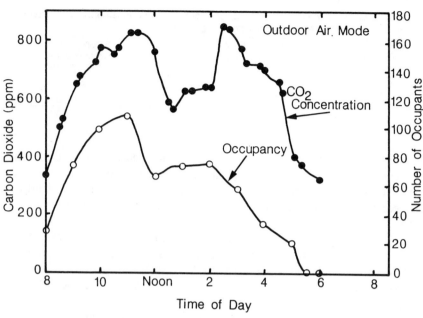

a

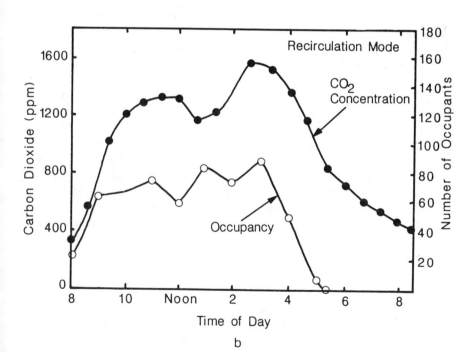

b

Figure 5.27.   Effect of occupancy on metabolic CO_2 levels in a San Francisco office building.[44]

Table 5.1. Recirculation Percentages of Contaminants in Supply and Return Air Corrected for Outdoor Concentrations[46]

% Recirculated Air	CO	Outdoor Organics	Indoor Organics	Strong Odor Components	CO_2
80	81	80	88	80	80
50	60	71	48	61	47
20	33	57	33	40	24

measurements to HVAC system operation, CO_2 levels can be used to control the ventilation percent recirculation air rate. In studies of Sodegren,[45] a CO_2 sensor in the main return air duct was connected to a computer, which adjusted the percentage of recirculated and outdoor air to maintain a CO_2 level of 700 ppm.

The management of ventilation rate based on measured CO_2 levels has the advantage that ventilation is provided on a basis of building occupancy and ventilation need. The sensitivity of the HVAC system in providing ventilation relative to changing occupancy levels has the potential for considerable energy savings.

As described above, ventilation requirements in mechanically ventilated large buildings are, in a major way, based on CO_2 and occupancy levels, which are indicators of human bioeffluents. By implication, ventilation rates based on these criteria should also reduce concentrations of other contaminants as well, ideally to acceptable levels. But how are other contaminants affected by the ventilation rate?

Berglund et al.[46] studied the effect of mechanical ventilation on the concentrations of CO_2 and a variety of organic compounds as a function of the percentage of recirculated air. As can be seen in Table 5.1, CO_2 levels are strongly correlated to the percent recirculation air rate. However, concentrations of CO and organic compounds were seen to be transferred to supply air from return air to a larger extent than predicted from the transfer of CO_2 or calculated flow rates. Changes in the percentage of recirculated air affected the concentration of indoor contaminants differently for different compounds. These differences may be due to the fact that emission rates of some compounds (e.g., formaldehyde) are coupled with ventilation rates, with increased emissions occurring at higher ventilation rates.

Smoking Considerations

In ASHRAE Standard 62-1981, a higher ventilation rate was specified for smoking areas (Table 5.2). In general office environments where smoking is permitted, ASHRAE recommended a ventilation rate of 20 CFM (10 L/sec) per person, a rate four times that of nonsmoking areas. This increased rate was based on the need for reducing levels of suspended particulate matter associated with cigarette smoking. Assuming a total suspended matter generation

Table 5.2   Ventilation Requirements for Selected Building Environments Under ASHRAE 62–1981[41]

Space Type/Use	Ventilation Rate (CFM/person)	
	Smoking	Nonsmoking
Restaurant—dining	35	10
Bars/cocktail lounges	50	10
Hotel bedrooms	30	15
Hotel lobbies	15	5
Office space	20	5
Retail stores	25	5
Ballrooms and discos	35	7
Theater lobbies	35	7
Classrooms	25	5

rate of 31.9 mg/cigarette, an occupant smoking rate of two cigarettes/hr by 30% of the occupants, and a 24-hour National Ambient Air Quality Standard (NAAQS) of 0.260 mg/m^3, an outside air requirement of 41 CFM (20 L/sec) per person would be needed to meet the ambient particle standard in a smoking area. Because this ventilation rate did not seem reasonable in light of energy concerns, the ASHRAE recommendation represented a compromise between health needs and energy costs. Higher ventilation rates were also recommended for smoking areas where occupant density is high (e.g., auditoriums) or where smoking density is expected to be high (e.g., bars).

Recent chamber studies designed to determine the effect of smoking on indoor particle levels have shown that when chamber data are replotted for real-world conditions, an approximate ventilation rate of 35 CFM (17.5 L/sec) per person is needed to meet the 24-hour NAAQS for suspended particulate matter.[47] This corresponds to the 1981 ASHRAE guideline for auditorium lobbies. Based on studies of smoking-related odor, a ventilation rate of 35 CFM (17.5 L/sec) per person would be acceptable to 75% of occupants.[48] Smokers, therefore, would account for 90% of the ventilation demand during smoking occupancy, even though they constitute about 30–35% of the adult population. The relationship between acceptability and ventilation rate per cigarette is illustrated in Figure 5.28.

Air Quality Option

ASHRAE Standard 62-1981 introduced an alternative procedure to specifying ventilation requirements. The Air Quality option was deemed a desirable alternative to the Ventilation Rate Procedure as it would permit and encourage innovative, energy-conserving solutions to the problem of building ventilation.[24] In the Air Quality option, a building designer or manager could use any amount of outside air as long as it could be shown that specific contaminants would be maintained below guideline levels. These guidelines are summarized in Table 5.3.

The introduction of the Indoor Air Quality option concept into building

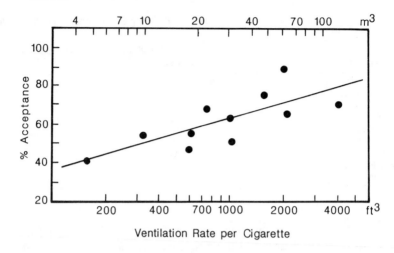

Figure 5.28. Relationship between ventilation rates and acceptability of indoor air associated with tobacco smoking.[48]

ventilation design provoked a great deal of confusion and controversy. The Air Quality option in practice is a complicated one, placing a considerable burden on ventilation engineers who are unfamiliar with sources of contaminants and methods of measurement. Because of this, it has rarely, if ever, been used.

Controversy arose in regard to the indoor air quality guideline for formaldehyde. Manufacturers of mobile homes and suppliers of formaldehyde-releasing building components challenged the guideline as being unreasonably stringent and unattainable in mobile home manufacture.[24] As a consequence of this challenge and controversy, ANSI declined to approve ASHRAE Standard 62-1981, which at this writing is still in a process of review and revision. In the forthcoming revisions, air quality guidelines used to support the Air Quality option in ASHRAE Standard 62-1981 are anticipated to be deemphasized by removing them from the text of the standard and placing them in an appendix.

The Air Quality option, though apparently never used in building ventilation system design or operation, is significant in one major way. It is, in essence, the only real attempt to specify indoor air quality guidelines for a

Table 5.3   ASHRAE Guidelines for Air Contaminants of Indoor Origin[24]

Contaminant	Concentration	Exposure Time
Carbon dioxide	4.5 mg/m3	continuous
Chlordane	5 μg/m3	continuous
Formaldehyde	120 μg/m3	continuous
Ozone	100 μg/m3	continuous
Radon	0.01 WL	annual average

variety of contaminants by any organization that has significant stature and professional credibility in the United States.

Acceptability of Outdoor Air

The use of outdoor air for ventilation assumes that outdoor air is reasonably free of contaminants and will not adversely affect the health and well-being of building occupants. Both ASHRAE Standards 62-73 and 62-1981 recommended that outdoor air used for ventilation meet ambient air quality standards for carbon monoxide, nitrogen dioxide, oxidants (ozone), sulfur dioxide, hydrocarbons, and particulate matter, or be treated so that it does. Though this recommendation has not been subject to challenge, it has been rarely, if at all, followed.

HEAT EXCHANGE VENTILATION SYSTEMS

General ventilation (as opposed to local ventilation) can be used effectively to maintain acceptable indoor air quality on a routine basis or to reduce levels of specific problem contaminants. However, the mechanical introduction of ventilation air to attain or maintain acceptable air quality may be limited by the increased energy requirements (and associated costs) of temperature-conditioning cold or warm outdoor air. In such instances, building managers and home owners may conclude that increased energy costs associated with ventilation are not affordable. This would be particularly true for the use of mechanical ventilation in residential buildings.

An obvious approach to the dilemma of unacceptable costs vs poor indoor air quality is to reduce the energy needs of conditioning outdoor air used for ventilation. These needs can be reduced by extracting warmth or "coolth" from the exhaust air and transferring it to air drawn from the outside. This describes the principle of heat exchange. Devices designated to condition air by this principle are called heat exchangers. Since energy is transferred from one air stream to another, they are described as air-air heat exchangers.

Heat Exchangers

In a heat exchanger, heat is transferred from the exhaust to the supply air stream during cool and cold weather and from the supply to the exhaust air stream when outdoor temperatures exceed indoor temperatures. The two air streams come into thermal contact but do not mix. The part of the system where heat is transferred is called the core. The heat exchange core can be made from a variety of materials, with metals, plastics, and treated paper being the most common. It may be in the form of flat sheets or concentric cylindrical shells. Commonly, air movement through the core is either crossflow or counterflow (Figure 5.29).

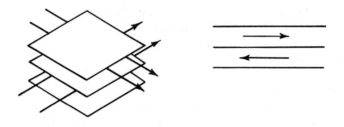

Figure 5.29. Air flow through heat exchanger cores.[49]

In most cases only sensible heat is transferred. Sensible heat exchangers typically have transfer efficiencies in the range of 65-75%.[49,50] Heat exchangers may also transfer latent heat, i.e., heat which changes liquids to gases or solids to liquids. Water vapor, for example, contains latent heat. Transfer of latent heat can occur (from one air stream to another) if the core is permeable to water vapor. Heat exchangers that transfer both sensible heat and latent heat are called enthalpy exchangers. In general, enthalpy exchangers have a relatively high heat transfer efficiency, commonly 75-85%.[49]

In addition to the transfer core, the typical exchanger is equipped with two blower fans — one supply, the other exhaust. Exhaust and supply air flows are balanced so that there is no effect on internal pressure. The unit may also have one or more dust filters. Blower units, filters, and the core are housed in an insulated box (Figure 5.30).[51]

Rotary heat exchangers (Figure 5.31) differ in design from what is described as typical. They employ a slowly rotating heat wheel through which two side-by-side countercurrent air streams must pass.[50,52] Heat wheels as used in residential applications are typically 9 inches (23 cm) in diameter and 2.5 inches (6.4 cm) thick. They contain thousands of air passages which are parallel to one another and the rotor axis. Passages in the wheel accept heat from exhaust air and then, as they rotate, release heat into the supply air stream. Major advantages of rotary exchangers include a large heat transfer area per core volume and the potential transfer of latent heat. When a hygroscopic material such as lithium chloride is used to coat the passages, water vapor is first removed from the exhaust air and then released into the supply air. Moisture does not pass through the core material as is the case with other enthalpy-type exchangers.[49,50]

Heat exchangers vary in size depending on ventilation needs. In residential applications, design air flows are in the range of 50–500 CFM (25–250 L/sec). Smaller units (50–80 CFM; 25–40 L/sec) are usually window- or wall-mounted. These low design flows limit their applicability to providing ventilation to a single room. Larger units (100–500 CFM; 50–250 L/sec) are designed to provide whole-house ventilation. They are normally installed with

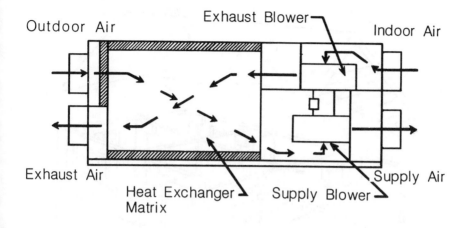

Outdoor Air

Exhaust Blower

Indoor Air

Exhaust Air

Heat Exchanger
Matrix

Supply Blower

Supply Air

Figure 5.30. Common air-air heat exchanger design.[51]

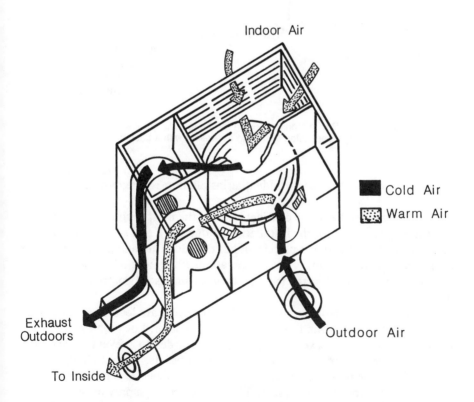

Indoor Air

Cold Air

Warm Air

Exhaust
Outdoors

Outdoor Air

To Inside

Figure 5.31. Rotary air-air heat exchanger design.[52]

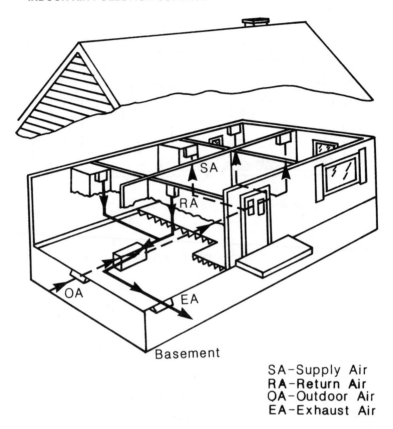

SA–Supply Air
RA–Return Air
OA–Outdoor Air
EA–Exhaust Air

Figure 5.32. Recommended design for the installation of an air-air heat exchange ventilation system in a dwelling with a basement.[53]

a system of ducts. A typical residential installation is diagrammed in Figure 5.32.

Basement installation is considered to be the most desirable, as the basement is easy to access for the installation of ductwork and subsequent maintenance of the unit. Basement placement has the added advantage of minimizing back drafts when the unit is not in operation. Placement of a heat exchange system in an unheated space is to be avoided unless all components can be insulated.

In a good installation, exhaust and intake ports are located at least eight feet apart, with one around the corner from the other. Location of supply air intakes upwind of the exhaust is necessary in order to prevent re-entry (as much as possible). To minimize stagnant air pockets, supply air should be delivered into the residence as far away from the exhaust as is practical so that supply air will flow through the building structure. If supply air is delivered through an existing duct system providing heating, cooling, or both, it should

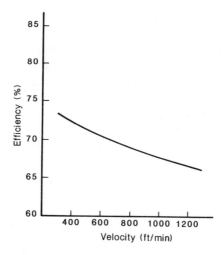

Figure 5.33.   Relationship between heat transfer efficiency and air velocity.[53]

be delivered upstream of the furnace or cooling unit. Exhaust ducts must be installed separately from the existing duct system so that short-circuiting is minimized.

Performance

The efficiency of heat recovery is a function of flow rate;[53-55] as flow rate increases, heat transfer decreases (Figure 5.33). Heat transfer efficiency is also affected by available core surface area; heat transfer efficiency increases with surface area available.

In addition to design features, thermal performance is also affected by operating parameters including leakage of air between the exchanger and its surroundings, imbalances in air flow, and freezing of condensate water.[54,55] Leakage of air from heat exchangers appears to be a common phenomenon. Such leakage appears to be on the order of 5–8%.[56] Imbalances in air flow may cause the ventilated space to be under negative pressure, thus increasing the entry of unconditioned air by infiltration.

Freezing of Exchanger Cores

In cold climates, performance can be significantly diminished by the freezing of condensate water in the exchanger core.[57-60] Under normal operating conditions, water vapor cooled below the dewpoint is collected in a condensate pan and drained from the system. Freezing of condensate in the exchanger core can cause reductions in the rate of heat transfer and the airstream flow rate. Ice clogging of the passages of the core can result in system imbalance.

The onset of freezing in residential heat exchangers depends on the temperature of the exhaust and supply airstreams and the humidity level inside the ventilated space. Studies of Fisk et al.[57] indicate that freezing occurred in the range from –6 to –12°C for a typical range of relative humidities. In other studies the onset of freezing was not observed until temperatures were much lower. Ruth[58] in Manitoba reported, for example, that freezing was initiated in rotary exchangers at supply air temperatures of –16 to –26°C with exhaust relative humidities of 25–30%. Ruth suggested that supply temperatures as low as –26°C could be tolerated at an exhaust air temperature of 24°C and a relative humidity below 22%. In other studies, freezing onset varied linearly from –23 to –9°C with exhaust and inlet relative humidities of 60% and 30%, respectively. Reported differences in freezing onset may have been due to how the onset of freezing was determined. Fisk et al.[60] determined it visually; Ruth[58] and Sauer[59] measured pressure drop through the unit.

The effect of freezing on thermal performance may be substantial. Fisk et al.[60] reported a 1.5–13% reduction in efficiency per hour in a cross flow exchanger at an unconditioned supply air temperature of –12°C and an exhaust relative humidity of 30%. In a counterflow exchanger, the loss of efficiency was less (0.6–2%/hr.). Fisk et al. estimated that freezing and periodic defrosts reduced the thermal efficiency of the crossflow and counterflow exchangers by 15% and 10%, respectively.

Freezing can be prevented by preheating the supply air before it reaches the exchanger core. Such heating will, of course, reduce the overall thermal efficiency. Once freezing has occurred, the unit must be defrosted. Defrosting can be accomplished by shutting off the supply fan while the exhaust fan continues to operate. The exhaust air flow warms the core and causes the ice to melt. Within an hour the supply fan can again be engaged for normal operation.

In Situ Thermal Efficiency

Thermal efficiency as measured under laboratory conditions may not be achieved under conditions of use. For example, Persily[61] studied a wall-mounted crossflow heat exchanger which had an efficiency (measured by heat balance) of 70%. Measured in situ, heat recovery was less, on the order of 55–60%, and as low as 50%, if fan power requirements were considered. This loss of thermal performance was apparently due to the incomplete mixing of supply air in the ventilated space. Side-by-side placement of supply and exhaust vents in the window-mounted heat exchange unit apparently resulted in short-circuiting of air flows.

Ventilation Efficiency

Though good to excellent thermal performance is a desirable feature of such systems, it is important that they effectively reduce indoor contaminant levels as well. The report of Persily[61] described above suggests that because of the side-by-side placement of exhaust and supply vents in window or wall-mounted units, ventilation efficiencies are considerably less than 100%, providing less contaminant control than expected and desired.

Turiel et al.[62] used tracer gases to determine ventilation efficiency of two different wall-mounted units applied for the ventilation of a single room. They reported that the Mitsubishi VL-1500 (operated at a fan speed of 65 CFM; 32.5 L/sec) had an average ventilation efficiency of 50% with a range of 36-65%. A Sharp GV-120 operated at a fan speed of 56 CFM (28 L/sec) had an average ventilation efficiency of 50% with a range of 44-56%.

Wall-mounted units are attractive because of their low cost and ease of installation. They appear to be particularly suitable for use in all-electric houses since no ductwork is needed. However, because of their low design flows, they are only suitable for ventilation of small air volumes, such as a single closed room. Their ability to effectively ventilate even such small spaces appears to be limited.

Central or whole-house heat exchange ventilation systems should in theory provide sufficient ventilation air to effectively reduce contaminant levels. However, as seen in Table 5.4,[63] the actual exchange rate as determined from tracer studies is significantly less than the systems are designed to provide. Data in Table 5.4 include relatively good (based on manufacturers' specifications) as well as poor system installation. In all installations, both supply and exhaust vents delivered or extracted air at floor level. This apparently resulted in short-circuiting similar to that reported for ceiling-mounted supply and exhaust vents in general ventilation used in large buildings.

The apparently low ventilation efficiency observed in Table 5.4 may not have been due only to short-circuiting of intake and exhaust flows. In studies of 227 residential whole-house heat exchanger systems by the Bonneville Power Administration (BPA),[64] actual or delivered air flows averaged only about 55% of design values (range 48-82%). Significant differences were observed among heat exchanger types. Differences between design flows and those delivered were thought to have been due to significant pressure drops associated with the use of lengthy runs of high-resistance flexible ducting.

Cross Contamination

In sensible heat–type exchangers, supply and exhaust air streams are separated by an impermeable barrier preventing the cross contamination of supply and exhaust air. In enthalpy-type exchangers, the heat exchange matrix is permeable to water vapor and other water-soluble gases such as formaldehyde.

Table 5.4 Ventilation Performance of Whole-House Heat Exchange Ventilation Systems[63]

Residence Type	Residence Volume (m³)	Average Air Change Rate Without Mechanical Ventilation (ACH)	Theoretical Air Change Rate (ACH)	Actual Air Change Rate (ACH)	Actual as % of Theoretical
Conventional (unoccupied)	152[a]	0.20	1.20	0.48	40
			1.60	0.67	42
Mobile home (unoccupied)	164[a]	0.19	1.50	0.79	52
			3.10	1.33	42
Conventional (occupied)	269[a]		1.60	0.51	32
Condominium (occupied)	470[b]		1.50	0.43	25

[a]Installed per vendors' recommendations.
[b]Installed with supply and exhaust into existing ductwork.

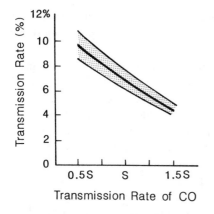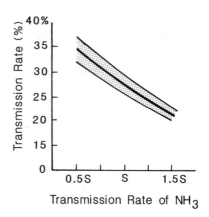

Figure 5.34. Contaminant transmission across a treated-paper enthalpy-type heat exchanger matrix. S = unit capacity (CFM).[65]

Transmission rates for several gases through the Mitsubishi Lossnay treated paper enthalpy-type exchange matrix are illustrated in Figure 5.34.[65]

In studies conducted by Offerman et al.,[66] they observed no reduction in indoor formaldehyde levels associated with the use of enthalpy-type heat exchange ventilation units, whereas a 30% reduction was observed for the sensible heat type, suggesting that formaldehyde was being transferred along with water vapor from the exhaust to the supply airstream. Fisk et al.,[56] studying the problem in detail, observed that only 7–15% of formaldehyde in the exhaust stream was transferred to the supply stream. They concluded that formaldehyde transfer did not seriously compromise performance of the enthalpy-type heat exchanger. However, they also concluded that transmission of water vapor by enthalpy-type heat exchangers may significantly diminish their ability to lower formaldehyde levels as a result of the coupling between indoor relative humidity and release of formaldehyde from source materials.

Sizing

If a heat exchange ventilation system is to provide ventilation sufficient to maintain or attain acceptable indoor air quality in a general way, or reduce specific problem contaminants to acceptable levels, it must be properly sized, that is, it must have sufficient capacity to provide the desired volumetric flow rate and, in the case of a residence, the desired air exchange rate. What exchange rate should be used? There is no easy answer to this question. It is claimed that most experts agree that an air exchange rate of 0.5 ACH is adequate.[50,67] This recommendation, unfortunately, is not based on any systematic evaluation of ventilation needs. Rather, it is an intuitive, best-guess approximation. It is based in part on the assumption that typical houses do not have indoor air quality problems and that such houses have infiltration rates of approximately 0.5 ACH or greater. There is also an implicit assumption that residences with infiltration rates less than this are likely to have poor indoor air quality.

As seen from empirical measurements presented elsewhere in this chapter, reduction in ventilation rates can result in increased levels of a variety of contaminants. It cannot, however, be assumed that 0.5 ACH is a sufficient level of ventilation. For contaminants such as formaldehyde and radon, ventilation needs cannot be prescribed in a generic way. For these contaminants, source strength is the primary determinant of indoor levels.[68,69] Ventilation needs will therefore be determined by the magnitude of emissions, resultant concentrations, and acceptable levels of exposure.

What are acceptable levels of exposure? Acceptable levels may be those recommended by some governmental body or some standard-recommending group. The EPA, for example, recommends that 4 picocuries/L be the action level for radon remedial measures.[70] For formaldehyde, ASHRAE has recommended a maximum exposure level of 0.10 ppm;[44] the Canadian government has recommended 0.10 ppm as an action level for corrective action with a target level of 0.05 ppm.[71] Ventilation needs, and therefore system sizing, can be targeted to achieve these levels.

Acceptable levels of exposure may also depend on whether building occupants are experiencing symptoms or not. In urea-formaldehyde foam–insulated (UFFI) houses, peak levels of exposure average approximately 0.10 ppm;[72] these levels meet the ASHRAE guidelines. However, it is apparent from several studies that hypersensitive individuals experience acute symptoms at 0.10 ppm and below, a fact recognized in the Canadian guidelines.[71]

In sizing a ventilation system to control specific contaminants, it is essential that the level of contamination be known. Because of variability in contaminant levels, it is advisable to make multiple measurements to provide confidence that measured values reveal the true nature of the contamination problem.

Based on the studies of Broder et al.,[73] there appears to be no relationship between formaldehyde levels at an average concentration of 0.03 ppm and acute symptoms, whereas there does appear to be a relationship when levels average 0.045 ppm. Let us assume, therefore, that an acceptable level of exposure to formaldehyde is 0.03 ppm and the average peak levels in a UFFI house are 0.10 ppm. If a heat exchange ventilation system were to be used to reduce formaldehyde to an acceptable level of 0.03 ppm, an air exchange rate capable of providing this must be known. In UFFI house studies, a minimum air exchange rate of 0.85 ACH was needed to reduce levels from 0.12 to 0.03 ppm.[74]

Many UFFI houses are small bungalow-style buildings with an average area of 1000–1500 ft^2 (92–138 m^2). Let us assume that one or more residents are affected by formaldehyde exposures at average peak levels of 0.12 ppm (peak levels defined at 23°C and 60% relative humidity) and that the purpose

of introducing a heat exchange ventilation system is to reduce exposures sufficiently to mitigate observed health problems.

If a UFFI house has an area of 1400 ft^2 (129 m^2), an 8-ft (2.44 m) ceiling,
and a total volume of 11,200 ft^3 (314 m^3), what is the delivery rate of outside
air that is needed to provide 0.85 ACH in this house? Let us address this
question using the following equation:

$$Q = \frac{A \times V}{60} \qquad (5.21)$$

where   Q = CFM (or L/sec)
         A = air exchange rate, ACH
         V = building volume, ft^3 (or m^3)
        60 = conversion of hours to minutes

Solving:

$$Q = \frac{0.85 \times 11,200 \text{ ft}^3}{60} \qquad (5.22)$$

$$= 160 \text{ CFM (or 75 L/sec)}$$

Therefore, in order to achieve an air exchange rate of 0.85 ACH, clearly
one would need to install a heat exchange ventilation system with a delivery
capacity near 160 CFM (75 L/sec). Because of problems of short-circuiting of
intake and exhaust air, and the relatively low delivered:design flows ratio
reported by BPA,[64] it is essential that heat exchange systems be oversized by a
factor of two.

Cost-Effectiveness

Although the use of heat exchange can significantly reduce the energy
penalty associated with providing ventilation air, it is, nevertheless, a fact that
such systems will increase heating or cooling costs because they typically are
only 75% efficient. In addition, there are capital costs associated with purchase and installation. However, ventilation with heat recovery imposes a
smaller heating load than ventilation due to infiltration (assuming air flow
rates are equal), as heat recovery systems preheat incoming air.[55]

A major question is raised when the use of heat exchangers is considered
for residential use: "Are they cost-effective?" In most instances, that question
is addressed in the context of new, energy-efficient houses, with an assumption
that such systems must pay for themselves lest they be judged not worthy of
use or a poor investment.

Let us discuss cost-effectiveness based on the premise that the system must
entirely pay for itself in energy savings in order for it to be cost-effective.

Let us assume that one has a choice of constructing a very tight energy-efficient house with an average air exchange rate of 0.1 ACH as compared to a more conventional house with an average air exchange rate of 0.5 ACH. In our fancy, we assume that the latter provides a reasonably healthful environment and the former does not. In order to provide an environment equivalent to 0.5 ACH in the 0.1-ACH house, we decide to introduce a mechanical ventilation system with heat recovery, so that we have both an energy-efficient house and reasonably good air quality. We assume that this choice is going to result in a significant reduction in energy usage over the conventional house, and this reduced energy requirement will result in savings that will balance out the costs of system purchase and installation. The number of years required to achieve this balance is called the payback period. The payback period is highly variable, being affected by the cost of the system, climate, type of heating fuel, heat exchange performance, and ventilation rate. As a result of extensive studies, Fisk and Turiel[55] concluded that if low ventilation rates (< 0.5 ACH) were deemed adequate, heat exchangers in energy-efficient houses would only be cost-effective in the coldest climates (such as Chicago and Minneapolis), and then only when high-performance heat exchangers were used. Because much less energy is used for air conditioning than heating, the use of heat exchange ventilation systems in conjunction with air conditioning would not be cost-effective. Fisk and Turiel further concluded that, from a homeowner's point of view, investing in a tightly constructed house and installing a heat exchange ventilation system would, in most instances, not be considered economical since the payback period would be too long, e.g., greater than 10 years.

Let us return to the point of what ventilation is for. In theory, it is to provide a healthy, habitable environment at a reasonable cost. In most cases it provides both. A ventilation system, like any other appliance, provides a service which imposes a cost on the purchaser. One purchases appliances for the services they provide and does not require that they pay for themselves. Why should a ventilation system be any different?

EFFICIENCY OF VENTILATION
IN RESIDENTIAL CONTAMINANT CONTROL

Intuitively, one would assume that the use of mechanical ventilation in the form of heat exchange ventilation systems or by some other means would be effective in controlling or reducing levels of exposure to specific contaminants in residences. Beyond the intuitive, how effective is mechanical ventilation in residential contaminant control?

Offerman et al.[66] studied the effect of mechanical ventilation on indoor levels of radon, formaldehyde (HCHO), NO_2, and relative humidity in nine occupied houses (Table 5.5). In the absence of mechanical ventilation, air exchange rates in the nine houses were low, ranging from 0.22 to 0.5 ACH.

Table 5.5   Effect of Mechanical Ventilation on Air Exchange and Contaminant Levels in 9 Occupied Rochester, New York Houses[66]

House No.	Ventilation Mode	Air Exchange Rate (ACH)	Radon (pCi/L)	% Reduction	HCHO (ppb)	% Reduction	NO₂ (ppb)	% Reduction	NO₂ (ppb) Outdoor	Relative Humidity %	RH % Reduction
1	off	0.22 ± 0.09	0.4	75	36	47	1	-67	7	47	19
	on	0.47 ± 0.19	0.1		19		3		10	38	
6	off	0.38 ± 0.09	1.6	56	29	24	4	0	16	30	13
	on	0.66 ± 0.16	0.7		22		4		13	26	
10	off	0.30 ± 0.09	1.2	42	7	+29	1	-80	9	35	14
	on	0.61 ± 0.02	0.7		>5		5		25	30	
33	off	0.38 ± 0.15	0.3	100	33	42	23	10	9	42	7
	on	0.78 ± 0.16	0.0		19		20		12	39	
45	off	0.37 ± 0.10	2.1	57	28	-3	2	-67	14	44	11
	on	0.61 ± 0.23	0.9		29		6		29	39	
49	off	0.42 ± 0.11	0.1	-50	30	3	11	-30	15	25	-8
	on	0.64 ± 0.17	0.2		29		16		11	27	
52	off	0.28 ± 0.10	1.1	64	64	3	3	67	12	41	-2
	on	0.73 ± 0.13	0.4		62		1		11	42	
56	off	0.50 ± 0.13	0.2	100	18		10	10	15	36	8
	on	0.61 ± 0.21	0.0				9		18	33	
60	off	0.33 ± 0.10	2.2	37	57	26	12	-8	18	52	13
	on	0.52 ± 0.06	1.6		42		13		18	45	

The increase in air exchange induced by mechanical ventilation varied from 22 to 114%. Except for one anomalous outcome (house #49), mechanical ventilation was observed to decrease indoor radon levels by 37–100%. In general, even relatively small increases in air exchange rates resulted in significant reductions in indoor radon levels. Results for formaldehyde were inconsistent, varying from 0–44%. Ventilation was ineffective in reducing NO_2 levels. In general, mechanical ventilation was observed to reduce relative humidity by –8 to +14%.

Inconsistencies observed in the effectiveness of ventilation for formaldehyde control were apparently due to variations in environmental conditions (for example, temperature and relative humidity), which significantly affect emission rates from formaldehyde-releasing materials. The ineffectiveness of mechanical ventilation on indoor NO_2 levels was apparently due to the fact that outdoor levels were higher than those indoors. The use of mechanical ventilation increased the average indoor NO_2 level by 2 ppb.

Offerman et al.[66] studied the effect of mechanical ventilation on inhalable (< 10 μm) and respirable (< 2.5 μm) particles in two houses (Table 5.6). In one house, an increase in the air exchange rate from 0.38 to 0.66 ACH resulted in a reduction in the concentrations of inhalable and respirable particles by 35% and 43%, respectively. In the second house, an increase in air exchange rate from 0.42 to 0.64 ACH resulted in only 17% and 21% reductions. Both series of evaluations were conducted during days when outdoor particle levels were relatively low.

Evaluations of the effectiveness of mechanical ventilation in reducing formaldehyde concentrations in residences under controlled indoor conditions of temperature and relative humidity are summarized in Table 5.7. Studies of Jewell[75] and Godish and Rouch[74] indicate that mechanical ventilation can significantly reduce indoor formaldehyde levels. Studies of Matthews et al.[76] indicate somewhat lower levels of formaldehyde control associated with mechanical ventilation.

From the studies of Godish and Rouch, it is evident that when initial air exchange is low (around 0.2 ACH), very high ventilation rates (0.85–1.4 ACH) are required to achieve significant formaldehyde reduction (70% or more). This suggests that ventilation will not be the preferred method of control in many circumstances. On the other hand, when control options are very limited, as is the case with UFFI houses, mechanical ventilation at the reported rates does appear to be both economically viable and effective in reducing levels sufficiently to mitigate health problems.

It is evident from studies of Jewell,[75] Godish and Rouch,[74] and Matthews et al.[76] that the effect of mechanical ventilation on formaldehyde levels is not linear and that as the ventilation rate increases, its effectiveness in reducing formaldehyde levels diminishes. This phenomenon is most likely a result of the interaction between formaldehyde source emissions and formaldehyde levels above/around formaldehyde sources, particularly those sources that are comprised of or that utilize urea-formaldehyde resins. The effect of increased

Table 5.6   Effectiveness of Mechanical Ventilation in Reducing Inhalable and Respirable Particulate Matter in Two Occupied Houses in Rochester, New York[66]

House No.	Ventilation Mode	Air Exchange Rate (ACH)	Inhalable Particle Fraction (<15μm) ($\mu g/m^3$)			Respirable Particle Fraction (<2.5μm) ($\mu g/m^3$)		
			Indoor	% Reduction	Outdoor	Indoor	% Reduction	Outdoor
6	off	0.38 ± 0.09	76	35	28	54	43	19
	on	0.66 ± 0.16	49		13	31		9
49	off	0.42 ± 0.11	52	17	20	38	21	14
	on	0.64 ± 0.17	43		6	30		6

Table 5.7 Effectiveness of Mechanical Ventilation in Reducing Formaldehyde Levels in Residential Structures

Investigator(s)	Residence Type	Air Exchange Rate (ACH)	Formaldehyde Concentration (ppm)	Reduction (%)
Jewell[75]	Mobile home	0.56	0.21	
		1.20	0.11	48
		2.00	0.08	62
	Mobile home &	0.56	0.28	
	added particleboard	1.20	0.15	46
		2.00	0.13	54
Godish & Rouch[74]	Mobile home			
	Series #1	0.20	0.20	
		0.45	0.11	45
	Series #2	0.17	0.19	53
		0.77	0.09	
	Mobile home & added particleboard			
	Series #1	0.19	0.18	67
		1.22	0.06	
	Series #2	0.21	0.21	57
		1.33	0.09	
	Urea-formaldehyde foam– insulated (UFFI)			
	Series #1	0.20	0.25	
		1.45	0.06	76
	Series #2	0.18	0.10	
		1.36	0.03	70
	Series #3	0.20	0.12	
		0.85	0.03	70
	UFFI & particleboard subflooring			
	Series #1	0.21	0.22	
		0.70	0.10	55
	Series #2	0.24	0.19	58
		0.48	0.08	
Matthews[76]	Conventional— Particleboard subflooring			
	Series #1	0.20	0.091	
		0.50	0.085	7
		0.94	0.059	35
	Series #2	0.25	0.099	
		0.46	0.082	17
		0.98	0.053	47

ventilation is to reduce formaldehyde levels, which in turn contributes to increased source emission rates. The effect of formaldehyde levels on emission rates can be seen in Figure 5.35. The coupling effect between emission rates and formaldehyde levels results in a decrease in the effectiveness of ventilation for formaldehyde control.

Matthews et al.[68] have developed a single-compartment model to predict

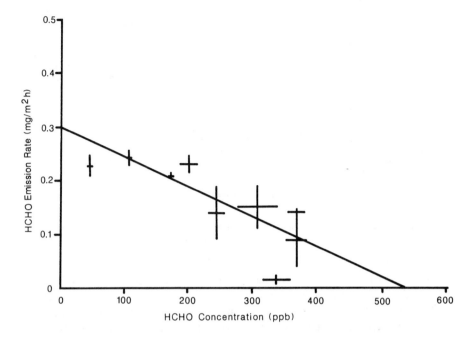

Figure 5.35. Effect of formaldehyde levels on emission rates.[68]

the effect of ventilation on formaldehyde levels in a conventional house with pressed wood products with a loading factor of 0.29 m²/m³, Equation 5.23.

$$Q_2 = Q_1X + 1.22X - 1.22 \qquad (5.23)$$

In this model Q_2 = the ventilation rate required to achieve an X-fold reduction in formaldehyde concentration measured at a ventilation rate of Q_1. The Q_1X term represents a reduction in formaldehyde levels which is proportionate to an increase in the ventilation rate. The second and third terms on the right adjust for the negative formaldehyde concentration dependence of the formaldehyde emission rate from pressed wood products based on molecular diffusion principles. The second and third terms are dependent on X but are independent of the initial ventilation rate Q_1. The terms Q_1 and Q_2 are approximately equal to Q_1X when the initial ventilation rate Q_1 is large. As a consequence, the decrease in formaldehyde levels associated with increased ventilation rates is insensitive to the formaldehyde concentration dependence of the emission rate from the pressed wood products at the low formaldehyde levels that are achieved with high air exchange rates.

While the use of mechanical ventilation in controlling formaldehyde levels in residential environments appears to be moderately effective, it is, on the other hand, highly effective in reducing radon levels. This is evident in the

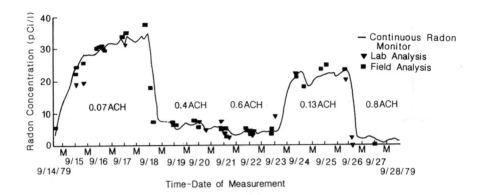

Figure 5.36. Effect of ventilation on radon levels in an energy-efficient house.[77]

studies of Offerman et al.[66] as summarized in Table 5.5. Levels of control ranged from 37 to 100% even for small absolute increases in air exchange rate (0.11–0.40 ACH). The effect of ventilation rates in an energy-efficient house with a significant radon source strength can be seen in Figures 5.36 and 5.37. In Figure 5.37, there appears to be a loglinear relationship between radon level and air exchange rate. Although radon levels are more significantly reduced by increased ventilation rates than are levels of formaldehyde and the relationship appears to be loglinear, as with formaldehyde, radon emission rates are cou-

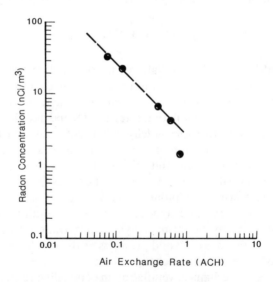

Figure 5.37. Relationship between building air exchange rates and radon concentrations.[78]

Table 5.8   Reduction in Radon Progeny Concentrations by the Balanced Ventilation of Basements[79]

House No.	Basement			First Floor		
	WL Before	WL After	% Reduction	WL Before	WL After	% Reduction
1	0.086	0.006	93	0.014	0.001	93
2	0.053	0.002	96	0.064	0.003	95
3	1.849	0.097	95	0.738	0.287	61
4	0.245	0.033	87	0.123	0.029	76
5	0.047	0.002	96	0.030	0.004	87
6	0.105	0.004	96	0.029	0.007	76
7	0.259	0.063	76	0.060	0.017	72
8	0.534	0.025	95	0.168	0.020	88
9	1.039	0.027	98	0.405	0.060	85
10	0.107	0.004	96	0.043	0.005	88

pled or linked to air exchange rates. However, in this case, coupling is associated with infiltration. Radon entry is coupled with infiltration because increased indoor/outdoor temperature differences cause a greater driving force for infiltration and greater depressurization in portions of a building where radon entry is likely to occur. Balanced mechanical ventilation apparently does not increase radon emission rates.[66,78] This is not surprising, since maintenance of an equal amount of both exhaust and supply air will not result in depressurization; only if there is an imbalance on the exhaust side will depressurization occur.

Use of mechanical ventilation for reducing residential radon levels to acceptable levels has been of particular interest in the Reading Prong area of Pennsylvania, where unusually high radon levels have been reported. In studies conducted by Wellford,[79] balanced ventilation appears to be a very effective radon mitigation measure. In Wellford's studies, 10 homes were initially ventilated in the basement area only, employing the rationale that a basement-only system would have a more limited volume and the basement would be a semi-conditioned space. Results are presented in Table 5.8. Ventilation systems (with a rated air delivery capacity of 0.10 inches static pressure) reduced basement radon progeny exposures an average of 93% (range 76–98%) and radon concentrations by 79%. The reduction of radiation exposure in the first floor living area averaged 82%. When the systems were modified to supply air directly to the first floor as well as to the basement, radon progeny exposure in the basement increased by 9%, while it decreased in the upstairs living area by 18%. Despite the initial compelling logic that basement ventilation would prove to be the most effective and economical, these studies indicate that a combined ventilation of both basement and living areas provides the more effective reduction of radiation exposures. Since the living space is ventilated, it would also reduce exposure to other indoor contaminants as well.

Table 5.9   Effectiveness of Retrofit House-Tightening Measures in the Reduction of Air Change Rates Due to Infiltration[82]

Measure Applied	% Reduction in Air Change Rates
Storm windows or window weather stripping	6
Door weather stripping	1
Caulking	3
Blown-in wall insulation	10
Sealing air ducts	9
Total	29

IMPLEMENTATION OF ENERGY CONSERVATION MEASURES: IMPACT OF REDUCED VENTILATION ON INDOOR AIR QUALITY

In numerous news stories in both the print and electronic media, indoor air pollution is almost without exception linked to energy conservation measures and the construction of energy-efficient buildings. Implicit in this linkage is that indoor air pollution is caused by energy conservation attempts. Also implicit in this notion is that older, leaky, energy-inefficient buildings are safe. Unfortunately, the world is not as simple as most make it or wish it to be. In the cases of both radon and formaldehyde, the dominant factor affecting levels is not ventilation but the magnitude of source emissions, that is, source strength. This is evident in the apparent lack of a relationship between radon levels and air exchange rates in a large population of houses[80] and the relatively small effect that incremental increases in ventilation rates have on formaldehyde concentrations (Table 5.7).

To what extent do energy conservation measures increase contaminant exposure levels and, implicitly, health risks? This is not an easy question, particularly if the effectiveness of energy conservation measures or practices is unknown. Studies conducted for BPA indicate that the sum of a variety of retrofit weatherization measures can be expected to reduce leakage or air infiltration on the order of 20–30%, with a range of 0–60%.[81] The effectiveness of different measures in reducing infiltration is summarized in Table 5.9.[82] The relatively small decreases in air exchange rates evidenced in Table 5.9 would be expected to increase contaminant levels and aggravate existing contamination problems, but apparently not substantially.

The contaminant which has created the most concern relative to the application of residential weatherization measures, particularly those supported by large utility authorities such as BPA, has been radon. Since such measures have the potential of reducing ventilation rates and thereby increasing exposure levels, residents may be at higher risk of developing lung cancer. The advocacy and support of weatherization measures by utilities brings into question their liability in any potential future lawsuits claiming that an individual's lung cancer was due to the increased exposure to radon associated with such measures. At this time, this is a question of considerable uncertainty both in a technical and a legal sense.

Reduced house ventilation associated with infiltration reductions of 20–30% may not necessarily result in significant increases in radon exposures. A reduction in infiltration can be expected to reduce house depressurization, the phenomenon primarily responsible for the inflow of radon-bearing soil gas. There is evidence that the coupling effect between reduced infiltration and reduced house depressurization can partially counterbalance increases in radon levels that would have occurred as a result of decreased natural ventilation rates.[83]

Nazaroff et al.[77] reported a positive correlation between air exchange rate and radon emissions in five of six houses studied in the northeastern United States. In three of the houses, peaks in radon source emissions corresponded to peaks in infiltration-driven air exchange rates. This corresponds to the fact that radon from soil gas enters building spaces by pressure-driven flows associated with the stack effect.

Burkhart et al.,[84] however, conducted studies in Switzerland in which they attempted to determine the effect of extensive weather stripping of windows and doors in pairs of dwellings matched for factors such as subsoil and climate. They assumed that the weatherization measures would, in theory, reduce air exchange in houses from an assumed average of 0.5 ACH to 0.1 ACH and conjectured that radon levels would increase by a factor of 5. Their matched pair analyses revealed an increase of radon concentration in weatherized houses by a factor of 1.8. Although the effect of the weatherization measures was significantly smaller than expected, they did result in an approximate twofold increase in radon levels.

Fleischer and Turner[85] measured radon levels in 21 energy-efficient and 14 conventional houses near Albany, New York. Energy-efficient houses (without heat storage mass) were observed to have radon levels in the range of 1.5–1.7 times higher in basement, first floor, and second floor areas. Energy-efficient houses with heat storage mass had radon levels averaging 3.1–4.9 times levels in conventional non–energy-efficient houses.

Studies by Burkhart[84] and Fleischer and Turner[85] indicate that energy conservation measures can increase radon levels in houses by 1.5- to 2-fold. This is considerably less than would be predicted from studies of air exchange based on the use of mechanical ventilation (Figure 5.36).

LOCAL VENTILATION

When indoor contaminants are known to be generated from specific pieces of equipment, appliances, or processes, it may be desirable to control them by local rather than general ventilation, as the former is likely to be more effective and economical than the latter. Local ventilation is widely used in industrial hygiene to reduce contaminants to acceptable levels in workrooms and laboratory environments. It is also widely used to exhaust combustion by-products and odors from cooking in cafeterias and restaurants, and odor from lavatories. Its use in controlling contaminants released from equipment or appliances in office environments and cooking appliances in homes is less consistent.

Local ventilation or exhaust systems capture contaminants at or near the source before they can become dispersed into the surrounding air. Such systems, if designed and sized properly, can reduce contaminant concentrations from intermittent sources by 90–95% in a matter of minutes. Exhaust is to the ambient environment, where it is presumed to be carried away from the building by air currents and diluted to very low concentrations. As was seen in the section on cross contamination and re-entry (pp. 199–207), this presumption may not always be valid.

A typical exhaust system consists of a hood, ducts, a blower fan, and an exhaust vent. Exhaust systems primarily intended for the control of gaseous contaminants may also have low-efficiency particulate filters, which serve to provide some protection for the blower fan or to avoid potentially hazardous deposits in the duct system (for example, cooking greases). If the exhaust air is to be recirculated within the building, it may pass through an air cleaner incorporated into the exhaust system.

Hoods are roomside openings into the ventilation system where contaminants are to be captured. They vary in size, shape, and application. A hood is the most important part of a local exhaust ventilation system. Hoods work in different ways. Some reach out and catch contaminants (capturing hoods), some catch contaminants which are thrown into the hood (receiving hoods), and some enclose the contaminant source entirely (enclosure hoods). The capturing hood is the most widely used in industrial hygiene practice. A kitchen range hood would be best described as a capturing hood when the fan is engaged and as the receiving type when it is not.

In using hoods of the capturing or receiving kind, it is important to locate the hood as close to the contaminant source as is possible. Even in well-designed exhaust systems, a capturing hood can generate an adequate capture velocity only up to two feet from its opening, as random drafts and other currents disperse contaminants into the room beyond this distance.[86] How well a hood works relative to exhausting contaminants depends on whether it develops an adequate capture velocity and how well the capture velocity is distributed relative to the source of contaminants. Capture velocities may vary from 50 to 2000 ft/min (16.4–656 m/min), with choice depending on exhaust needs, which are usually determined by contaminant toxicity and conditions in the local environment.

The capture zone (Figure 5.38) of an open duct can be described as a sphere. Contaminants inside the zone will be captured and exhausted; those outside the zone may escape. One can expand the capture zone by increasing the capture velocity (by increasing fan static pressure). Depending on the hood

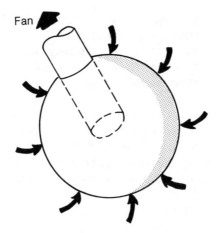

Figure 5.38.   Capture zone of an open
duct under negative pressure.[86]

type, however, increases in capture velocity beyond a certain maximum may result in little additional contaminant control.[86]

Ducts are the network of pipes which carry contaminants from the hood to the outdoors. Blowers or fans provide the energy necessary to draw contaminants into the hood by negative pressure. Fan size and speed will be determined by hood design and desired capture velocity.

The theoretical effectiveness of local ventilation as compared to general ventilation, with nominal air exchange rates of 4 and 10 ACH, is summarized in Table 5.10. Local exhaust ventilation is seen to be far more efficient in reducing dust levels from an intermittent source than is general ventilation.[87]

As air is exhausted from a space, it must be replaced by an adequate supply of make-up air. Make-up air may be supplied by the general ventilation system or drawn from adjacent spaces or the outdoors by infiltration. This inflow of make-up air results from the negative pressure generated as the exhaust system expels air to the outdoors. In some cases, make-up air will be insufficient; because of the added resistance, the blower fan may not work

Table 5.10   Reduction of Dust Concentrations by Local and General Ventilation in an Industrial Workroom[87]

	% Dust Reduction 1 min after Fan Induction
General ventilation	
4 ACH	6.0
10 ACH	15.0
Local ventilation	
4-inch pipe	96.0
6-inch pipe	99.2

properly. The reader may observe this phenomenon by closing the door to a small bathroom and engaging the exhaust fan.

Make-up air should be provided mechanically on a basis of predetermined needs rather than by infiltration. The make-up air inlets should also be located (as described previously) to minimize re-entry of contaminants from nearby exhausts.

In office environments, local ventilation may be applied to exhaust ammonia from blueprint machines, ozone from office copiers, methanol from spirit duplicators, and acetic acid, formaldehyde, and other vapors from photographic film processing. Considering the noxious gases produced by such equipment and processes, it is a point of irony that spaces utilized for them are typically the most poorly ventilated ones in the building.

The primary applications of mechanically driven local ventilation in residences are in the bathroom and kitchen. Bathroom ventilation is intended to prevent structural damage from mold, which is favored by the high humidity levels associated with bathing. In the kitchen, exhaust ventilation is deemed desirable to remove combustion gases from gas ranges as well as particulate matter and volatile gases generated in cooking.

Although most cooking ranges are equipped with hoods, the need for and use of such hoods is poorly understood by consumers and by individuals who design, manufacture, and sell them. A prime example of such poor understanding is the widespread availability and use of what are described as ductless hoods. These hoods have no ducts to the outside. They ostensibly remove contaminants by drawing air through a three-eighths-inch filter bed of activated charcoal. Such filters are so poorly designed and constructed that filtration efficiency would be nearly zero even if the charcoal had any affinity for the contaminants generated. Additionally, these filters are almost never replaced by consumers.

Hoods that are ducted to the outside are usually placed directly over the cooking surface. Taking into consideration the relatively low fan power and volume flow rates used (20–250 CFM; 9.3–116.7 L/sec), the rising air flow and particles produced from cooking, and the fact that, in most cases, users do not engage the fan (except after an episode of burned food!), kitchen range hoods are best described as receiving hoods. In addition to infrequent use (in the context of fan activation), kitchen range hoods are being increasingly replaced by microwave ovens which, because of space limitations, are often located directly over the cooking surface.

Assuming that a kitchen range hood is in place and is available for venting cooking exhausts to the outside, how well do they perform in practice?

Because of differences in hood configurations, fan flow rates, hood location, and environmental factors, the performance of kitchen range hoods would be expected to vary significantly. Rezvan[88] has conducted studies of kitchen range hood efficiency in a two-room test chamber using SF6 tracer gas. He observed that a strong linear relationship existed between contaminant removal and fan volume flow rate when the tracer gas or contaminant was

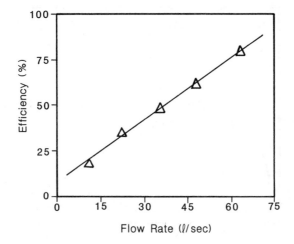

Figure 5.39. Range hood contaminant removal efficiency
as a function of the volume flow rate.[88]

heated (Figure 5.39). Over a range of flow rates from 10 to 60 L/sec, the highest measured efficiency was 77%. With unheated tracer gas, effectiveness was highly dependent on environmental conditions, particularly free convection.

Traynor et al.[89] studied the effect of a kitchen range hood on kitchen NO_x concentration as compared with that obtained with whole-house ventilation. They observed that a kitchen range hood operated at a flow rate sufficient to provide a whole-house air exchange rate of 1.5 ACH (approximately 108 L/sec) reduced NO_2 levels in the kitchen to about one-sixth of those observed in a whole-house mechanical ventilation experiment (0.83 ACH).

Traynor et al.,[90] using a 27-m³ environmental chamber, have also reported on the effects of range hood operation on NO_2 levels associated with the use of a gas-fired oven for a series of air exchange rates. These results are presented in Figure 5.40. Note that at the highest air exchange rate of 7.0 ACH (normal operating mode), the range hood was very effective in preventing buildup of NO_2. The 1.0, 2.0, and 7.0 ACH experimental room air exchange rate would have been equivalent to hood volumetric flow rates of 7.5, 15, and 52.5 L/sec, respectively, assuming that all air exchange under those conditions was solely due to hood fan operation. The effectiveness of local ventilation by use of a range hood would, in these experiments, be expected to be higher (as compared to a whole house), since contaminants were confined to a single room.

Studies described here suggest that effective control of gas range and oven combustion by-products can be achieved by using a kitchen range hood with a volumetric flow rate of 106–235 CFM (50 to 110 L/sec). This, of course, assumes good hood design as well as installation close to the source.

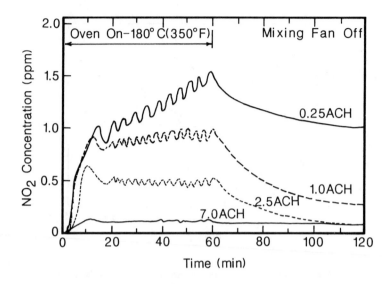

Figure 5.40. Effect of range hood air exchange rates on NO_2 levels in a kitchen.[90]

A special application of local exhaust for the control of indoor air contamination is the use of sub-slab and crawl space ventilation as a radon mitigation measure. Because such ventilation is directed at controlling radon at its soil gas source before it enters building airspaces, it would be more appropriate to describe it as a source control technique. As such, it has been treated extensively in Chapter 2.

REFERENCES

1. American Society of Heating, Refrigerating and Air-Conditioning Engineers. 1985. "Natural Ventilation and Infiltration." 22.1–22.18. *ASHRAE Handbook. Fundamentals.* American Society of Heating, Refrigerating and Air-Conditioning Engineers. Atlanta.
2. Allen, C. 1984. "Wind Pressure Data Requirements for Air Infiltration Calculations." Technical Note AIC 13. Air Infiltration Centre, Berkshire, Great Britain.
3. Sharo, C.V. and W.C. Brown. 1982. "Effect of a Gas Furnace Chimney on the Air Leakage Characteristic of a Two-Story Detached House." In: *Proceedings of the Third IEA Conference.* Air Infiltration Centre, Berkshire, Great Britain.
4. Tamura, G.T. and A.G. Wilson. 1967. "Pressure Differences Caused by Chimney Effect in Three High Rise Buildings." *ASHRAE Trans.* 73. Pt.2.
5. Wang, F.S. and C.F. Sepsy. 1980. "Field Studies of the Air Tightness of Residential Buildings." 24–35. In: C.M. Hunt et al. (Eds.). *Building Air Change Rate and Infiltration Measurements.* ASTM STP 719. American Society for Testing and Materials. Philadelphia.

6. Grimsrud, D.T. et al. 1983. "Calculating Infiltration: Implications for a Construction Quality Standard." Presented at American Society of Heating, Refrigerating and Air-Conditioning Engineers Conference on Thermal Performance of Exterior Envelopes of Buildings II. LBL-9416. Lawrence Berkeley Laboratory, Berkeley, CA.

7. Sandia National Laboratory. 1982. *Indoor Air Quality Handbook for Designers, Builders, and Users of Energy Efficient Residences.* Sand UC-11. 82-1773.

8. Hunt, C.M. 1980. "Air Infiltration: A Review of Some Existing Techniques and Data." 3-23. In: C.M. Hunt et al. (Eds.). *Building Air Change Rate and Infiltration Measurements.* ASTM STP 719. American Society for Testing and Materials, Philadelphia.

9. Grimsrud, D.T. 1984. "Tracer Gas Measurements of Ventilation in Occupied Spaces." In: *Ann. ACGIH: Evaluating Office Environmental Problems.* 10:69-76.

10. Allen, C. 1985. "Leakage Distribution in Buildings." Technical Note AIC 16. Air Infiltration Centre, Berkshire, Great Britain.

11. American Society for Testing and Materials. 1982. "Standard Practice for Measuring Air Leakage by the Fan Pressurization Method." 14-84-1493. In: *Annual Book of ASTM Standards — E779-81.* Part 18. American Society for Testing Materials, Philadelphia.

12. Sherman, M.H. and D.T. Grimsrud. 1982. "Energy Efficient Domestic Ventilation Systems for Achieving Acceptable Indoor Air Quality." In: *Proceedings of the Third IEA Conference.* Air Infiltration Centre, Berkshire, Great Britain.

13. Sherman, M.H. 1986. "Infiltration Degree-Days: A Statistic for Infiltration-Related Climate." *ASHRAE Trans.*

14. Sherman, M.H. 1984. "Description of ASHRAE's Proposed Air Tightness Standard." LBL-17585. Lawrence Berkeley Laboratory, Berkeley, CA.

15. Godish, T. et al. 1988. "Formaldehyde Building-Related Illness Survey" (unpublished data).

16. Godish, T. 1986. "Residential Formaldehyde Control Measures." Indoor Air Quality Note #2. Ball State University, Muncie, IN.

17. Melius, J. et al. 1984. "Indoor Air Quality — The NIOSH Experience." In: *Ann. ACGIH: Evaluating Office Environmental Problems.* 10:77-92.

18. McNall, P.E. and A.K. Persily. 1984. "Ventilation Concepts for Office Buildings." In: *Ann. ACGIH: Evaluating Office Environmental Problems.* 10:49-58.

19. Benard, J.M. et al. 1986. "Relating Health and Environmental Studies in a Tight Building Syndrome Investigation." 23-34. In: D.S. Walkinshaw (Ed.). *Transactions: Indoor Air Quality in Cold Climates.* Air Pollution Control Association. Pittsburgh.

20. Godish, T. et al. 1986. "Ventilation System Performance in a New Classroom Building Assessed by Measurements of Carbon Dioxide Levels." 603-610. In: *Proceedings of IAQ '86: Managing Indoor Air for Health and Energy Conservation.* American Society of Heating, Refrigerating and Air-Conditioning Engineers. Atlanta.

21. Sandberg, M. 1981. "What is Ventilation Efficiency?" *Building and Environ.* 16:123-135.

22. Janssen, J. et al. 1982. "Ventilation for Control of Indoor Air Quality: A Case Study." *Environ. Int.* 8:487-496.

23. Janssen, J.E. 1984. "Ventilation Stratification and Air Mixing." 43-48. In: *Pro-

ceedings of the Third International Conference on Indoor Air Quality and Climate. Swedish Council for Building Research. Stockholm. Vol. 5.

24. Janssen, J. 1983. "Ventilation for Acceptable Indoor Air Quality: ASHRAE Standard 62-1981." In: *Ann. ACGIH: Evaluating Office Environmental Problems.* 10:49–58.

25. Woods, J.E. 1984. "Measurement of HVAC System Performance." In: *Ann. ACGIH: Evaluating Office Environmental Problems.* 10:77–92.

26. Gorman, R.W. 1984. "Cross Contamination and Entrainment." In: *Ann. ACGIH: Evaluating Office Environmental Problems.* 10:115–120.

27. Bearg, D.W. and W.A. Turner. 1986. "Building Assessment Techniques for Indoor Air Quality." 276–283. In: D.S. Walkinshaw (Ed.). *Transactions: Indoor Air Quality in Cold Climates.* Air Pollution Control Association. Pittsburgh.

28. Morey, P.R. et al. 1984. "Environmental Studies in Moldy Office Buildings: Biological Agents, Sources, and Preventative Measures." In: *Ann. ACGIH: Evaluating Office Environmental Problems* 10:21–35.

29. Reible, D.D. et al. 1985. "The Effect of the Return of Exhausted Building Air on Indoor Air Quality." 62–71. In: D.S. Walkinshaw (Ed.). *Transactions: Indoor Air Quality in Cold Climates.* Air Pollution Control Association. Pittsburgh.

30. Wilson, D.J. 1983. "A Design Procedure for Estimating Air Intake Contamination from Nearby Exhausts." *ASHRAE Trans.* 89. Pt. 2.

31. Wilson, D.J. 1977. "Dilution of Exhaust Gases from Building Surface Vents." *ASHRAE Trans.* 83. Pt.1.

32. Smith, D.E. 1978. Doctoral dissertation. Harvard School of Public Health, Cambridge, MA.

33. Wilson, D.J. and E. Chui. 1985. "Influence of Exhaust Velocity and Wind Incidence Angle or Dilution from Roof Vents." *ASHRAE Trans.* 90. Pt.3.

34. Wilson, D.J. 1976. "Contamination of Air Intakes from Roof Exhaust Vents." *ASHRAE Trans.* 82. Pt.1.

35. Wilson, D.J. 1985. "Ventilation Intake Contamination by Nearby Exhausts." 335–347. In: D.S. Walkinshaw (Ed.). *Transactions: Indoor Air Quality in Cold Climates.* Air Pollution Control Association. Pittsburgh.

36. Sagendorf, J.F. et al. 1979. "Near Building Diffusion Determined from Atmospheric Tracer Experiments." *Proceedings of the Fourth Symposium on Turbulent Diffusion and Air Pollution.* American Meteorologic Society. Boston.

37. National Research Council. 1981. *Indoor Pollutants.* National Academy Press, Washington, D.C.

38. Yaglou, C.P. et al. 1936. "Ventilation Requirements." *ASHVE Trans.* 42:133–162.

39. Klauss, A.K. et al. 1970. "History of the Changing Concepts in Ventilation Requirements." *ASHRAE J.* 12:51–55.

40. ASHRAE Standard 62-73. 1977. "Standards for Natural and Mechanical Ventilation." American Society of Heating, Refrigerating and Air-Conditioning Engineers. Atlanta.

41. ASHRAE Standard 62-1981. 1981. "Ventilation for Acceptable Indoor Air Quality." American Society of Heating, Refrigerating and Air-Conditioning Engineers. Atlanta.

42. Woods, J.E. 1929. "Ventilation, Health and Energy Conservation: A Status Report." *ASHRAE J.* 21:23–27.

43. Huber, G. and H.W. Wanner. 1983. "Indoor Air Quality and Minimum Ventilation Rate." *Environ. Int.* 9:153–156.

44. Turiel, I. et al. 1983. "The Effects of Reduced Ventilation on Indoor Air Quality in an Office Building." *Atmos. Environ.* 17:51–64.

45. Sodegren, D. 1982. "A Carbon Dioxide Controlled Ventilation System." *Environ. Int.* 8:483–486.

46. Berglund. B. et al. 1982. "The Influence of Ventilation on Indoor/Outdoor Air Contaminants in an Office Building." *Environ. Int.* 8:395–399.

47. Leaderer, B.P. et al. 1984. "Ventilation Requirements in Buildings. II. Particulate Matter and Carbon Monoxide from Cigarette Smoking." *Atmos. Environ.* 18:99–106.

48. Cain, W.S. et al. 1983. "Ventilation Requirements in Buildings. I. Control of Occupancy Odor and Tobacco Smoke Odor." *Atmos. Environ.* 17:1183–1197.

49. Shurcliff, W.A. 1983. *Air-to-Air Heat Exchanger for Houses.* Brick House Publishing Co., Andover, MA.

50. Reef, D. 1983. "Should You Install an Air-to-Air Heat Exchanger?" *Mech. Illus.* December, pp. 70, 74, 78, 80–81.

51. Des Champs Laboratories Incorporated. 1983. Bulletin #EZV-284.

52. Nutone, Inc. 1983. Bulletin.

53. Des Champs Laboratories Incorporated. 1983. Bulletin 3–75C.

54. Fisk, W.J. et al. 1981. "Test Results and Methods: Residential Air-to-Air Heat Exchangers for Maintaining Indoor Air Quality and Saving Energy." LBL-12280. Lawrence Berkeley Laboratory, Berkeley, CA.

55. Fisk, W.J. and I.J. Turiel. 1983. "Residential Air-to-Air Heat Exchangers: Performance, Energy Savings and Economics." *Energy and Buildings* 5:197–211.

56. Fisk, W.J. et al. 1984. "Formaldehyde and Tracer Gas Transfer Between Airstreams in Enthalpy-Type Air-to-Air Heat Exchangers." LBL-18149. Lawrence Berkeley Laboratory, Berkeley, CA.

57. Fisk, W.J. et al. 1984. "Onset of Freezing in Residential Air-to-Air Heat Exchangers." LBL-18025. Lawrence Berkeley Laboratory, Berkeley, CA.

58. Ruth, D.W. et al. 1975. "Investigations of Frosting in Rotary Air-to-Air Heat Exchangers." 410–417. *ASHRAE Trans.* 87. Pt.1.

59. Sauer, H.J. et al. 1981. "Frosting and Leak Testing of Air-to-Air Energy Recovery Systems." 211–234. *ASHRAE Trans.* 87. Pt.1.

60. Fisk, W.J. et al. 1984. "Performance of Residential Air-to-Air Heat Exchangers During Operation with Freezing and Periodic Defrosts." LBL-18024. Lawrence Berkeley Laboratory, Berkeley, CA.

61. Persily, A. 1982. "Evaluation of an Air-to-Air Heat Exchanger." *Environ. Int.* 8:453–459.

62. Turiel, I. et al. 1983. "Energy Savings and Cost Effectiveness of Heat Exchanger Use as an Indoor Air Quality Mitigation Measure." *Energy.* 8:323–335.

63. Godish, T. 1988. Unpublished data.

64. Reiland, P. et al. 1985. "Preliminary Air-to-Air Heat Exchangers Testing Results for the Residential Standards Demonstration Program." Bonneville Power Administration, OR.

65. Mitsubishi Electric Industrial Products. Mitsubishi Lossnay Air-to-Air Enthalpy Heat Exchanger Bulletin.

66. Offerman, F.J. et al. 1982. "Low Infiltration Housing in Rochester, New York: A Study of Air-Exchange Rates and Indoor Air Quality." *Environ. Int.* 8:435–446.

67. Flower, R.G. and F.S. Langa. 1984. "Fresh Air Without Frostbite." *New Shelter* January, pp. 58–66.

68. Matthews, T.G. et al. 1983. "Formaldehyde Release from Pressed Wood Products." In: T. Maloney (Ed.). *Proceedings of the WSU Seventeenth International Particleboard/Composite Materials Symposium.* Washington State University. Pullman, WA.
69. Nero, A.V. 1983. "Indoor Radiation Exposures from 222-Rn and Its Daughters: A View of the Issue." *Health Physics* 45:277–288.
70. *Fed. Reg.* 1980. April 22. 45:27366.
71. Armstrong, V.C. and D.S. Walkenshaw. 1987. "Development of Indoor Air Quality Guidelines in Canada." 553–557. In: *Proceedings of the Fourth International Conference on Indoor Air Quality and Climate.* Institute for Water, Soil and Air Hygiene. West Berlin. Vol. 3.
72. Godish, T. 1983. "Interpretation of One-Time Formaldehyde Sampling Results from Measurements of Environmental Variables." 463–467. In: *Proceedings: Measurement and Monitoring of Noncriteria (Toxic) Contaminants in Air.* Air Pollution Control Association. Pittsburgh.
73. Broder, I. et al. 1986. "Health Status of Residents in Homes Insulated with Urea-Formaldehyde Foam." 155–166. In: D.S. Walkinshaw (Ed.). *Transactions: Indoor Air Quality in Cold Climates.* Air Pollution Control Association. Pittsburgh.
74. Godish, T. and J. Rouch. 1988. "Residential Formaldehyde Control by Mechanical Ventilation." *App. Ind. Hyg.* 3:93–96.
75. Jewell, R. 1984. "Reducing Formaldehyde Levels in Mobile Homes Using 29% Aqueous Ammonia Treatment or Heat Exchangers." Weyerhaeuser Corp., Tacoma, WA.
76. Matthews, T.R. et al. 1986. "Preliminary Evaluation of Formaldehyde Mitigation Studies in Unoccupied Research Homes." 389–401. In: D.S. Walkinshaw (Ed.). *Transactions: Indoor Air Quality in Cold Climates.* Air Pollution Control Association. Pittsburgh.
77. Nazaroff, W.W. et al. 1981. "Measuring Radon Source Magnitude in Residential Buildings." LBL-1284. Lawrence Berkeley Laboratory, Berkeley, CA.
78. Nazaroff, W.W. et al. 1981. "The Use of Mechanical Ventilation with Heat Recovery for Controlling Radon and Radon Daughter Concentrations in Houses." *Atmos. Environ.* 15:263–270.
79. Wellford, B.W. 1986. "Mitigation of Indoor Radon Using Balanced Mechanical Ventilation." 122–124. In: Proceedings of the APCA Specialty Conference: Indoor Radon. Air Pollution Control Association, Pittsburgh.
80. Nero, A.V. et al. 1983. "Radon Concentrations and Infiltration Rates Measured in Conventional and Energy-Efficient Houses." *Health Physics* 45:401–405.
81. Turiel, I. et al. 1983. "Energy Savings and Cost Effectiveness of Heat Exchanger Use as an Indoor Air Quality Mitigation Measure in the BPA Weatherization Program." *Energy* 8:323–335.
82. Bonneville Power Administration. 1984. "Home Weatherization and Indoor Pollutants." DOE/BP-310.
83. Arvela, H. and K. Winqvest. 1987. "A Model for Indoor Radon Variations." 310–315. In: *Proceedings of the Fourth International Conference on Indoor Air Quality and Climate.* Institute for Water, Soil and Air Hygiene. West Berlin. Vol. 2.
84. Burkhart, W. et al. 1984. "Matched Pair Analysis of the Influence of Weather-Stripping on Indoor Radon Concentrations in Swiss Dwellings." *Rad. Prot. Dos.* 7:299–302.

85. Fleischer, R.L. and L.G. Turner. 1984. "Indoor Radon Measurements in the New York Capital District." *Health Physics* 46:99–101.

86. McDermott, H.J. 1985. *Handbook of Ventilation for Contaminant Control,* 2nd ed. 402. Butterworth Publishers, Boston, MA.

87. Goldfield, J. 1985. "Contaminant Reduction: General vs Local Exhaust Ventilation." *Heating/Piping/Air Cond.* February, pp. 47–51.

88. Rezvan, K.L. 1984. "Effectiveness of Local Ventilation in Removing Simulated Pollutants from Point Sources." 65–75. In: *Proceedings of the Third International Conference on Indoor Air Quality and Climate.* Swedish Council for Building Research. Stockholm. Vol. 5.

89. Traynor, G.W. et al. 1982. "The Effects of Ventilation on Residential Air Pollution Due to Emissions from a Gas-Fired Range." *Environ. Int.* 8:447–452.

90. Traynor, G.W. et al. 1982. "Technique for Determining Pollutant Emissions from a Gas-Fired Range." *Atmos. Environ.* 16:2979–2987.

6

AIR CLEANING

Air cleaning is a form of pollution control widely applied to particulate matter, gases, and vapor generated by industrial sources. It is one of the primary means by which pollutant sources comply with local, state, and federal ambient air quality requirements. Air cleaning is also widely used as a part of local exhaust and ventilation systems where the capture and elimination of contaminants is practiced for worker health protection, protection of sensitive electronic equipment, and nuisance control.

The application of air cleaning to nonindustrial environments such as office buildings, schools, and commercial establishments, as well as single- and multi-family residential buildings, has been a more recent occurrence. Its application here is in some ways similar to the industrial case, in that the development and use of air cleaning technologies has centered on the control of particulate matter, commonly referred to as dust. This reflects in part the relative ease of controlling dust and the perception that dust needs to be controlled to prevent damage to equipment and maintain a physically clean environment. Unlike the industrial case, the use of air cleaning to control indoor contaminants does not proceed from any regulatory requirement. Its use is entirely voluntary.

PARTICULATE MATTER CONTROL

Particulate matter, or dust, is a problem sufficient in most buildings to justify some level of air cleaning. Dust may be generated indoors (tobacco smoke, fabric lint, etc.) or drawn in from the outdoors through ventilation systems and/or by infiltration. In large buildings with mechanical heating, ventilation, and air conditioning (HVAC) systems, and in residences with forced-air heating and/or cooling units, a minimal level of air cleaning is engineered into such systems to collect lint and large dust particles to prevent damage to blower fans. Dust stop filters of low efficiency (10–15%) are typically used for this application.

Beyond this minimal need, use of air cleaning equipment in a given building depends upon the awareness of architects and building owners (or managers) of the need for dust control and on decisions relative to the level of control desired.

Dust and a variety of other particles can be removed from contaminated indoor air by the application of relatively simple techniques adapted from use in industrial gas cleaning. In contrast to industrial flue gases, which may have dust loadings in the range of 200–40,000 mg/m^3, dust contents of indoor air seldom exceed 2 mg/m^3 and are usually less than 0.2 mg/m^3.[1] Fibrous filtration and electrostatic precipitation are the two primary particle collection principles employed in removing particulate matter from nonindustrial indoor air.

Filtration

Filters receive the widest application in cleaning particulate materials from indoor air. Typically they consist of a mat of fine fibers oriented perpendicular to the direction of air flow. Filter fibers are often cellulose, glass, or some form of plastic, and vary in size from < 1 to 100 μm. The filter mats themselves vary in density, with porosities in the range of 70–99%.[2] Because of these porosity/density differences, filters vary considerably in efficiency.

Contrary to the conventional wisdom that particle collection by filters occurs by straining or sieve action, fibrous filtration is relatively complex, with particle deposition occurring as a consequence of a variety of processes. These include interception, impaction, diffusion, and electrostatic deposition.[3]

Deposition by interception occurs when a particle follows a streamline within one radius of a fiber (Figure 6.1a). A particle collides with the perimeter of a fiber, loses velocity, and is captured. Interception is an important deposition mechanism in the size range of minimum collection efficiency (described on p. 251). Over a broad size range, interception is important in collecting particles with diameters in excess of 0.1 μm (Figure 6.2). The efficiency of collection by interception increases with increasing filter density; efficiency is, however, independent of particle velocity.

Impaction occurs as relatively large particles collide with a fiber (Figure 6.1b). Due to their inertia, particles are incapable of adjusting quickly to abrupt changes in gas streamlines in the vicinity of the fiber. Impaction is governed by the Stokes number, defined as the ratio of particle stopping distance to fiber diameter. The efficiency of particle deposition by impaction increases with increasing particle size, particle velocity, and Stokes value. It is an important deposition mechanism only for particles greater than 1 μm (Figure 6.2).

Particles smaller than 1 μm are subject to forces of diffusion and Brownian motion. The random Brownian motion of particles associated with diffusion processes increases the likelihood that a small particle will move into an intercepting streamline and be deposited on a collecting fiber (Figure 6.1c).

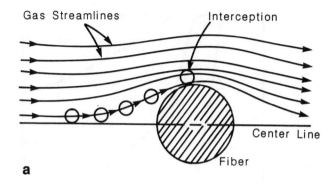

a

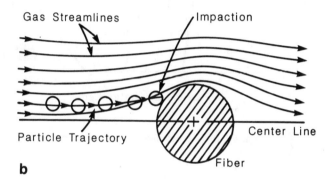

b

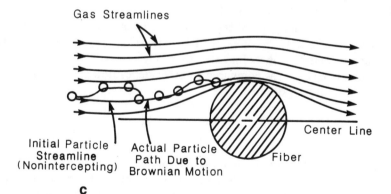

c

Figure 6.1.   Particle deposition processes on filter fibers.[1]

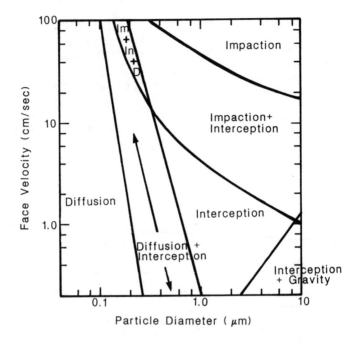

Figure 6.2. Fibrous filter filtration characteristics.[1]

Diffusion is the only important collection mechanism for particles less than 0.1 μm in diameter. It is notable that the efficiency of deposition by diffusion increases with a decrease in particle size.

Lastly, particles may be collected in a filter by electrostatic deposition. In this collecting process, particles, which naturally carry a small electrical charge, are attracted to fibers that are oppositely charged. The relative role of electrostatic deposition (as compared to other deposition processes) in filtration has not been well defined.

Filter Performance

Filter performance or efficiency is determined by a variety of factors. The most important of these are particle size, filter thickness, fiber diameter, filter packing density, and air flow rate. Filtration efficiency for monodisperse (particles all the same size) aerosols increases as filter thickness increases. As can be seen in Equation 6.1, particle penetration decreases exponentially with increasing filter thickness.

$$P = e^{-rt} \qquad\qquad (6.1)$$

where   P = particle penetration, %
         e = natural log base
         r = fractional capture per unit thickness
         t = filter thickness, mm

The value of r depends on particle size, packing density, fiber size, and face velocity.[2]

Particles that are relatively easily collected are deposited in the first few layers of filter fibers. As particle-laden air passes through the filter mat, particle size distribution changes. For each filter type there is a particle size, usually between 0.05–0.5 μm, for which collection is minimal (Figure 6.3). The minimum collection efficiency for a given particle size and filter is affected by face velocity (cm/sec), i.e., the ratio of volumetric flow rate to cross-sectional area of the filter. The effect of face velocity on particle collection efficiency as a function of particle diameter is illustrated in Figures 6.3 and 6.4. As can be seen, increasing the face velocity decreases the collection efficiency until a minimum value is reached. Collection due to diffusion at this point is at a minimum. Collection efficiency due to impaction increases significantly as face velocities are increased beyond those of the minimum collection values (Figure 6.3).

The performance of a filter depends on its thickness; as thickness increases, collection efficiency also increases. However, an increase in filter

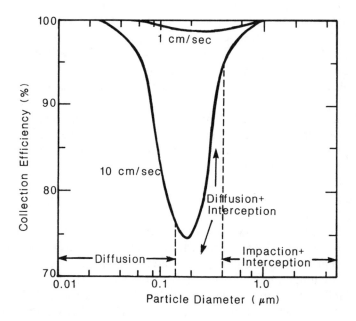

Figure 6.3. Relationship between particle size and collection
efficiency at two different face velocities.[1]

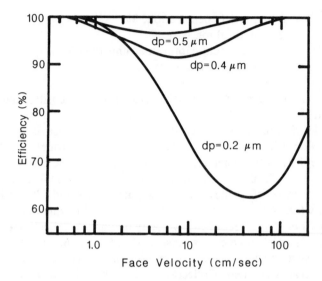

Figure 6.4. Effect of face velocity on the collection efficiency of particles of different aerodynamic diameters. (dp = particle diameter.)[1]

thickness results in an increase in resistance to air flow, with a concomitant drop in pressure and decrease in air flow. This pressure drop is directly proportional to filter thickness and inversely related to fiber diameter.[2] The effect of an increase in pressure drop is to reduce the actual rate of air flow through the filter, and thus the volume of air treated or cleaned. To compensate for the pressure drop, it is necessary to either extend the surface of the filter medium or employ fans/blowers with higher horsepower ratings and energy consumption.

The effect of three fiber diameters (0.5, 2, and 10 μm) on filter efficiency as a function of particle size can be seen in Figure 6.5. Note that by decreasing fiber diameter, the minimum efficiency particle size decreases, while overall collecting efficiency increases.

Fibrous Filtration Applications

Fibrous filters are used in a variety of indoor air cleaning applications. The most common of these is the filtration of air streams driven by air handling equipment in large buildings and residences. Such in-duct filtration may have the minimum objective of protecting the blower fan of the air handling unit (AHU) or may, in addition, be designed to reduce particle levels in the building for aesthetic purposes or to improve physical cleanliness of indoor surfaces. Fibrous filters also find application in modular free-standing units designed to clean the air of small spaces such as rooms of residences. The latter

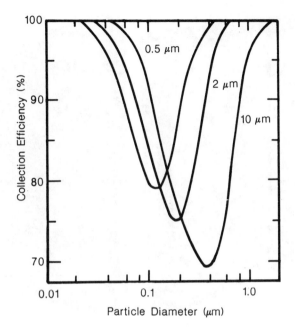

Figure 6.5. Effect of fiber size on the collection efficiency
of different particle sizes.[1]

have become very popular in recent years and represent an annual market of
over $150 million.[4] Special fibrous filters of very high performance are used
for applications in which a very high degree of particle control is required,
typically in nuclear power plants, industrial clean rooms, and surgical
theatres.

Filter Types

Three basic kinds of fibrous media filters are available for indoor air
cleaning applications. These are dry, viscous impingement, and charged media
filters. They may be single-use panel filters or they may be renewable. In the
latter, a fresh filter surface from a roll is introduced into the air stream as
needed to maintain acceptable air resistance levels and cleaning efficiency.
Examples of panel and renewable filters are illustrated in Figures 6.6 and 6.7.

Dry-type panel filters. These filters have high porosities and low efficiencies.
Their typical application is as dust stop filters to protect mechanical equipment
in AHUs and home furnaces, and as prefilters for higher-efficiency filters.
Dry-type panel filters collect large particles such as fabric lint by impaction
and interception. Typical air velocities range from 200 to 700 ft/min (FPM)
(60.9–213.3 m/min [MPM]). Initial air flow resistances are low (0.05–0.25

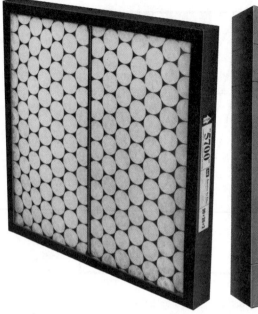 

Figure 6.6. Dry panel-type filters. (Courtesy of American Air Filter Co.)   **Figure 6.7.** Renewable media filter system. (Courtesy of American Air Filter Co.)

inches H_2O; 12.4–62 pascals) and units are operated to 0.50–0.75 inches H_2O (124–186 pascals) before filters are replaced. Dry panel filter media may consist of open-cell foams, textile cloth nonwovens, paperlike mats of glass or cellulose fibers, wood fill, etc.[5]

Viscous media panel filters. These are filters comprised of coarse fibers coated by a viscous, oily material to which particles adhere when they impact/impinge on a medium fiber. Like the dry type described above, these filters have a high porosity and are characterized by low resistance to air flow. They have low efficiencies over the particle range common to indoor air, but are very efficient in collecting fabric dust or lint.[5]

Typical operating velocities range approximately from 300 to 600 FPM (91.4–182.8 MPM) and filter change is required when the operating resistance reaches 0.5 inches H_2O (124 pascals). Limiting factors in service life are the consumption of available viscous material by collected dust and, to a lesser extent, increased resistance because of dust loadings.[5]

The viscous material used will, in part, determine filter efficiency and dust-holding capacity. Desirable characteristics include (1) low volatility, (2) minimal variation in viscosity, (3) ability to wet and retain particles, (4) high flash point, and (5) low odor.

Renewable media filters. When panel filters reach their maximum permissible loading (as determined by pressure drop or some predetermined schedule) they must be discarded and replaced by clean filters. To minimize labor costs for maintenance, filter systems are available that automatically provide a fresh surface when the previous surface has become clogged to the point that acceptable pressure drops have been exceeded. Renewable media filters are of both the viscous and dry types.[5] They consist of an automatically moving curtain of fiber or metal medium that advances in response to a signal from a pressure sensor or timer. As the dry-type medium becomes soiled, it moves to a takeup roll on the bottom of the filter system. When the fresh roll has been completely expended, it must be replaced. Typically, when the roll is about to be exhausted, a signal is activated to alert maintenance staff that the filter requires replacement. Dry, renewable fibrous-media filters have weight arrestance efficiencies of 60–90% and atmospheric dust spot efficiencies of 10–30%. (Arrestance and dust spot efficiencies are discussed later in the chapter — see "Performance Testing.")

In viscous media renewable filters, the traveling curtain passes through a viscous reservoir where the dust load is released and a new coat of viscous material is taken up. The media curtain, normally formed of metal, is continually cleaned of dust and renewed with fresh viscous collecting agent. Accumulated dust must be removed from the reservoir and the reservoir refilled periodically.

Extended surface dry-type filters. Filters discussed in previous paragraphs can be described as low-efficiency, being used primarily to protect mechanical equipment and remove large nuisance particles such as fabric dust. One of the most effective ways of increasing dust collecting efficiency is to increase filter thickness or density. Increasing filter thickness or density, however, significantly increases resistance to air flow, causing a drop in pressure and flow of air through the filter. Filters can be designed to overcome this problem. The principal approach is to extend the surface area by pleating the filter medium. Such pleating increases the medium surface relative to the face area of the filter. This reduces the velocity of air through the filter and decreases overall resistance to air flow; pressure drop is reduced to acceptable levels, despite the increased overall thickness and/or density of the filter medium.[6]

Extended media filters are available in a variety of designs and performance levels. They vary in media thickness and density, fiber size, media materials, number of pleats per unit area, and filter depth. The filter medium typically is in the form of a random (in terms of fiber orientation) mat or blanket. Common media fibers include cellulose, bonded glass, wool felt, and a variety of synthetics.

Extended media filters can also be constructed in the form of bags, similar in design and concept to those used in industrial applications (baghouse filters). A bag-type extended media filter is illustrated in Figure 6.8.

In the pleated-type filter, the medium is usually supported in a panel

Figure 6.8.   Bag-type filter. (Courtesy of American Air
Filter Co.)

frame or box (Figure 6.9). The V-shaped plates may vary from 2 to 36 inches
(5.1 to 91.4 cm) in depth, or even more in special applications. The filter
medium may be rigid enough to be self-supporting or may be held in place by a
combination of rigid metal spacers inserted between the pleats and an adhesive
boundary at the edges of the filter frame.

The performance or efficiency of extended surface media filters is usually
much higher than for other dry-type filters. Depending on the medium
employed and the filter design, performance levels can vary widely (Table 6.1).
(Note: Dust cleaning ratings used in Table 6.1 are discussed in detail in the
section titled "Performance Testing" later in this chapter.) Extended surface
filters, characterized by their efficiency or level of performance, are described
as medium-, high-, and very high–efficiency. In intermediate-efficiency types,
air velocities through the filter medium range from 6 to 90 FPM (1.8 to 27.4
MPM), although the initial face velocity of the filter may be as much as 750
FPM (228.5 MPM). Filter depth may vary from 2 to 36 inches (5.1 to 91.4
cm).[5]

The most efficient type of extended media filter commonly available is the
HEPA (High-Efficiency Particulate Absolute) filter. HEPA filters are characterized by efficiencies in excess of 99.97% at a minimum particle diameter of
0.3 μm. They were originally designed for applications in the nuclear power
industry and have seen widespread use in industrial and military clean rooms.
A HEPA filter is illustrated in Figure 6.9.

HEPA filters are effective in capturing particles as small as 0.01 μm. Their

Figure 6.9. Extended surface media filters. (Courtesy of American Air Filter Co.)

Table 6.1   Performance Levels of Dry Media Filters[5]

Filter Media Type	Arrestance (%)	Dust Spot Efficiency (%)	DOP Efficiency (%)	Dust Holding Capacity (g/1000 CFM cell)
Fine open foams and textile denier nonwovens	70–80	15–30	0	180–425
Thin paperlike mats of glass fibers, cellulose	80–90	20–35	0	90–180
Mats of glass fiber, multi-ply cellulose, wool felt	85–90	25–40	5–10	90–180
Mats of 5- to 10-μm fibers, 1/4 to 1/2 inches (6–12 mm) thickness	90–95	40–60	15–25	270–540
Mats of 3- to 5-μm fibers, 1/4 to 1/2 inches (6–12 mm) thickness	>95	60–80	34–40	180–450
Mats of 1- to 4-μm fibers, mixture of various fibers	>95	80–90	50–55	180–360
Mats of 0.5- to 2.0-μm fibers (glass)	NA	90–98	75–90	500–1000
Wet laid papers of mostly glass fibers, <1-μm diameter (HEPA)	NA	NA	95–99.999	500–1000
Membrane filters (cellulose, acetate, nylon) having holes ≤1 μm in diameter	NA	NA	~100	NA

Table 6.2 Air Flow Rates and Static Pressure Levels for Two Blower Fans

Static Pressure (H$_2$O) (resistance to air flow)	Air Flow Rate (CFM)	
	Fan #1	Fan #2
free air	125	125
0.1	120	115
0.2	115	105
0.3	110	95
0.4	105	85
0.5	100	—

small fiber diameter and packing density favor the collection of very small particles by Brownian motion.

Electrostatic forces are very important in particle collection by HEPA filters. They act to cause particle agglomeration and adherence to the surface of the HEPA filter medium. Van der Waal's forces act to capture and retain particles without blow-off. Larger particles may be retained on the filter surface by the force or velocity of moving air. In some HEPA applications, the filter surface is coated to discharge the static build-up. It can then be renewed or cleaned by a pneumatically operated pulse cleaning mechanism.[7]

Resistance to Air Flow

One of the main problems associated with fibrous filters is the increase in static pressure and resistance to air flow associated with soiling and clogging of the fibrous media. If fans used with such filtration systems are inadequately sized, they will not develop sufficient static pressure to overcome the resistance to air flow and maintain desired air velocity as filters become soiled. The pressure loss (drop), or resistance to air flow, is directly proportional to the air flow rate.[6] The relationship between static pressure and air flow is illustrated for different fans in Table 6.2. Note that the first fan can accommodate relatively higher static pressures than the second without much change in air flow. As an illustration, consider a system in which air flow of 100 CFM (47 L/sec) is needed and filters are changed when their resistance to air flow reaches 0.5 inches H$_2$O. In this case, fan #1 will still be moving the required 100 CFM, while the air flow from fan #2 will be substantially below 100 CFM. Note, however, that both fans would have been more than adequate when filters were clean (low static pressure). The two fans described in Table 6.2 differ significantly in their horsepower rating and energy consumption. Fan horsepower required to force air through a filter at constant velocity increases with increased resistance to air flow. Higher-horsepower fans consume more energy and therefore increase the cost of operation.[8] However, an adequately sized fan is necessary for proper system performance.

Filter manufacturers usually provide resistance values for each of their products. Two resistance values are typically given: (1) for a clean filter at its rated air flow and (2) a final resistance value after which the filter should be replaced or reconditioned.

How does one know when a filter has reached the end of its useful life relative to acceptable air flow and fan operation? The best approach is to use a differential gauge that gives the static pressure as the filter becomes soiled. In some systems an alarm is tripped when a predetermined static pressure or resistance value is exceeded. This alerts maintenance personnel to replace or renew filters.[5] Most fibrous filter systems are not equipped with pressure sensing gauges and alarms, however. Because of the variability in indoor dust loadings, it is difficult to know when the end of the useful life of a filter has been reached. In such circumstances it is a matter of service personnel's judgement. Alternatively, the dirty filters can be routinely replaced on a predetermined schedule. Because dirty filters can result in decreased air flows and increased operating costs, it is important that operators of fibrous filtration systems develop and implement a rational service program.

Electronic Air Cleaners

Electrostatic air cleaners remove particles by electrostatic forces similar to electrostatic precipitators used in industrial gas cleaning. There are three basic designs: the ionizing plate type, the charged-media non-ionizing type, and the charged-media ionizing type.

Ionizing Plate Type

In most cases, electronic air cleaning refers to the ionizing plate type. Ionizing plate devices are widely employed in residential, commercial, and office buildings. They may be used as modular, free-standing units, modular units that can be suspended from a ceiling or mounted on a wall, in-duct units installed in residential heating/cooling systems, or large units placed in the HVAC systems of large buildings. Electronic air cleaners used in residential applications are illustrated in Figure 6.10. An electronic air cleaner used to clean air in a HVAC system is illustrated in Figure 6.11.

Operation of electronic air cleaners is based on the principle that particles moving in an air stream can be charged and then subsequently collected on plates of opposite charge. An electronic air cleaner draws particle-laden air past a series of ionizing wires that produce positive ions; this air then passes through channels between a series of negatively charged collection plates which attract the positively charged particles.[9,10]

Two-stage electronic air cleaning is illustrated in Figure 6.12. In the first stage, a high electrical potential (12,000 V) is applied to thin tungsten wires strung vertically in the air stream. Electrons present in the air are accelerated toward the positively charged ionizing wires. These accelerated electrons strike air molecules, stripping them of electrons, creating positive ions and additional electrons. Liberated electrons strike other gas molecules, producing

Figure 6.10.   Residential electronic air cleaners. (Courtesy of Honeywell, Inc.)

more positive ions, which become vast in number. The ionizing process causes a corona discharge near the ionizing wires. (Ozone is produced in the process because of the high energies involved. To minimize ozone production, most electronic cleaners use positive coronas [as opposed to negative ones], since this polarity produces less ozone.)[10]

The positive ions produced attach to particles moving in the air stream, making them positively charged. The magnitude of electrical charge on a particle depends on the number of positive ions picked up as it passes through the ionizing field. Typically, the larger the particle, the more ionized it becomes and, therefore, the easier it is to collect.

After becoming ionized, charged particles flow with the air stream into a collector section consisting of a series of thin metal plates. Alternate plates are positively and negatively charged by means of a high dc voltage (6000 V). The positively charged particles are attracted to negatively charged plates. The magnitude of electrostatic force acting on a charged particle depends on the extent of a particle's charge, distance between the plates, and voltage applied.[9]

When particles hit the collecting plates, they lose their original charge and take on the charge of the collecting surface where they remain attached because of molecular adhesion and cohesion to other particles already collected. Collector plates may be coated with a special oil or adhesive to improve particle retention. However, most electronic air cleaners use no adhesive materials.

As particle buildup occurs, collection efficiency decreases as accumulated particles reduce the strength of electrostatic forces at the collector plates.

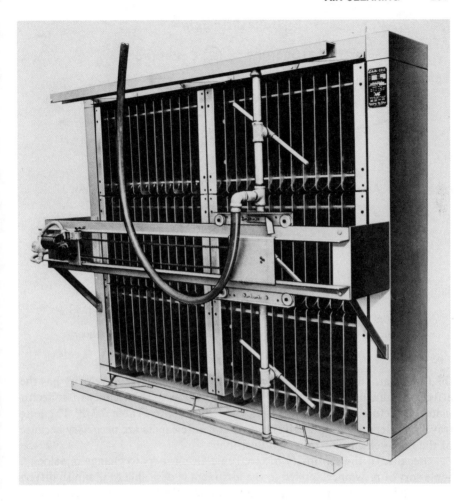

Figure 6.11. Electronic air cleaner used in large building.
(Courtesy of American Air Filter Co.)

Plates must then be cleaned of their accumulated dust load. Typically, such cleaning is done by washing with hot water; in some systems the washing is done automatically.

The efficiency of electronic air cleaners is determined by particle migration velocity, collection surface area, travel path length (distance through collection field) and air flow rate. Migration velocity is directly proportional to the charge on the particle and the strength of the electric field. In most air cleaners used for indoor dust cleaning, the travel path length for particles is relatively short (5–8 inches [12.7–20.3 cm]). In industrial electrostatic precipitators, the travel path length may be 20–25 ft (6.1–7.6 m) or longer. In most cases, collection efficiency will be increased with increased travel time through

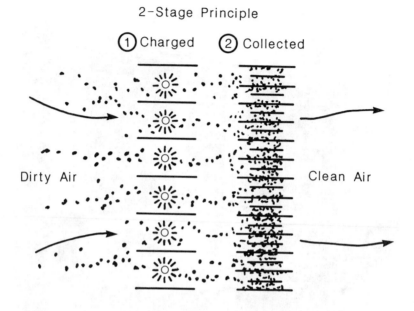

Figure 6.12. Principles of operation of a two-stage electronic cleaner.[9]

the filter, particularly at higher-velocity air flows. Increasing velocity has the effect of reducing the amount of time a particle is in the vicinity of collector plates; therefore, collection efficiency is reduced also (Table 6.3).[11] The long travel path lengths in industrial electrostatic air cleaners are necessary because of the high-velocity air flow and dust loads.

Because of the sensitivity of electronic air cleaners to change in velocity, some sort of resistance applied to the airstream is desirable to diminish differences in air velocity across the collection module. Such resistance may be in the form of prefilters, afterfilters, baffles, perforated plates, etc.[9]

The double-stage ionizing-type air cleaner has been described above. Ionizing units with wires placed between collecting plates are of the single-stage type. Because of the decreased particle migration path length, single-stage units are usually less efficient than their double-stage counterparts.

Electronic air cleaners have distinct advantages over fibrous filters. The most important is their very low resistance to air flow and attendant ability to operate at constant air velocity. They can be used with fan equipment that has

Table 6.3   Effect of Air Flow on the Collection Efficiency of an Electronic Air Cleaner[11]

Air Flow Rate (CFM)	Dust Spot Efficiency (%)
400	93
550	90
750	85

little ability to operate at constant static pressure, translating into lower fan power needs and operating costs. Maintenance is usually limited to cleaning collector plates. The frequency of cleaning depends on particle loading. To maintain high collection efficiency and decreased frequency of collector cleaning, it is important that prefilters be used. Prefilters typically are of the dust stop type.

Though relatively high efficiencies (dust spot values [see p. 264] of 80-95%) can be achieved by electronic air cleaners, they cannot be used in applications where very high efficiencies are required (dust spot > 95%). Additionally, they should not be used where sensible moisture (i.e., liquid) is likely to affect them.[8]

Charged Media Non-Ionizing Type

These cleaners combine characteristics of both electronic air cleaners and dry filters. They consist of a dielectric filtering medium made of a mat of glass fiber, cellulose, or some similar material supported on or in a gridwork of alternately charged (typically 12,000 V dc) or grounded members. A strong electrostatic field is formed through the dielectric filter medium. As particles approach the charged filter medium, they are polarized and drawn to it.[5] Particles are not ionized.[9] Since this filter is a media filter, resistance to air flow occurs, increasing as the filter becomes soiled. The filter must therefore be replaced.

Charged Media Ionizing Type

In this type of electronic air cleaner, dust in an air stream is initially charged (by passing it through a corona discharge ionizer) and then collected on a charged-media filter.[5]

Performance Testing

The performance of filters and filtration systems was discussed earlier in both a general and a theoretical way. Beyond this, however, lies the real-world need to describe the performance of the wide array of products available. Products vary in their collection efficiency for different size aerosol particles, resistance to air flow, service life, and dust holding capacity. Recognizing the need for a uniform and consistent approach to evaluating filter performance, a committee of the American Society of Heating, Refrigerating and Air-Conditioning Engineers (ASHRAE) has developed ASHRAE Standard 52-76, which describes methods of testing the performance of filters and/or filtration systems in removing dust from an air stream.[12] ASHRAE Standard 52-76 describes test procedures allowing comparison of performance characteristics of one air cleaning device with another type or with another device of the same type. ASHRAE test procedures can also be used as a basis for establishing product classifications and standards. Performance results obtained from the application of ASHRAE test methodology alone, however, cannot be used to quantitatively predict area cleanliness or length of service life. Performance as determined by the application of ASHRAE test methodology is described in the context of dust spot efficiency and/or arrestance, expressed as percentage.

The dust spot test is frequently applied to high-efficiency filters/filtration systems such as electronic units and fine fiber extended media filters. It measures the discoloration caused by deposition of particles as they pass through a glass fiber filter tape located both upstream and downstream of the filter undergoing evaluation. Discoloration is measured by changes in optical density read on a photometer. Filter efficiency is calculated from the following equation:

$$E = 100 \left[1 - \left(\frac{V_1}{V_2} \right) \left(\frac{T_{20} - T_{21}}{T_{10} - T_{11}} \right) \left(\frac{T_{10}}{T_{20}} \right) \right] \% \tag{6.2}$$

where

E = efficiency, %
V_1 = air volume drawn through upstream filter tape
V_2 = air volume drawn through downstream filter tape
T_{10} = initial light transmission, upstream filter tape
T_{11} = final light transmission, upstream filter tape
T_{20} = initial light transmission, downstream filter tape
T_{21} = final light transmission, downstream filter tape

Arrestance is the term applied to filter performance based on feeding a standardized synthetic dust into the air stream of the filter or filtration system. ASHRAE synthetic dust consists (by weight) of 72% standardized air-cleaner dust fines, 23% molacco black, and 5% No.7 cotton linters ground in a Wiley mill. It is designed to take into account the large variability in particle size distribution and composition that normally passes through general ventilation systems. Arrestance is the weight of the synthetic dust collected on the filter expressed as a percentage of the initial weight of the synthetic dust injected upstream of the test filter. Arrestance testing is typically associated with panel filters that find application as prefilters, with filters protecting mechanical equipment, and where low-efficiency air cleaning is acceptable.

Dust spot efficiency and arrestance measure different performance characteristics of filters and air cleaners. The synthetic dust used to measure arrestance is, on the whole, coarser than typical atmospheric dusts. Arrestance describes a filters's ability to remove large particles, and as a consequence gives little indication of its ability in removing smaller particles. Because most of the weight is associated with large particles, collection efficiency expressed as arrestance is exaggerated relative to the total percentage of particles actually collected.

The dust spot efficiency test was developed as a result of the limitations of the arrestance method described above. It is designed to measure an objectionable characteristic (ability to soil interior building surfaces) of finer dust parti- cles. Because they measure different performance characteristics, values based on one test cannot be used as a relative measure of the other.

ASHRAE Standard 52–76 includes a method for determining dust hold- ing capacity, which is defined as the amount of a particular kind of dust that a filter can hold to some maximum resistance value. Dust holding capacity is the weight of synthetic dust fed to an air stream flowing into the test device multiplied by (1) its average arrestance prior to reaching the maximum resist- ance or (2) two consecutive arrestance values that are less than 85% of the maximum arrestance values.

As is typically the case for filters, efficiency and arrestance increase with increased dust loading. This increased efficiency is associated with the buildup of a particle mat which itself is involved in particle deposition. The effect of increasing particle load on arrestance and dust spot efficiencies can be seen in Figure 6.13a; the effect on air resistance is shown in Figure 6.13b.

The ASHRAE protocol is designed primarily for testing filters and air cleaning devices to be used in general ventilation systems, those to be set into an air flow path described by a duct. The protocol provides specific details on how performance characteristics are to be measured. Filters and devices are installed and operated according to manufacturer's recommendations. In addi- tion to performance testing, resistance is measured at various flow rates.

The ASHRAE dust spot method is not intended to rate the performance of filters with efficiencies greater than 98%. For high-efficiency filters of the HEPA type, particle collection efficiency is determined by the use of the di- octyl phthalate (DOP) smoke penetration method.[9] In this performance evalu- ation method, a smoke of essentially uniform droplets is produced by the vaporization and subsequent condensation of DOP, a high–boiling point liq- uid. The particle diameter is 0.3 μm, with a cloud concentration of 80 mg/m^3. The 0.03-μm particle size is significant here in that it is the size range of minimal filter collection efficiency. Any of the DOP smoke penetrating the filter or leaking through or around gaskets passes downstream where it is mixed and measured by a light-scattering photometer. Filter penetration is calculated from Equation 6.3.

$$P = 100 \frac{C_2}{C_1}$$ (6.3)

where P = penetration, %
           C_1 = upstream concentration
           C_2 = downstream concentration

Because HEPA filters have efficiencies near 100% (e.g., 99.97%), pene- tration rather than efficiency is specified. Efficiency and penetration are related, however, and efficiency can be expressed in the form:

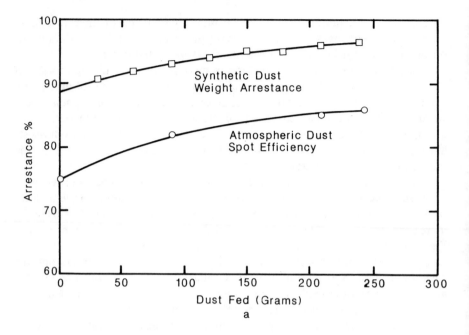

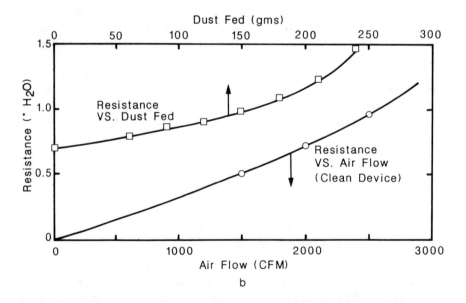

Figure 6.13. Effect of particle loading on (a) filter arrestance and (b) resistance.[12]

$$E = 100 - P \qquad (6.4)$$

where E = efficiency, %

Selection of Filters/Filtration Systems

The choice of a filtration system depends on a variety of factors. These include collection efficiency desired, initial costs, operating costs, and space limitations. The availability of sufficient space for filters/filter systems depends on the nature of air handling equipment. For example, fan coil and packaged terminal air conditioning units installed in confined spaces or as room consoles are usually limited to the use of simple panel filters. Central station AHUs and penthouse units generally have sufficient space for a variety of filter types. It is important that careful attention be given to the selection of filters when a variable air volume HVAC system is used.[13] It is also important that both the initial and operating costs of filters and filtration systems be taken into account when filtration system selection is being made.

Residential Air Cleaners

The selection of a particle or dust cleaner for use in residential environments poses more of a problem for homeowners than it does for owners/managers of large buildings with HVAC systems. In the latter case, filtration systems are usually selected by architectural engineers or individuals with technical expertise. Such individuals can be expected to have some technical knowledge about the installation and use of filters and filtration systems. The need for air cleaning in mechanically ventilated buildings is well established, and stable market conditions for product manufacture and distribution exist. The provision of reliable technical information by manufacturers' representatives is at a level equal to other well-established products.

The air cleaner market aimed at average consumers is much less mature and represents a wide array of products, many of which are of unknown efficacy. These include small- to medium-sized desktop units with small panel or charged media filters; ion generators; portable units that rest on floors or cabinet surfaces, or are mounted on ceilings or walls; and in-duct devices. In-duct devices have the potential for cleaning the air of an entire building. At best, portable cleaners are designed to clean the air of a single room. Residential air cleaners represent a rapidly growing market with annual sales of several hundred million dollars.

Most popular in terms of unit sales are the small desktop cleaners ranging in price from $10 to $30 (Figure 6.14). These basically consist of small panels of dry, loosely packed, low-density fiber filters located upstream of a high-velocity fan, all contained in a plastic housing. Some units may utilize an electrostatically charged electret filter. Electret filters are thin plastic materials, in either fibrillated or planar form, treated to imprint high-voltage charges of

Figure 6.14. Small panel filter units. (Courtesy of Rodale Press.)

opposite polarity on opposite sides of the film.[14] A scented medium is often included for odor control.

Negative ion generators have also developed a wide popularity as portable desktop-type air cleaners. Ion generators produce a large stream of negative ions that flow out into a space, negatively charging airborne dust. A space charge builds up, driving charged dust to the surrounding walls and surfaces.

A variety of negative ion generator–type air cleaners is available commercially (Figure 6.15). The simplest and least expensive generate negative ions that attach to particles and plate out on room surfaces. More advanced models are theoretically designed to eliminate the "dirty wall effect." They generate negative ions that flow out into a space, causing particles to become charged. The charged particles are then drawn back into the dust cleaner by a blower fan, where they are deposited on an electrostatically charged panel filter.[15] In other ionizers, a stream of negative ions is generated in pulses and negatively charged particles are drawn passively back to the ionizer, which contains a positively charged cover or sleeve.[16]

Negative ion generators used as air cleaners are similar to those marketed in the United States in the 1950s and early 1960s as health-promoting devices based on theories that negative ions promote human well-being. They had been forced off the market by the Food and Drug Administration because health claims could not be proven.

Another major class of residential air cleaners includes the larger second-generation models made by the manufacturers of the small fan/filter types and products that have been commercially available for many years. This class of larger portable air cleaners is considerably more effective in cleaning particles from room air than those previously described. Such cleaners usually use an extended surface filter (such as HEPA) or electrostatic deposition. High-efficiency filters in such cleaners are normally preceded by low-efficiency prefilters to remove lint and very large particles. Typically, costs range from

Figure 6.15. Negative ion generators. (Courtesy of Ion Systems, Inc. and Bionaire, Inc.)

$100 to $500. Portable room air cleaners of the electronic type can be seen in Figure 6.10.

A third class of residential air cleaners includes those designed to be placed in the ducts of central heating and/or air conditioning systems. These typically are extended surface fibrous media filters or two-stage electrostatic cleaners (Figure 6.11), the latter being the most popular. In theory, such systems are designed to provide whole-house cleaning, and in fact have the potential to do so. However, based on common installation and use practices, air cleaning only occurs when the blower fan is activated by heating and cooling demands. When the heating/cooling fan is not activated, no air cleaning takes place. Filtration systems are normally installed in the cold air return immediately upstream of the blower fan.

Air Cleaner Performance Under Actual Use Conditions

Air cleaner performance was discussed previously in theoretical terms and in the context of standard testing protocols. From the consumer's perspective, there is the larger question: "How effective is a given air cleaner under actual use conditions?"

Fan/filter units and negative ion generators. Because of their popularity, let us first address the performance of small desktop fan/filter air cleaners. Performance tests of 12 fan/filter models have been reported by the staff of *Rodale's New Shelter* magazine.[17] Tests were conducted in a 1200-ft³ (33.6-m³) room in which cigarette smoke was generated by a smoking machine. Nine of the twelve models reduced smoke levels by less than 24%, compared to a 17% reduction (by settling) that occurred in the absence of an air cleaner. The researchers concluded that effectiveness of small fan/filter air cleaners in removing smoke particles was marginal at best, and differed little from using no device at all. Similar results were reported by Offerman et al.[4]

Tests conducted by *Rodale's New Shelter* staff showed that negative ion generators performed well in reducing smoke levels. Cleaning effectiveness of 96% was observed for four models, and 71% for a fifth, during a 4-hour test period.[17]

Portable electronic and HEPA air cleaners. *Rodale's New Shelter* reported significantly more effective performances when researchers tested the larger HEPA and/or electronic filter–type portable air cleaners (15 different models made by 11 different manufacturers). The time required to reduce smoke levels by 95% varied from 26 to 120 minutes.[18] Product types and test results are summarized in Table 6.4.

Offerman et al.[4] tested a variety of portable air cleaners using tobacco smoke as a source of particles. The test protocol involved turning cleaners on for a 3- to 5-hour period following a period of decay and mixing followed by a 6- to 8-hour period of natural decay; results of a test of a HEPA-type cleaner are illustrated in Figure 6.16. They expressed cleaning performance in the context of an effective cleaning rate (ECR) calculated from differences in the observed smoke particle decay rate (with and without an air cleaner operating), multiplied by the chamber volume. The ECR is the flow rate of particulate-free air required to produce the observed decay rate of cigarette smoke. The ECR value can be used to estimate cleaner effectiveness in rooms of various sizes. System efficiency is the ECR divided by actual flow rate of the device. Values of system efficiency were used to compare performances of different air cleaners or to evaluate the performance of a specific air cleaner as a function of particle size. Effective cleaning rates and system efficiencies for a variety of portable air cleaner types are summarized in Table 6.5. Note that the highest efficiency was observed in the product utilizing a HEPA filter.

Kimmel[19] studied the performance of electronic air cleaners suspended from the ceilings of seven Finnish restaurants. Reduction in respirable suspended particles generated primarily from cigarette smoke ranged from 5 to 56%. This relatively low performance was believed to have been due in part to the loss of volatile components from the collected particle mass on cleaner plates. Kimmel found that cleaning performance declined in two of these seven case evaluations over a 2- to 3-week period. Performance degradation occurred in only the most heavily polluted restaurants. Because of the odor

Table 6.4   Smoke Removal Performance of Room Air Cleaners[18]

Product/Model	Cleaning Principle	95% Smoke Removal Time (min)
Air Techniques		
Cleanaire 150	HEPA filtration	56.5
Cleanaire 300	HEPA filtration	27.0
Associated Mills		
Pollenex 1801	negative ion	37.5
Bionaire Corp.		
BT-1000	negative ion, electret filtration	43.0
Five Seasons		
Air Duster 585A1	electronic, negative ion	79.0
General Time		
Ecologizer 98005	HEPA filtration	>120
Ecologizer 99005	HEPA filtration	54.7
Honeywell		
F56A1003	electronic	25.5
Lasko		
9152	negative ion	94.0
North American Phillips		
Cam 50	electronic	110.0
Oster		
402-06	electronic	>120
404-06	electronic	42.5
Teledyne Water-Pik		
AF3-W	electronic	57.3
Trion		
Table Top	electronic	49.0
Console	electronic	27.5

that emanated from collection plates, Kimmel recommended that plate cleaning be conducted weekly.

In-duct devices. Installation of an air cleaning device in a HVAC system in a large building or the return air plenum of a residential central heating/cooling system provides an opportunity for both building-wide air cleaning and, by recirculation of building air numerous times through the cleaning unit, the potential for more effective cleaning performance.

Sutton et al.[20] and McNall[21] have evaluated both theoretically and experimentally the performance of electronic air cleaners installed in the central heating/cooling ducts of a large room (1400 ft^2, 127.6 m^2) and a residence. They have developed model equations to predict smoke particle concentrations under steady-state conditions.

At steady state, indoor particle concentrations can be predicted from the following equation:

$$C_s = \frac{Q_t C_o + Q_r C_o (1 - E) + N_p}{Q_t E + Q_e + K} \tag{6.5}$$

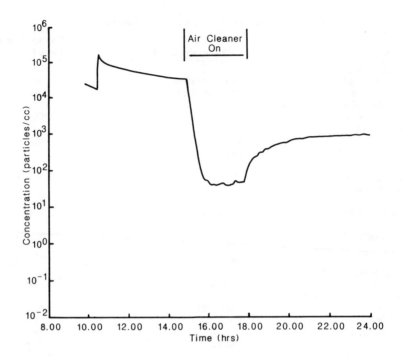

Figure 6.16. Performance test of a HEPA-type portable air cleaner on tobacco smoke.[4]

where C_s = indoor concentration, $\mu g/m^3$
C_o = outdoor concentration, $\mu g/m^3$
Q_v = outside air for ventilation, m^3/sec
Q_1 = infiltration air, m^3/sec
Q_r = recirculated air, m^3/sec
Q_e = exhaust air $(Q_v + Q_1)$, m^3/sec
N_p = rate of contaminant production, $\mu g/hr$
E = filter efficiency, %
K = particle removal rate on internal surfaces, m^3/sec

If the K term is primarily due to settling, then $K = AV$, where A is the floor area of the space (m^2) and V is the average settling velocity (m/sec). With very fine particle diameters, the effect of settling may be very small.

The significance of the ventilation (Q_v) and infiltration (Q_1) terms depends on the amount of this air compared to (1) the product of recirculation rate and filter efficiency (Q_rE) and (2) the quantity of contaminants entering from the outdoors. The Q_rE term is of major importance in predicting the effect of air cleaning in residential and small commercial and office environments. If ventilation rate is high and the recirculation rate is low, Q_rE becomes less and less

Table 6.5 Performance Tests of Portable Air Cleaning Units[4]

Model	Filter Type	Flow Rate (CFM)	System Efficiency	Effective Cleaning Rate	Time for 98% Smoke Removal (hr)
Rush Hampton 7305	foam panel	10	0 ± 1	0 ± 1	<8
Norelco HB 1920	electret panel	29	11 ± 1	3 ± 1	<8
Pollenex 699	electret panel	21	16 ± 1	3 ± 1	<8
Neolife-Consolaire	electret with negative corona charging	17	39 ± 11	7 ± 2	<8
Bionaire 1000	electret extended surface filter with negative ion generator	66	86 ± 9	57 ± 2	~1.8
Hepanaire	HEPA	157	115 ± 13	180 ± 8	7.5
Trion Console	electrostatic	215	57 ± 11	122 ± 19	~0.7
Micronaire P-520	electrostatic	200	58 ± 6	116 ± 5	~0.7
ISI-Orbit	negative ion generator	0	—	1 ± 1	<8
Zestron Z-1500	negative ion	0	—	30 ± 1	~2.5

important even with high-efficiency filters. For maximum effectiveness of air cleaning, it is desirable to have both a high-efficiency filter and a high recirculation rate. The effect of recirculation rate and air cleaner efficiency on indoor particle concentrations can be seen in Figure 6.17.

Equation 6.5 can be solved for various practical applications of air cleaning. In the case where no production of contaminant occurs indoors ($N_p = 0$), where ventilation is limited to infiltration ($Q_v = 0$), and where settling of particles is negligible ($K = 0$), Equation 6.5 reduces to

$$C_s = \frac{Q_i C_o}{Q_r E + Q_i} \qquad (6.6)$$

Though this case approximates a residence heated with a warm air furnace, the assumption of no particle generation indoors is at variance with real-world conditions.

Equation 6.7, derived from Equation 6.5, represents a case where N_p, the particle generation rate, is due to cigarette smoking within a residence. It has the following form:

$$C_s = \frac{Q_i C_o + N_p}{Q_r E + Q_i} \qquad (6.7)$$

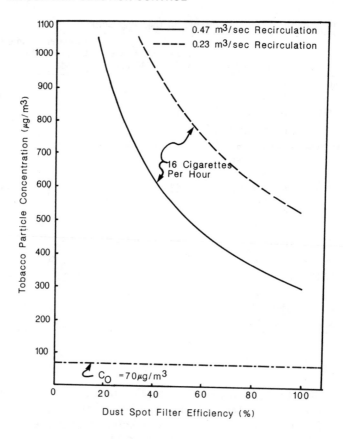

Figure 6.17. Effect of filter efficiency and recirculation rate on tobacco particle concentrations.[21]

McNall[21] applied this equation to the experimental conditions in a 425-m³ house,

where Q_1 = 0.06 m³/sec
Q_r = 0.35 m³/sec
C_o = 60 μg/m³
E = 95% dust spot efficiency (85% air cleaner, 10% furnace filter)
N_p = 40,000 μg/cigarette × smoking rate of 12 cigarettes/hr

Predictions based on this equation and results of actual measured values can be seen in Figure 6.18. The predicted and experimental values are seen to be in good agreement. These results indicate that the use of an electronic air filter in a central air circulating system can produce significant continuous

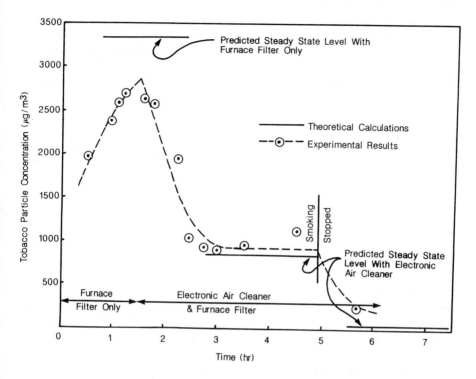

Figure 6.18. Measured and predicted levels of tobacco smoke particles in residence with electronic air cleaner in the central heating and cooling system.[21]

reductions in smoke particle levels. In the term Q_rE, both Q_r and E are of equal importance. A high-efficiency term is of little value where the recirculation rate is low or even zero (when the heating/cooling system blower fan is not activated), as is the case in residences when heating or cooling requirements are reduced.

Lefcoe and Inculet[22] measured the performance of electronic air cleaners installed in the central air circulation system in two residences and in three rooms of a chronic disease hospital. When experiments were conducted under rigorously controlled (no occupant activity) indoor conditions, particle counts in all measured size ranges decreased by an order of magnitude or more when the filtration system was activated. However, when system performance was studied over long periods of time under unrestricted occupant activity, particle counts decreased by only about 40–60%. Lefcoe and Inculet concluded that the number of occupants and their activities at any one time were significant in reducing filtration effectiveness.

Hermance[23] studied the performance of 95% NBS-type pleated filters on nitrate levels in indoor telephone switching environments. The use of such filters reduced nitrate concentrations by 73–87% and nitrate deposition on

Table 6.6 Bacterial Contamination Levels (Colony Forming Units/m³) in an Operating Room[25]

		Operating Room Status	
Position Location	Clean Air System Status	Empty	Surgery in Progress
	Off		
A		85	254
B		64	· 350
C		67	300
D		78	223
E		71	346
	On		
A		7	4
B		4	42
C		7	81
D		14	50
E		—	0

nickel-brass specimens by 96–98%. This reduction was achieved by filtering outdoor air used for ventilation. (The ambient [outdoor] environment is the major source of nitrate, which leads to the failure of telephone relay equipment by stress corrosion of nickel-brass switches.)

Control of Microbial Particles

Margard and Logsdon[24] evaluated the performance of a variety of filters on removing bacteria (*Bacillus globigii* and *Staphylococcus aureus*) introduced into the air stream of a test duct. Highest collection efficiencies were observed for a conventional 2-stage electronic air cleaner (91.5%) and a dry-type pleated filter (99.6%). Other filters had observed efficiencies less than half of these. Retention of bacteria upon the filter materials varied with the test organism employed. *B. globigii* was retained in all filters in uniform quantities during a one-month test period. Filters claiming to have bacteriostatic amendments displayed little if any bacteriostatic action.

Bacterial contamination of air in hospital operating rooms is of critical importance because of the potential for infection of surgical wounds or cuts. The primary source of such bacteria (primarily *S. aureus*) is the shedding of bacteria and skin particles by surgical staff. To reduce bacterial infection, hospitals often attempt to reduce particle concentrations by the use of HEPA filtration and very high laminar flow rates. Nelson[25] has reported results of the effect of a clean air system (HEPA filter and laminar air flow at 200 ACH) on bacterial counts in an empty operating room and during surgery. These results are summarized in Table 6.6. Note that bacterial counts are higher when surgery is being performed as compared to the empty operating room. Also note that the clean air system achieved a significant reduction in bacterial count levels compared to the conventional operating condition (air flow rate of

Table 6.7   Effect of Central Electrostatic Air Filtration on Mold Levels (Colony Forming Units/m³) in Residences[27]

| | | Air Cleaner Status | | | |
| | | Intermittent Operation | | Continuous Operation | |
	None		% Difference		% Difference
Mean	687	344	50	155	77
Minimum	106	125	0	36	66
Maximum	5,984	1038	83	757	87
Homes/category	40	8	—	13	—

12 ACH). Though Nelson's clean air system significantly reduced bacterial levels in operating rooms, there was no statistical evidence of a reduction in infection rate.

Solomon[26] conducted studies on mold levels in 21 residences that had electronic air cleaners installed in the return air plenum of forced-air heating systems. Mold counts averaged only 16% less than those in a group of structures matched for bedroom relative humidity but lacking an air cleaner. Kozak et al.[27] reported that the effectiveness of electronic air cleaners in reducing mold levels was significant under continuous operation and less so under intermittent operation (Table 6.7).

Alleviation of Allergy and Asthma Symptoms

One of the most common applications of air cleaners is to reduce symptoms in individuals who have atopic allergy or asthma. Such symptoms are usually induced by inhalation of particulate allergens such as pollen grains, mold spores, dusts containing mite fecal pellets, cockroach allergen, animal danders, etc.

Friedlander and Friedlander[28] evaluated the performance of portable electronic air cleaners on reducing symptoms of allergy patients with severe seasonal pollenosis. They observed excellent clinical relief of symptoms in 9 of 12 patients, and moderate improvement in the others. Symptoms abated within 10–30 minutes after patients entered an air-cleaned room, and reoccurred within relatively short periods after the patients left the room.

Zwemer and Karibo[29] evaluated the performance of a clean air headboard (consisting of a HEPA filter and laminar air flow) on symptoms in 18 symptomatic asthmatic children. All 18 patients improved, with 11 patients reporting symptom relief (reduced wheezing, coughing, and attendant disturbed sleep) as good to excellent. Scherr and Peck[30] reported a strong trend in reducing the incidence and severity of nighttime attacks in asthmatic children at summer camp using HEPA filtration units.

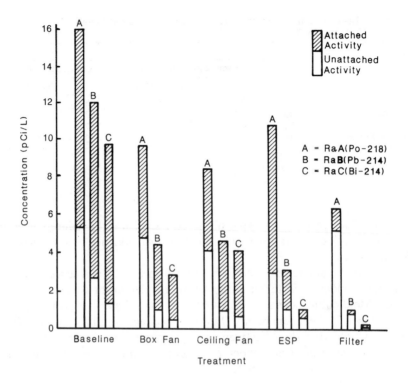

Figure 6.19. Effect of various air treatments on attached and unattached radon decay product levels.[31]

Radon Control

Hinds et al.[31] tested the effectiveness of portable electronic air cleaners and HEPA filters on radon progeny levels in a 78-m³ chamber at air change rates in the range of 0.2–0.8 ACH. Additionally, they evaluated the effectiveness of both box and ceiling fans. Radon progeny (attached fraction) reduction efficiencies were approximately 70% for the electronic cleaner and 89% for the HEPA filter. Radon progeny reductions were significant (57–65%) for the two fan systems evaluated as well, indicating that part of the air treatment effectiveness was due to enhanced deposition on chamber surfaces. Use of the HEPA filter resulted in a three-fold increase in unattached progeny activity. Although the HEPA filter caused a substantial reduction in RaB (Pb-214) and RaC (Bi-214) radon decay products, there was a limited effect on RaA (Po-218), one-third of which was present in the unattached state. The electronic air cleaner was observed to be more effective in reducing RaA levels than the HEPA filter. The effects of the different air treatments on both attached and unattached radon decay products can be seen in Figure 6.19.

Niles et al.[32] also conducted chamber measurements to determine the effectiveness of electronic air cleaners in reducing radon levels. They observed that this type of air cleaning reduced radon progeny activity by factors of 6–19, depending upon occupant smoking and the condensation nucleus concentration entering the room. Increasing condensation nuclei by lighting cigarettes increased radon progeny activity levels even when the electronic air cleaner was operating.

Lloyd and Mercer[33] tested the effectiveness of air cleaning devices on radon progeny concentrations in small apartments. When three 100- to 150-CFM (47-to 70-L/sec) electronic air cleaners were operating simultaneously in an apartment of 3100 ft³ (86.8 m³) volume, radon progeny concentrations decreased by 52% of their initial level in the first half-hour of evaluation. In a second apartment with an air volume of 5600 ft³ (156.8 m³), electronic air cleaners with a total volumetric air flow of 300 CFM (140 L/sec) were able to reduce the working level (WL) by 49% (from 0.043 to 0.022 WL). Reduction in an upstairs area was smaller (35%). Furnace filters were observed to reduce working levels by approximately 30% in two upstate New York houses.[34]

Use Problems of Residential Air Cleaners

The data of Kozak[27] on the effect of intermittent vs continuous operation of electronic air cleaners installed in the return air plenum of residential forced-air systems demonstrate the need for continuous operation of such units to achieve and maintain high mold particle collection performance, and, by analogy, performance in the collection of other particles as well. Based on the author's experience, continuous operation is the exception rather than the rule. In most residential installations, air cleaning only occurs when the forced-air system is activated to meet heating and cooling needs. When heating and cooling needs are low in the spring and autumn months, the forced-air system is only occasionally activated, and, therefore, little air cleaning actually occurs. In most instances, blower fans are wired in such a way that they cannot be engaged to operate continuously without shutting off the heating or cooling system. They can, of course, be wired to provide continuous operation of the blower fan and satisfy heating/cooling requirements. The need for such installation is not usually apparent to either the consumer or the installer. As previously described by Sutton et al.[20] and McNall,[21] performance of centrally installed air cleaners is dependent on both the recirculation rate and filter efficiency. When the recirculation rate is low, as often occurs with an intermittently operated air cleaner, its performance will be significantly less than capacity and represents a loss to consumers, particularly allergy/asthma patients who could achieve considerable symptom relief if design performance were achieved.

Portable electronic and HEPA air cleaners are limited in cleaning performance by low design air flows. They are typically designed for use in single rooms. Performance decreases as the air volume to be cleaned increases, that is, as the recirculation rate decreases. Most consumers view such devices generically and expect them to provide whole-house filtration.

For best performance, portable air cleaners should be used in a closed room with forced-air registers closed. To further maximize performance, the air cleaner should be placed in the center of the room where obstructions to air flow are minimized. For allergy patients, such operation is essential if significant symptom relief is desired.

As a general rule, the average consumer is notoriously poor in recognizing and acting on the need for equipment maintenance. HEPA and other dry-type filters must be replaced periodically in order to maintain design flow rates. Clogged filters decrease air flow rates and therefore air cleaner performance. Electronic air cleaners, unlike dry-type filters, do not have the problem of diminished air flow rate with increased filter soiling. However, excessive particle loads on collector plates will decrease performance; they therefore require frequent cleaning. Frequency of cleaning will depend on particle levels and dust loading of collector plates. Households with smoking residents should frequently replace dry-type filters in air cleaners and clean electronic air cleaner collection plates.

Ozone Production

Electronic air cleaners and negative ion generators are high voltage devices, and as such are capable of producing ozone. Because of ozone's toxicity at very low concentrations, the use of an air cleaner that itself produces a toxic contaminant is of some concern. As a consequence, all devices (to be used indoors) that produce ozone are regulated by the Food and Drug Administration.[35] The FDA ozone limit for electronic air cleaners and other devices is 0.05 ppm; that is, in normal use the device cannot produce ozone emissions that would exceed an indoor air concentration of 0.05 ppm.

Clean Rooms

Clean rooms are a special application of particle air cleaning in which elaborate attempts are made to limit the generation of airborne particles and to control their amount by air cleaning.[36-38] Clean rooms are widely used in the aerospace industry, industries making high-tech products, the military, and for various hospital applications where patients have a high risk of infection. Their history dates back to the 1950s, when HEPA filters were first developed and the space age began. In most instances, clean rooms are used to manufacture equipment and various components that would be rendered inoperative if particles of a critical size landed on a critical area (such as on an electrical contact).

Clean rooms achieve a high degree of particle control by the following means: (1) using high-efficiency filters to provide particle control, (2) purging the room of particles by changing the air, (3) limiting the generation of particles by special lint-limiting clothing and room materials, (4) protecting products from impact and settling of particles, and (5) providing an area for the cleaning of parts and personnel.[37]

In so-called first-generation clean rooms used in the 1950s, particle control was primarily achieved by HEPA filters installed in the air handling unit. Such filters had minimum efficiency ratings of 99.97% for particle diameters of 0.3 μm, particles considered to be most difficult to control by fiber filters. In 1962, clean rooms with laminar air flow were developed. A laminar flow pattern is induced when air is introduced uniformly at low velocities into a space confined on four sides and through an opening equal to the cross sectional area of the space. The laminar flow causes the air to be stratified so that minimum cross-stream contamination occurs.[37] In theory, there will be no transfer of energy from one streamline to another, and suspended particles will tend to stay in the streamline until removed.

Unlike conventional rooms in which particles falling out on room surfaces can be reentrained by foot traffic and other movements, laminar air flow in a downflow or crossflow room follows a predictable path. However, laminar flow patterns are disturbed by the presence of people and objects in the room. But this breakdown will not result in the degradation of room cleanliness. When a streamline is broken by an object, it reforms some distance downstream. Additionally, if some of the streamlines are broken, those that are not carry the contaminant away.[37]

The advantage of using laminar flow patterns is that high purging air flows can be employed without the particle-disturbing turbulence normally associated with such air flows. Laminar air flow is designed to remove particles generated in the space by the presence of people and to prevent the resuspension of particles that have deposited on surfaces, particularly the floor.

As air flow rates increase to a critical level, they begin to resuspend more particles than they purge. Twenty ACH has been widely used in laminar air flow systems and is considered to be optimal.[32]

In an ideal design, the entire ceiling is a diffuser and the entire floor an exhaust. Because of a variety of factors, including cost, clean rooms vary from this ideal. Figure 6.20 shows a clean room design that has been in common use. It has a wide, centrally placed diffuser in the ceiling and wall exhausts at the floor.

Clean rooms are usually limited in size since very large rooms are difficult to control environmentally. Because of long air travel distances and human traffic, air handling systems in large rooms tend to be less efficient in purging particles.

In clean rooms the air is recirculated through HEPA filters. The percentage of recirculated air may be on the order of 75%, with make-up air sufficient to provide adequate exhaust of human bioeffluents.

The HEPA filter or filters are located in the air handling unit on the discharge side of the blower fan (rather than the suction side) to prevent

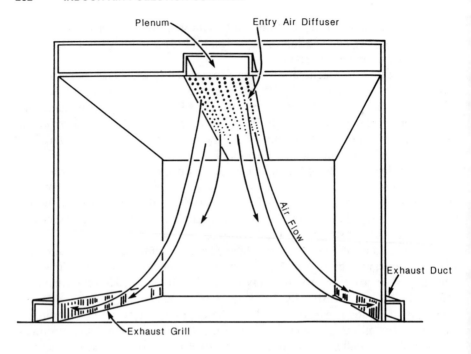

Figure 6.20. Laminar air flow patterns in a typical clean room.[37]

inward leakage of dust particles through inadequately sealed ductwork or filter gaskets. When the filter is on the discharge side, such leakage will be outward. The HEPA filter is the last unit of equipment downstream before the ductwork. This assures that all air entering the clean room is filtered and that there is no possibility of entry by unfiltered air. The installation of a filter to prevent leakage is critically important. The formation of an absolute seal, however, is difficult.

Pressurization is used in clean rooms to assure that all air is filtered prior to entering the room.[38] The maintenance of pressurization is facilitated by air locks used by employees to enter and leave the room.

As a consequence of modern needs, HEPA filters of ultra-high efficiencies have been developed. These efficiencies are on the order of 99.995%.[39]

REMOVAL OF GASEOUS CONTAMINANTS

In theory, the removal of gaseous contaminants from indoor air, or from outdoor air drawn into ventilation systems, can be achieved by the application of a variety of well-known principles. These include adsorption, catalytic oxidation or reduction, and absorption. Of these, adsorption has been the most widely applied and will, therefore, receive primary attention in this discussion.

Adsorption

Many gases, vapors, or liquids coming into contact with a surface will adhere to it to some degree. This adherence is a result of the same physical forces that hold atoms, ions, and molecules together in a solid state. Residual physical forces (Van der Waal's forces) at the surface of solids have the potential of binding molecules to the surface of the solid. This phenomenon is adsorption or sorption; the sorbed molecules are the adsorbate; the sorbing surface is the adsorbent.[40] The adsorbate can condense in the submicroscopic pores (capillary condensation) of the adsorbent.

Though adsorption is a chemical/physical phenomenon, no chemical reaction takes place. However, heat is released (heat of adsorption) and is approximately equal to that liberated when the adsorbed gas or vapor undergoes condensation. The adsorbate is present on the collection surface as a liquid or semiliquid.

Although adsorption can occur on a variety of solid surfaces, only a relatively few materials have adsorptive characteristics sufficiently favorable to air cleaning and other filtration applications. These include activated carbons, molecular sieves, zeolites, porous clay minerals, silica gel, and activated alumina. Common to these materials are high surface area:volume ratios. They are typically comprised of vast labyrinths of submicroscopic pores and minute channels. Most adsorption occurs in pores with diameters of approximately 10–30 Angstrom units. Access to these adsorption pores is provided by transition or macropores of 1000–10,000 Angstrom units.[41]

Metal oxides and silicaceous- and active-earth–type sorbents have an asymmetric molecular distribution, making them polar. Since polar substances have a strong affinity for each other and since water is strongly polar, such sorbents retain water preferentially. Consequently, they cannot efficiently adsorb gases (other than water vapor) in the humid atmospheres common to most air cleaning applications. Activated carbons, on the other hand, are nonpolar. Having no particular affinity for water vapor, they preferentially adsorb organic gases and vapors.[42]

Activated Carbons

Activated carbons are the most commonly used sorbents for gas cleaning. They are produced in a two-step process in which the raw material (hardwoods, coals, fruit pits, and coconut and other shells) is heated in a neutral atmosphere and then oxidized by a high-temperature oxidizing process. The raw material is cleared of materials that cannot be carbonized, and pores with extended surfaces are produced.[42] These activated carbons differ in their structural properties (e.g., density, hardness, pore size) and thus vary considerably in their performance as sorbents.

Adsorbability. The adsorbability or the degree of attraction of activated carbon for a vapor or sorbate is related to the sorbate's critical temperature and boiling point. For example, gases with critical temperatures well below -50°C and boiling points below -150°C cannot be adsorbed at normal temperatures. The high mobility of such light gases causes them to easily escape the attractive forces of carbon. These gases include oxygen, nitrogen, hydrogen, carbon monoxide, and methane.[42]

Gases with critical temperatures between 0° and 150°C and boiling points between -100° and 0°C are moderately adsorbable, as they have a lesser escaping tendency from activated carbon. These low–boiling point gases include ethylene, formaldehyde, ammonia, hydrogen chloride, and hydrogen sulfide. Activated carbons without special impregnants are not suitable for the removal of such gases from contaminated indoor air.

Higher–boiling point vapors (above 0°C) tend to have a low escaping tendency from activated carbon; therefore, their adsorptivity is considerable. Vapors in this category include a large variety of aldehydes, ketones, alcohols, organic acids, ethers, esters, nitrogen and sulfur compounds, alkylbenzenes, halocarbons, etc.

In general, the adsorbability of gases and vapors increases with molecular size and weight, so that in an organic series each larger or heavier member of the series is adsorbed more effectively than those in the series preceding it. Vapors with fewer than three atoms independent of hydrogen are not adsorbable.[42]

Hardness and size. Activated carbons vary in their degree of hardness, which is a function of raw material and production processes used. Hardness is an important parameter of activated carbons, since structural integrity is essential in withstanding the shear, compressive, and impact forces encountered in handling and use. When air is flowing through activated carbon at high rates, it may cause individual granules to vibrate. The loss of carbon due to vibration and/or crushing will result in voids developing in thin-bed adsorption panels, causing a portion of the air flow to be channeled around the adsorption medium and thus reducing adsorption efficiency.

Activated carbons are produced in a variety of particle size ranges described by the U.S. Sieve Series standard mesh sizes. For example, an 8 × 14 mesh size describes carbon particles that are 2.36 by 1.4 mm. Mesh numbers increase with decreasing granule size. Size specifications for carbon granules for general purpose air cleaning are in the 6–14 mesh size range. Granule size is important, as gas cleaning efficiency is directly related to it; increased efficiency is associated with decreasing size. As the adsorption bed becomes more tightly packed, the distance that contaminant molecules must travel to some point on the surface is smaller; therefore, the transfer rate from vapor to carbon is higher.[40]

Though gas cleaning performance is enhanced by the use of activated carbon granules of small size, such small granules, when tightly packed, cause significant increases in resistance and pressure drop. The effect of mesh size on pressure drop as a function of air velocity can be seen in Figure 6.21.[43]

Adsorption capacity. Activated carbons differ in their ability to adsorb gases and vapors. Adsorption capacity, or the weight of adsorbate collected per unit weight of adsorbent, depends on (1) surface area of the adsorbent, (2) volume of pores capable of condensing adsorbed gases, (3) characteristics of gas molecules to be collected, and (4) environmental conditions, including temperature, relative humidity, and pressure.

Surface areas for typical activated carbons range from 500 to 1400 m^2/g.[44] The relationship between surface area and pore size can be seen in Figure 6.22.

The adsorption capacity of activated carbons is rated by reference to their ability to adsorb carbon tetrachloride (CCl_4) vapors. A standard weight of activated carbon is exposed at 20°C to a dry air stream previously saturated with CCl_4 at 0°C until it no longer increases in weight. The ratio of sorbed CCl_4 to the weight of the carbon is the activity, or maximum possible adsorption, of the gas. This adsorptive capacity is expressed as CTC%. The CTC% ranges from a low of 20% to a high of 90%. Adsorptive capacity varies as a function of the surface area relative to carbon mass.[45]

Though adsorption capacity of activated carbons is rated with reference to CCl_4, it must be recognized that the adsorption capacity for other chemical vapors will vary from that of CCl_4. A sense of this variation can be obtained from the data for 27 carcinogenic vapors summarized in Table 6.8. Observed adsorption capacity by activated carbon was highest for 1,2-dibromomethane (1.0201 g/g) and lowest for 1,1-dimethyl hydrazine (0.359 g/g). Intermediate between these two extremes was CCl_4 (0.741 g/g).[46]

Adsorption capacity can be significantly affected by environmental temperature, pressure, and relative humidity. Though elevated temperature and low pressures can result in a significant loss of adsorption capacity, these factors are insignificant within the context of air cleaning needs in large buildings and residential environments.

Elevated relative humidity levels (> 50%) have the potential for significantly reducing the adsorption capacity of activated carbons. Werner[47] studied the effects of relative humidity on the adsorption of trichloroethylene (TCE) by activated carbon. The effect of increasing relative humidity on TCE adsorption at four different concentrations is illustrated in Figure 6.23. The effect was more pronounced at the lower TCE levels employed. A significant effect of relative humidity (RH) on the adsorption of radon from the air of uranium mines has been reported. Kapitanov et al.[48] observed that at 100% RH, adsorption capacity for radon was decreased by 50% relative to a humid-

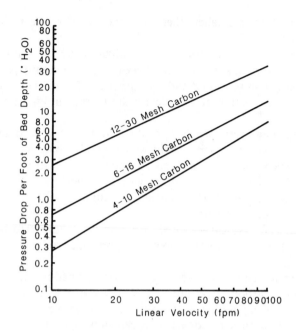

Figure 6.21.   Effect of carbon mesh size on pressure drop.[43]

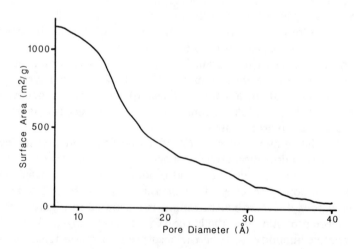

Figure 6.22.   Relationships between carbon pore size and surface area.[44]

Table 6.8   Adsorption Capacities of Activated Carbon for 27 Carcinogenic Vapors[46]

Vapor	Adsorption capacity g/g
Acetamide	0.494
Acrylonitrile	0.357
Benzene	0.409
Carbon tetrachloride	0.741
Chloroform	0.688
bis(Chloromethyl)ether	0.608
Chloromethyl methyl ether	0.480
1,2-Dibromo-3-chloropropane	0.992
1,1-Dibromomethane	0.962
1,2-Dibromomethane	1.020
1,2-Dichloroethane	0.575
Diepoxy butane (meso)	0.510
1,1-Dimethyl hydrazine	0.359
1,2-Dimethyl hydrazine	0.375
Dimethyl sulfate	0.615
p-Dioxane	0.475
Ethylenimine	0.354
Hydrazine	0.380
Methyl methane sulfonate	0.595
1-Naphthylamine	0.585
2-Naphthylamine	0.506
N-Nitrosodiethylamine	0.442
N-Nitrosomethylamine	0.458
B-Propiolactone	0.508
Prophylenimine	0.361
Vinyl chloride	0.404
Urethane	0.450

ity condition of 50% or less; Strong and Levins[49] reported that radon adsorption was reduced by up to 30% when entering gas was saturated with H_2O vapor.

Sleik and Turk[42] report that although activated carbon affinity for water is high, particularly above 40% RH, the adsorption of water by activated carbons is of little significance when carbon use is considered in the context of air cleaning in buildings. This conclusion is based on the fact that the retentivity of water vapor on activated carbon is very low. Because of low retentivity, sorbed gases and vapors cause sorbed water to leave the carbon and progressively reduce the carbon's sorptive capacity for water vapor. They conclude that, as a consequence, activated carbon applied to indoor air cleaning is equally effective under dry and humid conditions.

Retentivity. Retentivity is the maximum amount of vapor that can be retained by carbon when the vapor concentration in the air stream has been reduced to zero. The percent retentivity is less than the adsorption capacity.[6] It is measured by passing clean, dry air (at constant temperature and pressure) through a bed of activated carbon previously saturated by a specific gas or vapor. Air flow is discontinued when the carbon no longer loses weight. Retentivity is

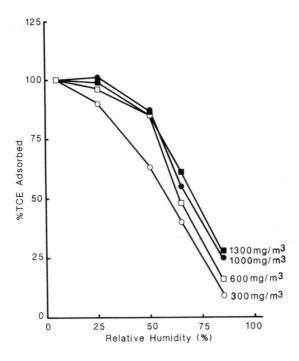

Figure 6.23. Relative humidity effects on trichloroethylene adsorption.[42]

expressed as the ratio of the weight of the retained substance to the weight of the carbon.[42]

Gas/vapor removal by a carbon bed. The effect of a carbon or sorbent bed on gases/vapors in a continuously flowing air stream is illustrated by a pattern of adsorption waves shown in Figure 6.24. Concentrations fall rapidly, reaching zero at some finite distance downstream on the bed face.[40] As adsorption continues with time, carbon granules near the face become partially saturated and the gases/vapors begin to penetrate more deeply (curve *b*). As adsorption continues, a portion of the carbon bed may become completely saturated (curve *c*). In the saturation zone a dynamic equilibrium is established between the saturated carbon granules and the gases/vapors in the incoming air stream. This is the equilibrium zone.[44] Downstream of the equilibrium zone, carbon granules are actively sorbing influent gases/vapors. This is the mass transfer zone. It is an area of the bed between saturation and zero concentration. The mass transfer zone moves progressively from one end of the carbon bed to the other with time. For each of the adsorption wave curves, the length of the mass transfer zone can be seen as the bed distance from the top to the bottom of the curve. As the adsorptive wave reaches the condition of curve *e*, breakthrough

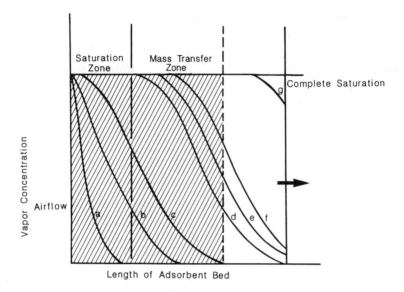

Figure 6.24.   Adsorption waves in a carbon bed.[40]

occurs, and the carbon bed reaches the end of its useful life. This is the breakpoint, and the carbon must now be replaced or reactivated.

Reactivation. In industrial applications, carbons are reactivated after they have become saturated. Such reactivation is accomplished by desorbing the collected gases/vapors by passing low-temperature steam or hot air through the bed. Because collected gases/vapors desorb differentially, reactivation usually does not result in complete desorption. Carbons in thin-bed filters used for indoor air cleaning are usually thrown away and reactivation is uncommon.

Residence time. Carbon beds used for solvent recovery and industrial air cleaning have depths on the order of 1–2 m. Because of the travel distances for gases/vapors in such systems, the residence time of contaminated air in the adsorbing medium is sufficiently long to effect high adsorption efficiencies. For indoor air cleaning applications, thick carbon beds are impractical, and thus thin-bed carbon filters are used. Because of the limited bed depth, the residence time in such carbon filters is critical to filter performance.

In a closely packed adsorbent thin-bed carbon filter, the half-life of a contaminant in a flowing air stream is approximately 0.01 seconds. In theory, four half-lives would be required to achieve 90% contaminant removal.[40] This would require a bed depth of approximately 0.4–0.5 inches (10–13 mm). Thin-bed carbon panels vary in bed depth depending on application and manufacturer. Thin-bed panels suitable for indoor air cleaning needs typically have bed depths of 1–1.25 inches (25–30 mm).

Service life. When the breakpoint is reached, the sorbent is nearing saturation and must be replaced. In practice it is difficult to know when the breakpoint has been reached. The service life of a carbon filter can, in theory, be calculated[42] from the following equation.

$$T = \frac{6.43 \ (10^6) \ S \ W}{E \ Q \ M \ C} \tag{6.8}$$

where T = duration of service before saturation, hr
         S = fractional saturation of adsorbent (retentivity)
         W = weight of adsorbent, lb
         E = fractional adsorption efficiency
         Q = rate of air flow, CFM
         M = average molecular weight of vapor adsorbed
         C = average vapor concentration, ppm

Let us apply this equation to the removal of toluene by a small air cleaning unit using these assumptions:

         E = 0.95
         C = 0.20 ppm
         Q = 100 CFM
         S = 0.30
         W = 6 lb

then

$$T = \frac{6.43 \times 10^6 \times 0.30 \times 6}{0.95 \times 100 \times 92 \times 0.20} \tag{6.9}$$
$$= 6621 \ hr$$
$$= 276 \ days$$

If we were to know the average concentration of contaminants in the air to be cleaned, we would calculate the service life from Equation 6.9 and thus determine when the carbon filter should be replaced. However, indoor air is contaminated with a large variety of adsorbable substances whose presence and concentration levels have not been well characterized. Equation 6.9, therefore, has little practical value for indoor air cleaning applications.

Schwabe[50] has proposed three different approaches to determining when a carbon filter applied for odor control should be replaced or renewed:

1. waiting until the filter ceases to function properly and odor is present
2. calculating from an equation, based on the assumption that activated carbon will adsorb 20% of its weight of odor
3. removing carbon granules from the filter at given times and determining their residual life in the laboratory

In the second option, filter life is determined from the following equation:

$$\text{filter life} = \frac{\text{carbon weight}}{5 \times \text{influent concentration} \times \text{flow rate}} \qquad (6.10)$$

The influent concentration is estimated from pilot studies using charcoal sampling tubes.

Schwabe suggested that the residual carbon life determinations of carbon granules removed from the filter could be carried out by some form of capacity determination or by determining the heat of wetting. The latter involves measuring in a calorimeter the heat of immersion of charcoal in a suitable wetting agent. The reduced value associated with the "dirty" activated carbon is then related to the degree of carbon contamination.

Turk et al.[51] investigated nondestructive methods for determining the degree of saturation of carbon beds using CCl_4 as a tracer gas. From analyses of test elements located upstream of carbon filters, they were able to determine the extent of saturation of a partially saturated carbon filter.

In a more recent report, Turk et al.[52] proposed that the most "practical" method for determining the degree of saturation of a carbon bed (utilized in indoor air cleaning) is by removing a sample and then assaying it by weighing before stripping it completely of adsorbed matter with superheated steam and by measuring the moisture content of a separate sample. Saturation S is calculated from the following equation:

$$S = \frac{\text{total mass of adsorbate} - \text{mass of adsorbed water}}{\text{mass of carbon}} \qquad (6.11)$$

Due to the physical disturbance of the filter medium and sophisticated procedures required, methods described above are for the most part not practical for determining service lives of carbon filters used in indoor air cleaning. Thus we are reduced to a choice of waiting until the odor problem returns (when the system is used to control odor) or to replacing filters at a predetermined time. Sleik and Turk[42] have suggested, as a rule of thumb, that carbon filters be replaced on an annual basis.

A potential approach that has not received systematic evaluation is the release of odoriferous chemicals such as isoamyl acetate (banana oil) or wintergreen on the upstream side of carbon filters. The presence of odor downstream would suggest that the carbon bed had reached or was approaching the end of its useful life. However, it could also indicate that a portion of the air stream was leaking around filter gaskets.

Activated carbons as catalysts. In addition to adsorption, activated carbons remove some gases/vapors by catalyzing their conversion to other less objectionable forms. Most notable is the catalysis of ozone (O_3) to oxygen (O_2). As a consequence, activated carbon can be used to remove O_3 from ambient air drawn into HVAC systems in such areas as southern California and may be applied to protect greenhouse crops from ozone injury.[53] In addition to O_3, activated carbons catalyze the destruction of ozonides, peroxides, and hydroperoxides.[40]

In the presence of free O_2, activated carbon functions as a catalyst, oxidizing hydrogen sulfide (H_2S) to elemental sulfur, which remains on the carbon surface. Activated carbons are used to control H_2S, one of the principal malodorants associated with sewage treatment.[41]

Activated carbons can catalyze the oxidation of sulfur dioxide to sulfur trioxide, which in the presence of water vapor is converted to sulfuric acid. Chlorine reacting with water can be catalytically converted to hydrochloric acid. In both cases strong acids are produced as a consequence of the catalytic action of activated carbons.

In addition to the natural catalysis of some compounds described above, activated carbons can be impregnated with catalysts for specific applications.

Chemisorption

The large internal surface area of sorbents provides a favorable medium for chemical reactions. These surfaces can be coated or impregnated with chemicals that will selectively react with or chemisorb molecules from a gas stream. The process in which an adsorbate chemically reacts with the adsorbent is called chemisorption.

Chemisorptive media are produced by the impregnation of materials such as activated carbons and activated alumina. Carbon impregnants include bromine, metal oxides, elemental sulfur, iodine, potassium iodide, sodium sulfide, etc. Bromine-impregnated carbon is used to chemisorb ethylene, a gas for which carbon has a low sorptive capacity. Ethylene and bromine react catalytically on the surface of activated carbon to produce ethylene bromide, which is adsorbable. Carbons impregnated with metallic oxides can be used to remove H_2S under low-oxygen conditions. H_2S is converted to a metallic sulfide. Elemental sulfur- and iodine-impregnated carbons are used to sorb mercury, producing stable mercuric sulfide and mercuric iodide, respectively. Activated carbons impregnated with potassium iodide are used to sorb radioactive Iodine-118 that might be released in a nuclear power accident. Activated carbons specifically developed to control radioactive krypton and xenon are used in nuclear power plant off-gas control systems.[41] Sodium sulfide and other impregnants are used where formaldehyde control is desired; they have been used in residential applications.

Activated alumina impregnated with potassium permanganate ($KMnO_4$) is widely used in air cleaning systems. It is best known by the trade names Purafil and Carasorb. Though activated alumina does not have the adsorptive power of most activated carbons, it can be used effectively for the control of a variety of contaminant gases, including formaldehyde and ethylene. Gaseous vapors are sorbed on the surfaces of the activated alumina, where the $KMnO_4$ oxidizes them in a film of water.[40,42] If oxidation is complete, organic vapors will be converted to carbon dioxide and water vapor. When the $KMnO_4$ reacts with adsorbate chemicals, it is reduced to manganous oxide ($KMnO_2$) and turns from pink to brown. As the activated alumina/$KMnO_4$ adsorbent turns brown, it is no longer effective and must be replaced. This can be determined by removing adsorbent pellets from the downstream side of the filter bed and crushing them.[53]

Activated alumina/$KMnO_4$ media have one potential drawback. They may produce by oxidation contaminants such as hydrochloric acid (HCl) when chlorinated hydrocarbons such as 1,1,1-trichloroethane, trichloroethylene, and tetrachloroethylene are oxidized. Because of the high solubility of HCl in water, it should effectively be removed on the water film of the oxidized alumina. This, nevertheless, points out a dilemma: under certain circumstances, new contaminant problems may be created while attempting to control the one at hand.

Adsorption/chemisorption equipment designs. Air cleaners based on adsorption/chemisorption are available in a variety of commercial designs.[55] These include units in which a variety of filter bed types and depths are employed (Figure 6.25). For HVAC systems these include:

- *Partial bypass*: Filters are loosely packed with numerous voids. They are of low efficiency and are used primarily for odor control. Bed depths available include 1, 2, and 4 inches (25, 50, and 100 mm). See Figure 6.25A.
- *Serpentine*: These contain thin-bed (0.375–0.5 inches [9.4–13 mm]) convoluted filters and are used in medium-duty recirculation applications. See Figure 6.25B.
- *Thin-bed tray*: These contain multiple filters of thin depth (0.5–0.625 inches [13–15.6 mm]) arranged in "Z" configuration in a module to achieve an extended surface area. See Figure 6.25C.
- *Intermediate bed depth trays*: These are similar to thin-bed trays except bed depth is on the order of 0.875–1 inch (23–25 mm). They are designed for higher efficiency and longer service life. See Figure 6.25D.
- *Intermediate bed depth V module*: Filters are in a "V" configuration, with a bed depth of 1–1.125 inches (25–28 mm). See Figure 6.25E.
- *Thick-bed tray*: This is a single filter with bed depth of 2 inches (50 mm) or more oriented perpendicular to the air stream. It has a high pressure drop and is used with low duct velocities. See Figure 6.25F.

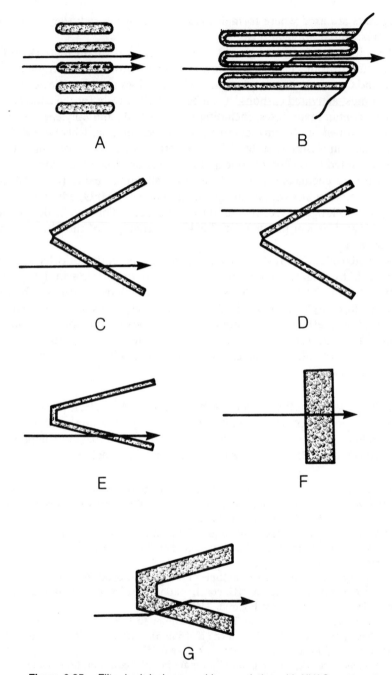

Figure 6.25.   Filter bed designs used in association with HVAC systems.[5]

Figure 6.26.   Adsorption module used in HVAC systems.
(Courtesy of American Air Filter Co.)

• *Thick-bed extended surface area module*: This unit has an average bed depth of 3 inches (75 mm) and attains low bed velocity by using extended surface area techniques. See Figure 6.25G.

The above equipment can be installed in side-access housing or other standardized equipment. It can also be installed in a self-contained cleaner unit.

Residential air cleaners utilize a more limited series of designs. In most instances, thin-bed filters oriented perpendicular to air flow are employed. These filters vary in depth from about 0.5 to 2 inches (12.5 to 50 mm). In-duct devices contain thin-bed filters housed in modules similar to those used in HVAC system applications (Figure 6.26). Some residential air cleaners employ cylindrical canisters to achieve a high surface area (Figure 6.27).

Performance Studies

Studies of the air cleaning effectiveness of adsorption/chemisorption systems designed and used for indoor applications have been very limited. It appears that judgments relative to performance have been based on subjective criteria. For example, if a carbon filtration system is installed in the HVAC system of a building to control odor and occupants perceive no objectionable

Figure 6.27. Residential air cleaner using a cylindrical canister. (Courtesy of E. L. Foust Co., Inc.)

odor, then performance would be considered to be excellent. Such assessments are anecdotal.

Richardson and Middleton[55] systematically tested the effectiveness of thin-bed carbon filters on smog-associated irritation levels experienced by employees in a Los Angeles building. When the air was filtered, significant irritation was reported on 5 of 109 sampling days, as compared to 25 of 109 sampling days for unfiltered air. Barely noticeable irritation was observed in a carbon filter–protected room on days when occupants of a nonfiltered room complained of high-moderate to severe irritation, indicating some penetration of carbon filters and/or the building envelope on days of intense smog. The degree of oxidant reduction was related to the residence time of air in the filter, but the degree of irritation was not (Table 6.9). The effect on NO_2 levels was quite variable.

Jewell[56] studied the effectiveness of a variety of air cleaning systems on controlling formaldehyde levels in a mobile home. He found that the most

Table 6.9   Effect of Carbon Filtration on Sensory Irritation and Contaminant Concentrations in a Los Angeles Office Building[55]

Residence Time (sec)	Irritation Ratio[a]	Oxidant Ratio[a]	NO$_2$ Ratio[a]
0.032	0.27	0.04	0.26
0.016	0.32	0.23	0.61
0.0075	0.39	0.38	2.07
0.0030	0.34	0.54	1.23
Particulate filter only	1.06	0.64	0.86

[a]Ratio of measurements in a filtered room to those in an unfiltered room.

effective control of formaldehyde levels could be achieved by thin-bed filters using an activated alumina/potassium permanganate medium. The control system utilized an 1800-CFM (849-L/sec) blower and 36 kg of activated alumina/KMnO$_4$. Formaldehyde reductions were on the order of 75%. The initial reduction, from approximately 0.5 ppm to 0.1 ppm, was maintained over a period of 140 days.

Godish[57] studied the effectiveness of portable thin-bed filter air cleaners in a mobile home and in a urea-formaldehyde foam–insulated (UFFI) house. The medium used was activated alumina/KMnO$_4$ and the air flow rate was 130 CFM (61 L/sec). Both building structures had comparable air volumes. Formaldehyde reduction in the mobile home was 25–30%; reduction in the UFFI house was 35–45%.

Activated carbon beds are used in conjunction with catalytic burners to reduce the level of organic compounds in submarine atmospheres.[59] The effect of carbon filtration as compared to catalytic burners on the removal rate of total hydrocarbon levels can be seen in Table 6.10. The carbon bed is used to remove odors and organic vapors that have a relatively high molecular weight. It precedes the catalytic burners, preventing them from being overtaxed.

Carbon filtration has been proposed as a means of reducing radon levels in uranium mine atmospheres[48,49,59,60] and indoor air. As mentioned previously, it is used extensively in the nuclear power industry to capture radioactive gases.

Table 6.10   Generation and Removal Rates of Organic Contaminants on Board the Submarine USS Sculpin[59]

Study Phase	Removal Unit	Hydrocarbon Equilibrium Concentration (mg/m³)	Hydrocarbon Removal Rate (g/day)	Hydrocarbon Generation Rate (g/day)
1	Burner #1	100	700	700
2	Burner #2	60	420	600
	Carbon bed		180	
3	Burner #1	40	280	680
	Burner #2		280	
	Carbon bed		120	

Because of its high adsorption capacity for radon,[49] activated carbon is widely used to measure radon levels in buildings.[61] Carbon filtration has been studied by Arthur D. Little, Inc.[60] as a potential alternative to ventilation for radon control in mines. The researchers concluded that carbon filtration for radon control in uranium mines, though technically effective, is not cost-competitive with ventilation.

Because of high adsorption capacities and relatively low absolute radon concentrations, activated carbon would appear to have a great ability to sorb and remove radon. Thus, filter life should be long. However, activated carbon is not specific to radon and can sorb a variety of vapors and gases present in much higher concentrations than radon. Water vapor in particular may be a problem. As discussed in a previous section, at a relative humidity of 100%, the adsorption coefficient for radon is decreased by a factor of two.[49] In residential environments, carbon filter life expectancy for radon is expected to be markedly reduced by the presence of a variety of organic molecules that compete with radon for binding sites. With these limitations in mind, a regenerative carbon filtration system for residential use has been proposed by Bocanegra and Hopke.[62]

Absorption

In industrial applications, a variety of contaminant gases (including SO_2 and HCl) are removed from waste gas streams by absorbing them in water or a reactive liquid reagent or slurry. The contaminant is removed by chemical reaction with the absorbing medium. The process is often referred to as scrubbing.

In theory, the principle of absorption or scrubbing can be applied to the removal of water-soluble gases from indoor air. The effect of absorption on the removal of formaldehyde from an air stream has been investigated by Pedersen and Fisk.[63] Formaldehyde removal efficiencies of two air washer units were 30–63%. No attempts were made to predict the impact of air washer operation on indoor formaldehyde levels.

Room Temperature Catalysts

Indoor air cleaning of certain contaminant gases can be achieved by the use of catalytic materials.[56,58,64,65] In submarine atmospheres, the use of catalytic burners is the principal means by which CO and a large variety of organic compounds are removed.[59] However, catalysis in submarines is associated with the use of relatively high temperatures (> 300°F). More conventional indoor air cleaning applications would require catalysts to be effective in the normal range of building temperatures.

The effect of catalysis by activated carbons on indoor air contaminants such as O_3 and H_2S has been previously described. The performance of automobile catalytic converters on formaldehyde levels in a mobile home was

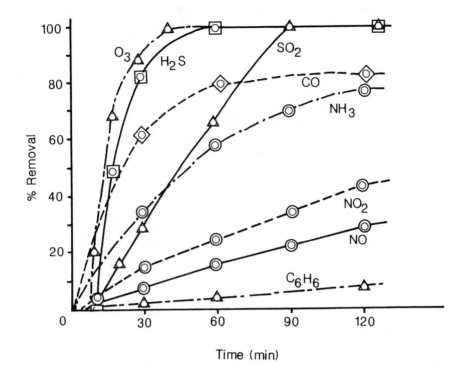

Figure 6.28. Performance tests of LTC catalyst in a sealed room.[64]

investigated by Jewell.[56] One catalytic converter type was totally ineffective. A second, containing platinum, palladium, rhodium, and cerium dioxide, reduced formaldehyde levels on an order similar to filter beds of activated alumina/KMnO$_4$. However, in operation, this converter produced an unknown irritant that Jewell speculated was formic acid. This irritant made the catalytic converter an unsuitable formaldehyde control technique. Low-temperature catalysts (LTC) have been developed by Teledyne Water-Pik for use in their Instapure residential air cleaning device. Performance studies of the Teledyne LTC have been reported by Collins.[64,65] This LTC is composed of copper chloride with small quantities of palladium chloride. A second copper salt is used to stabilize the active metal complex on a porous alumina medium. In the appliance, the LTC and activated carbon are mixed in 50:50 proportions in a tightly packed thin-bed filter (to minimize pressure drop). The flow rate is 300 CFM (142 L/sec), with a residence time of 0.02 seconds.

Performance studies of the Teledyne Instapure air cleaning appliance were conducted in a sealed 1152-ft^3 (32.3-m^3) room.[65] The effects of the LTC cleaning device on O$_3$, H$_2$S, SO$_2$, CO, NH$_3$, NO$_2$, NO, and C$_6$H$_6$ (benzene) can be seen in Figure 6.28. Removal of O$_3$ and H$_2$S was relatively rapid and

Table 6.11   Effect of Plants on Formaldehyde Removal from Laboratory Chambers[66]

Experiment	Formaldehyde Concentration (ppm)		
	0 hr	6 hr	24 hr
Controls without pots	17	17	17
Controls with pots	15	12	10
Scindopis aureus (golden pathos)	18	9	6
Syngonium podophyllium (nephthytis)	18	9	6
Chlorophytum elatum var. vittatum (spider plant)			
Set 1	14	2	<2
Set 2	37	8	<2

approached 100% within the first hour. Removal of benzene and the oxides of nitrogen was relatively poor. H_2S was observed to "poison" the LTC by reacting with the catalyst to form copper sulfide. Test results were based on a one-time infusion of a high concentration of test gases at the beginning of the test period.

Botanical Air Cleaning

A rather novel approach to removing gaseous contaminants from indoor air has been proposed by Wolverton et al.[66] In studies conducted for the National Aeronautics and Space Administration, plants, when exposed to formaldehyde (in chambers under laboratory conditions), were observed to significantly reduce formaldehyde concentrations. Most notable was the performance of spider plants (*Chlorophytum elatum* var. *vittatum*). The results of their studies, based on a one-time injection of formaldehyde into test chambers, are summarized in Table 6.11.

Godish et al.[67] have investigated the effectiveness of spider plants on removing formaldehyde from laboratory chambers in which formaldehyde levels were generated continuously (as occurs in a typical residential environment). They observed that only a modest reduction in formaldehyde levels occurred when plants were fully leaved, and that the greatest degree of reduction occurred when all leaves had been removed (Figure 6.29). Reductions in formaldehyde levels appeared to be associated with soil surfaces and soil water content. Comparisons of formaldehyde reductions under dark and illuminated conditions, when plants were at different metabolic states, revealed no differences in formaldehyde removal efficiencies. This suggested that formaldehyde removal was not due to metabolic consumption of formaldehyde by spider plants. The studies of Godish et al. did not support the conclusions of Wolverton et al. that spider plants are effective air cleaners for formaldehyde.

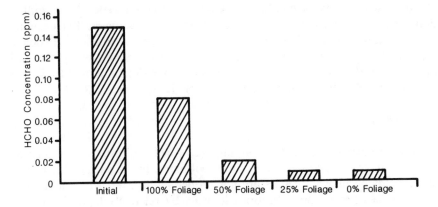

Figure 6.29. Formaldehyde reduction levels associated with spider plant foliage removal.[67]

Residential Air Cleaning of Gaseous Contaminants

Products available for removing gaseous contaminants from indoor air include small filter panel units, portable room models, and in-duct devices. Most such devices have not been tested under realistic, in-use conditions and their effectiveness is unknown. In-duct devices are usually produced by manufacturers who make systems for large buildings; such devices have a higher likelihood of effective performance. However, in most instances residential air cleaners, particularly the filter panel type and portable room models, are unlikely to be effective in removing gases and vapors.

Many products contain activated carbon filters. In most instances such filters are loosely packed, egg crate, filter frame–type affairs. Voids are so extensive that an individual can see through the filter. Since air travels in the path of least resistance, it will preferentially flow through the filter voids. Thus, the likelihood of effective contaminant removal is small.

Carbon filters are often used in conjunction with portable air cleaners, which are effective particle air cleaners. In the case of electronic air cleaners, such filters must have low resistance (high void space) to air flow to prevent damage to the axially driven fan and to maintain a high actual flow rate. The use of such filters appears to be intended only for consumer appeal.

The problems described above also hold for devices marketed as ductless hoods for kitchen stoves. These devices contain small, loosely packed, thin (0.375–0.5-inch [9.4–13-mm] depth) filter beds of activated charcoal. Because of voids and the relative ineffectiveness of activated carbons on combustion gases (e.g., CO and the oxides of nitrogen), such devices likely contribute to elevated contaminant levels, since they are used in lieu of a ducted hood.

To be effective in controlling gaseous contaminants, a residential air cleaner should be designed to use thin-bed filters similar to those used in

HVAC systems of large buildings (taking into account scale of the application). These devices are available from some manufacturers. Performance, based on this author's observations, is likely to vary considerably. Problems include poorly packed filters, undersizing the filter face relative to the static pressure that the blower fan can accommodate to maintain desired flow rates, and the use of a cleaner unit in an excessively large space.

Channeling

The occurrence of voids in an adsorptive bed can result in a significant decrease in cleaning performance, as air will move in the path of least resistance. Voids occur when filters are not properly filled and when attrition of filter media occurs from vibrations induced by relatively high air velocity. Reliable filter manufacturers use vibration techniques to attain dense packing of the media in filter frames to minimize void space and reduce the attrition of media by vibration.[42] The problems of poorly packed filters and channeling of air through voids (occurring in the periphery of the filter) are particularly acute in filter frames mounted vertically. In theory, the formation of voids and channeling should be less of a problem if filter frames are mounted horizontally.

Air Flow Rates

A well-designed and loaded adsorption media filter will produce a considerable resistance to air flow. The actual resistance will depend on the face surface area. High face surface area is desirable to maintain low pressure drop and desired air flow rates. The canister-type air cleaner in Figure 6.27 has a high face surface area. A "z" configuration (Figure 6.25) of filter panels that increases the effective surface area of the panels is required to maintain high air flow rates through in-duct devices. Product manufacturers who have developed air cleaners based on intuitive skills rather than technical understanding claim air flow rates based on the output of blower fans at zero static pressure without taking the air flow resistance into account. Actual flow rates or cleaner capacities are considerably less than those claimed. Consequently, such products clean less air than they profess to.

Cleaner Operation

Because many contaminants are generated continuously, air cleaner operation should also be continuous. Portable air cleaners are designed (but not necessarily marketed) to clean the air of a single room. Optimal performance would occur with the room air space isolated from the rest of the residence. This would require closure of doors and central air circulation inlets and outlets. In theory, the higher the recirculation rate, the more effective contaminant removal will be. No research studies are available to determine the actual distribution of contaminant levels in an isolated room as air cleaning is occurring. Lowest contaminant levels would be expected near the cleaning device, with increasing levels as a function of distance. Contaminant concentrations would be expected to depend on the mixing characteristics of the room and proximity to sources. As is the case for ventilation air in HVAC systems, some degree of short-circuiting may be expected where recently cleaned air is quickly returned to the intake of the air cleaning device. In such cleaners, air inlets and exhausts are often less than 12 inches (30 cm) from each other.

REFERENCES

1. DHEW. 1969. "Air Quality Criteria for Particulate Matter." Publication No. AP-44.
2. Hinds, W.C. 1982. "Filtration." 164–186. In: *Aerosol Technology—Properties, Behavior and Measurement of Airborne Particles.* Wiley-Interscience, New York.
3. Hinds, W.C. 1982. "Electrical Properties." 284–309. In: *Aerosol Technology— Properties, Behavior and Measurement of Airborne Particles.* Wiley-Interscience, New York.
4. Offerman, F.J. et al. 1983. "Control of Respirable Particles and Radon Progeny with Portable Air Cleaners." LBL-16659. Lawrence Berkeley Laboratory, Berkeley, CA.
5. American Society of Heating, Refrigerating and Air-Conditioning Engineers. 1983. "Air Cleaners." 10.1–10.12. *Equipment Handbook.* American Society of Heating, Refrigerating and Air-Conditioning Engineers. Atlanta.
6. Engle, P.M. and C.J. Bouder. 1964. "Characteristics and Application of High Performance Dry Filters." *ASHRAE J.* 6:72–75.
7. Multi-Fab, Inc. 1981. Bulletin No. M2-81-1.
8. American Air Filter. 1982. Bulletin No. AF-1-128D.
9. Honeywell, Inc. 1978. "Electronic Air Cleaner—Theory and Fundamentals." Publ. No. 70-9719.
10. Bevans, R.S. and J.H. Vincent. 1974. "An Experimental Grid-Type Residential Electrostatic Air Cleaner." 203–211. *ASHRAE Trans.* 80. Pt. 2.
11. Honeywell, Inc. 1985. Product literature.
12. ASHRAE Standard 52-76. 1976. "Method of Testing Air Cleaning Devices Used in General Ventilation for Removing Particulate Matter." American Society of Heating, Refrigerating and Air-Conditioning Engineers. Atlanta.
13. American Air Filter. 1982. Bulletin No. AF-1-172D.
14. Wallach, C. 1981. "Control of Indoor Air Pollution by Hypernegative Ionization and Passive Electret Filtration." In: Proceedings of the Second International Conference on Indoor Air Pollution, Health and Energy Conservation. Amherst, MA.
15. Bionaire Corporation. 1985. Product literature.
16. Ion Systems, Inc. 1986. Product literature.
17. Anonymous. 1982. "A Test of Small Air Cleaners." *Rodale's New Shelter* July/August. 47–49, 53–57. Rodale Press, Emmaus, PA.
18. Canine, C. 1986. "Clearing the Air." *Rodale's New Shelter* January. 64–66. Rodale Press, Emmaus, PA.
19. Kimmel, J. 1987. "Performance of Electrostatic Precipitators in Restaurants."

226–230. In: *Proceedings of the Fourth International Conference on Indoor Air Quality and Climate*. Institute for Water, Soil and Air Hygiene. West Berlin. Vol. 3.

20. Sutton, D.J. et al. 1964. "Performance and Application of Electronic Air Cleaners in Occupied Spaces." *ASHRAE J.* 6:55–62.

21. McNall, P.E. 1975. "Practical Methods of Reducing Airborne Contaminants in Interior Spaces." *Arch. Environ. Health* 30:552–556.

22. Lefcoe, N.M and I.I. Inculet. 1975. "Particulates in Domestic Premises. II. Ambient Levels and Indoor-Outdoor Relationships." *Arch. Environ. Health* 30:565–570.

23. Hermance, H.W. 1971. "Relation of Airborne Nitrate to Telephone Equipment Damage." *Environ. Sci. Technol.* 5:781–785.

24. Margard, W.L. and R.F. Logsdon. 1965. "An Evaluation of the Bacterial Filtering Efficacy of Air Filters in the Removal and Distribution of Airborne Bacteria." *ASHRAE J.* 7:49–54.

25. Nelson, C.L. 1979. "Environmental Bacteriology in the Unidirectional (Horizontal) Operating Room." *Arch. Surg.* 114:778–782.

26. Solomon, W.R. 1976. "A Volumetric Study of Winter Fungus Prevalence in the Air of Midwestern Homes." *J. Allergy Clin. Immunol.* 57:46–55.

27. Kozak, P.P. et al. 1979. "Factors of Importance in Determining the Prevalence of Molds." *Ann. Allergy* 43:88–94.

28. Friedlander, S. and A.S. Friedlander. 1954. "Effectiveness of a Portable Electrostatic Precipitator in Elimination of Environmental Allergies and Control of Allergic Symptoms." *Ann. Allergy* 12:419–428.

29. Zwemer, R.J. and J. Karibo. 1973. "Use of Laminar Control Device as Adjunct to Standard Environmental Control Measures in Symptomatic Asthmatic Children." *Ann. Allergy* 31:284–290.

30. Scherr, M.S. and L.W. Peck. 1977. "The Effects of High Efficiency Air Filtration System on Nighttime Asthma Attacks." *W. Va. Med. J.* 73:144–148.

31. Hinds, W.C. et al. 1983. "Control of Indoor Radon Decay Products by Air Treatment Devices." *JAPCA* 33:134–136.

32. Niles, J.C.H. et al. 1980. "The Effect of Domestic Air Treatment on the Concentration of Radon-222 Daughters in a Sealed Room." *Royal Soc. Health J.* 3:82–85.

33. Lloyd, L. and D. Mercer. 1981. "Testing of Air Cleaning Devices to Determine Their Capabilities to Reduce Radon Progeny Concentrations in Homes." Montana Department of Health and Environmental Services.

34. Offerman, F.J. et al. 1982. "Low Infiltration Housing in Rochester, NY: A Study of Air Exchange Rates and Indoor Air Quality." *Environ. Int.* 8:435–446.

35. U.S. Food and Drug Administration. 1975. "Maximum Acceptable Levels of Ozone." CFR 21.801:401–420.

36. Kranz, P. 1963. "Ventilation Effectiveness for Indoor Air Cleaning." *ASHRAE J.* 5:27–31.

37. Austin, P.R. and S.W. Timmerman. 1965. *Design and Operation of Clean Rooms*. Business News Publishing Co., Detroit.

38. Looney, R.J. 1962. "Controlling Contaminants for Air Cleaner Effectiveness." *ASHRAE J.* 4:22–26.

39. American Air Filter. 1986. Bulletin No. AF-1-110-G.

40. Turk, A. 1977. "Adsorption." 329–363. In: A.C. Stern (Ed.). *Air Pollution, Vol. IV. Engineering Control of Air Pollution*. 3rd ed. Academic Press, New York.

41. Calgon Corporation. 1975. "Air Purification with Activated Carbon." Bulletin No. 23–25.
42. Sleik, H. and A. Turk. 1953. "Air Conservation Engineering." Connor Engineering Corporation, Danbury, CT.
43. Calgon Corporation. 1981. "Effective Odor Control with Calgon Granular Activated Carbon Systems." Bulletin No. 27–46.
44. Doig, I.D. 1980. "Activated Carbon in Air Pollution and Odor Control." *Clean Air* November. 55–62.
45. Furniss, J.E. and A.J. Weir. "Granulated Activated Carbons for Air Treatment." Sutcliffe Speakman. CW 28304.
46. Sansone, E.B. and L.A. Jonas. 1981. "Prediction of Activated Carbon Performance for Carcinogenic Vapors." *Am. Ind. Hyg. Assoc. J.* 42:688–691.
47. Werner, M.D. 1985. "The Effects of Relative Humidity on the Vapor Phase Adsorption of Trichloroethylene by Activated Carbon." *Am. Ind. Hyg. Assoc. J.* 46:585–595.
48. Kapitanov, Y.T. et al. 1967. "Adsorption of Radon on Activated Carbon." *Int. Geo. Rev.* 12:873–878.
49. Strong, K.P. and D.M. Levins. 1978. "Dynamic Adsorption of Radon on Activated Carbon." 627–639. In: Proceedings of the 15th Department of Energy Nuclear Air Cleaning Conference. CONF-780819.
50. Schwabe, P.H. 1971. "Gas and Vapor Filtration with Special Reference to Activated Carbon." In: *Proc. Filtration Soc.* 8:153–157.
51. Turk, A. et al. 1964. "Tracer Gas Nondestructive Testing of Activated Carbon Cells." *Mat. Res. and Stand.* 9:24–26.
52. Turk, A. et al. 1984. "Assessing the Performance of Activated Carbon in the Indoor Environment." *Am. Ind. Hyg. Assoc. J.* 45:714–718.
53. Godish, T. 1983. "Indoor Air Cleaning of Gaseous Contaminants." *Arch. Clin. Ecol.* 2:35–42.
54. American Society of Heating, Refrigerating and Air-Conditioning Engineers. 1984. "Control of Gaseous Contaminants." 30.1–30.8. *ASHRAE Handbook— Systems.* American Society of Heating, Refrigerating and Air-Conditioning Engineers. Atlanta.
55. Richardson, N.A. and W.C. Middleton. 1959. "Evaluation of Filters for Removing Irritants from Polluted Air." *ASHRAE Trans.* 65:401–416.
56. Jewell, R.A. 1980. "Reduction of Formaldehyde Levels in Mobile Homes." 10–107. In: *Wood Adhesives— Research, Applications and Needs.* Forest Products Research Society. Madison, WI.
57. Godish, T. 1987. "Control of Formaldehyde Levels by Air Purification Systems" (unpublished studies).
58. Carhart, A. and J.K. Thompson. 1975. "Removal of Contaminants from Submarine Atmospheres." U.S. Naval Research Lab, Washington, DC.
59. Adams, R.E. et al. 1959. "Containment of Radioactive Fission Gases by Dynamic Adsorption." *Ind. and Eng. Chem.* 51:1467–1470.
60. Arthur D. Little, Inc. 1975. "Advanced Techniques for Radon Gas Removal." USBM Contract Report No. 2300.
61. Cohen, B.L. and E.S. Cohen. 1983. "Theory and Practice of Radon Monitoring with Charcoal Adsorption." *Health Physics* 45:501–508.
62. Bocanegra, R. and P.K. Hopke. 1986. "Optimal Parameters for the Adsorption of

Radon on Charcoal." In: Proceedings of the 79th Annual Meeting of the Air Pollution Control Association. Minneapolis.

63. Pederson, B. and W.J. Fisk. 1984. "The Control of Formaldehyde in Indoor Air by Air Washing." LBL-17381. Lawrence Berkeley Laboratory, Berkeley, CA.

64. Collins, M.F. 1986. "Room Temperature Catalyst for Improved Indoor Air Quality." 859–870. In: H.D. Goodfellow (Ed.). *Ventilation '85*. Elsevier Science Publishers, Amsterdam.

65. Collins, M.F. 1986. "Characterization of the LTC Catalyst. Performance Against Common Pollutants." In: Proceedings of the 79th Annual Meeting of the Air Pollution Control Association. Minneapolis.

66. Wolverton, B.C. et al. 1983. "Foliage Plants for Removing Indoor Air Pollutants from Energy Efficient Homes." *Econ. Bot.* 38:224–228.

67. Godish, T. 1988. "Botanical Air Purification Studies Under Dynamic Chamber Conditions." In: Proceedings of the 81st Annual Meeting of the Air Pollution Control Association. Dallas.

7

POLICY AND REGULATORY
CONSIDERATIONS

The identification of environmental quality problems in the United States and other western countries, and responses to them, have followed a predictable pattern of expanding governmental involvement and the institution of regulatory programs. We tend to look to government to solve environmental problems, or at least to participate in their solution.

Federal governmental authorities in the United States have been reluctant to become significantly involved in the problem of indoor air quality. This reluctance was due in part to the anti-regulatory philosophy of the Reagan administration. There has also been concern that new programs, such as indoor air quality control, would demand the allocation of additional federal resources or compete for the limited resources available for other environmental programs administered by federal agencies, such as the U.S. Environmental Protection Agency.

It would be overly simplistic to conclude that a lack of action by federal agencies in defining policy and becoming involved in a regulatory way was due to prevailing political philosophies and preoccupation with federal spending. It takes time for a health concern or problem to become recognized. Indoor air quality is the "new kid on the block." As a research activity, it is scarcely a decade old. The scientific data base that identifies and defines specific problems is only now beginning to mature. It is difficult to define and implement governmental policy and action on any problem in the absence of scientific data to support it. Lack of governmental involvement, particularly at the federal level, has been due in part to the problem's own immaturity. It has needed time to grow up.

There are also other factors in the slow pace of federal involvement. On a philosophical and practical basis, it is relatively easy to deal with problems such as ambient air pollution, water pollution, and hazardous wastes. These are problems involving the contamination of media external to private and public buildings. Risks associated with these are involuntary, and their control is seen as a public good.[1] Control burdens are placed only on sources that release particular contaminants to the environment. On a population-wide basis, regulation of such problems is relatively nonintrusive. Few individuals are actually burdened with regulatory requirements.

Air in buildings is another matter. There is a practical question of respecting property rights, particularly of individual homeowners. Public policy as it relates to indoor air quality must address questions touching on one of the most cherished of individual possessions and on the equally cherished right of privacy. The question of property rights is not an insignificant one.

Direct federal or other governmental involvement in regulation of indoor air quality is more likely to occur with those problems affecting public access buildings. Buildings such as offices, schools, restaurants, hotels, and the like are commonly open to the general public. Since health risks would be involuntary, control of indoor air quality in public access buildings by means of regulatory activities may be seen as in the public's interest. Numerous precedents, particularly at the state and local levels, have already been set in regulating various aspects of such buildings. For example, most public buildings must meet building, fire, and safety codes, both during and after construction.

The imposition of regulatory requirements that are intended to protect and enhance the quality of indoor air in public access buildings would not be breaking new ground. Indeed, in specific cases such requirements already exist. In many states and localities, ventilation standards are incorporated into building codes.[2] Regulation of smoking in public access buildings is widespread, widely accepted, and growing.[3] A precedent for direct federal involvement in regulating an indoor air quality problem in public access buildings involves asbestos in public and private schools. Based on EPA regulatory requirements promulgated in 1982,[4] all public and private schools had to be inspected for asbestos. More recent regulations[5] require that all schools be inspected by certified individuals using an EPA-approved protocol, and that schools develop an asbestos control program. At this writing, it appears that this requirement will be extended in a few years to all public access buildings.

RESEARCH AND TECHNICAL ASSISTANCE

In the early history (pre-1970) of ambient air pollution control, federal involvement focused on research to define the nature of the problem, technical assistance, and money for developing state and local air pollution control programs. Present federal involvement in indoor air quality control differs only in that no federal money is available for state and local programs. What the future may bring is unknown, but it is likely that some support of state and local activities will be needed.

Given budget constraints, federal support for indoor air quality research has been limited but growing. A variety of federal agencies have conducted or are supporting indoor air quality research activities. These include the EPA, the Department of Energy (DOE), the Department of Housing and Urban Development (HUD), the Consumer Product Safety Commission (CPSC), the National Institute of Occupational Safety and Health (NIOSH), the Bonneville Power Administration (BPA), the Tennessee Valley Authority (TVA), and others. Interestingly, research funding by the private sector has been significant, including the Electrical Power Research Institute, the Gas Research Institute, the American Tobacco Institute, and the Formaldehyde Institute, among others. Major research activities are also being funded by federal governments in Canada and Western Europe.

The technical assistance aspect of federal involvement is most clearly demonstrated in EPA's response to the radon problem.[6] In the area of radon, EPA sees its role primarily in the context of information development and delivery. EPA senior management, from various headquarters and regional offices, serves on the Radon Management Committee. This committee's responsibilities include:

1. identifying cost-effective mitigation technologies for existing homes
2. assisting states in developing programs
3. developing information materials
4. promoting good practices in radon measurement
5. assisting states in designing and conducting surveys to identify high-radon areas
6. conducting a national survey to determine the distribution of indoor radon levels
7. sharing with states and the private sector all technical knowledge
8. identifying cost-effective control techniques for new houses
9. developing measurement protocols
10. providing a quality assurance program for radon measurement

In practice, EPA has responded to public requests for information by developing and disseminating technical bulletins (specifically designed for laymen and free of charge) that describe the nature of the problem, associated risks, methods of air testing, and effective mitigation techniques. It has conducted surveys to assess radon levels nationwide and provides considerable technical assistance to state radiation programs. EPA has developed a program that approves testing laboratories that provide analytical services for passive radon samplers.

REGULATORY STRATEGIES

Air Quality Standards

Air quality standards form the basis for most regulatory ambient air pollution activity. In the air quality standard approach, numerical limits are placed on air contaminant levels. The limit is set, in theory, low enough to

Table 7.1   Formaldehyde Standards and Guidelines

	Level (ppm)	Status
Occupational		
USA (OSHA)	1.00	Promulgated
USA (NIOSH)	0.10	Recommended
Residential		
USA (HUD—mobile homes)	0.40 (target)	Recommended
USA (ASHRAE)	0.10	Recommended
USA (California)	0.05	Recommended
Denmark	0.12	Promulgated
Netherlands	0.10	Promulgated
Sweden	0.20	Promulgated
West Germany	0.10	Promulgated
Finland	0.12	Promulgated
Italy	0.10	Promulgated
Canada	0.10 (action level); 0.05 (target)	Promulgated
Other		
USA (submarines—U.S. Navy: 90-day continuous)	0.50	Promulgated
USA (NASA: 6-month continuous space travel)	0.10	Promulgated
USA (ambient air—AIHA)	0.10	Recommended
USSR (ambient air)	0.03	Promulgated

protect public health with a legislative requirement of an adequate margin of safety. In the United States, health-based air quality standards are developed on the assumption that there is some threshold dose (concentration/time) below which no adverse health effects occur. Difficulties arise in setting such standards because the scientific data base is usually inadequate to clearly reveal threshold values. Even the theory of threshold values is open to challenge. As a consequence, the standard setting process is plagued with uncertainties. Standards, whether they are applied to ambient air or the air of industrial environments, include scientific judgement as well as political and economic considerations.

Standards have the force of the law. If they are exceeded, the regulated entity is out of compliance and must implement measures to achieve compliance or face legal penalties. This is true for ambient air quality standards as well as standards for occupational health.

In theory, the air quality standard concept could be applied to regulate and control air quality in public access buildings and residences. It has been applied by the states of Minnesota and Wisconsin to control formaldehyde levels in new mobile homes.[7,8] Standards for formaldehyde have also been promulgated for residences by a number of Western European governments. Formaldehyde standards for occupational environments and for residences, as well as recommended health guidelines, are summarized in Table 7.1.

The development and setting of air quality standards and other regulatory activities associated with toxic contaminants has been a slow, difficult administrative process. Proposed regulatory activities are usually preceded by quantitative assessments that attempt to estimate health risks.[9]

The risk assessment process typically includes (1) hazard identification, (2) dose-response assessment, (3) exposure assessment, and (4) risk characterization. Hazard identification and dose-response assessment attempt to determine whether human exposure to a toxic contaminant is causally linked to observed health effects. In exposure assessment, human exposures to the toxic contaminant under real-world conditions are characterized. Risk characterization involves quantifying the magnitude and uncertainty of the health risks of a particular contaminant. Because of the nature of the process, federal responses to a toxic hazard associated with indoor air can be expected to be very slow, particularly if specific indoor air quality standards are to be promulgated.

In addition to the slow, arduous standard setting process, the application of the air quality standards in controlling indoor air pollution would be fraught with significant problems. One of the most notable of these is the compromising between health protection and political and economic considerations often involved. As a consequence, standards may not be sufficiently stringent to protect public health. Minnesota, for example, attempted to set an indoor air quality standard of 0.1 ppm formaldehyde in new mobile homes but had to settle on 0.5 ppm in a compromise with state legislators.[7] In another instance, the Department of Housing and Urban Development used a target level of 0.4 ppm as a de facto standard, claiming that it afforded reasonable health protection to occupants of new mobile homes,[10] even though health effects data indicated that formaldehyde exposures well below 0.4 ppm were sufficient to cause serious health effects. The target level primarily reflected the existing ability of the wood products industry to produce particleboard and hardwood plywood with relatively low formaldehyde emissions.

Another problem associated with the use of standards is that they carry with them a sense of implicit safety. The lay perception is that numerical values are absolute; that is, exposures below the standard are safe and exposures above are not. These perceptions result in a false sense of security in the former case and excessive fear in the latter. The true nature of the standard is often obscured.

In the case of ambient air quality standards, compliance is determined by monitoring community air and modeling specific sources. Monitoring ambient air quality is a significant undertaking in terms of time, personnel, and resources, but it has proven to be manageable; so too is the use of monitoring to assess compliance with occupational standards.

Assessing compliance with indoor air quality standards would be more difficult, both in philosophical and practical terms. Based on precedents established in regulatory programs for ambient air and occupational exposures, some measure of monitoring activity would be desirable to assess compliance. Such monitoring would not be practical in most cases. By its very nature, effective monitoring would be intrusive, particularly applied to residences.

Many homeowners would not be receptive to it. Passive monitors would be less intrusive; however, the quality of such test results would depend on the care and integrity of those placing them in their homes. In addition to the intrusiveness of monitoring and legitimate questions of privacy and property rights, it would be physically impossible to test the approximately 82 million residences in the United States for even one contaminant. Monitoring public access buildings would pose fewer problems from a standpoint of access, property rights, and privacy. In addition, such buildings are considerably less numerous than residences. Nevertheless, monitoring public access buildings for compliance with an air quality standard would be an enormous task. The burden of monitoring would have to be placed on building owners.

Air quality standards promulgated for formaldehyde levels in new mobile homes by the states of Minnesota and Wisconsin[7,8] were not enforceable because no resources were made available to determine compliance. Based on the Minnesota experience, Oatman[7] concluded that use of an air standard was not a reasonable approach to solving formaldehyde problems in Minnesota homes.

Emission Standards

Emission standards are used as both a strategy and a tactic in achieving the objectives of ambient air pollution control programs. As a strategy, emission standards are applied to sources in a source category, irrespective of existing air quality. These include New Source Performance Standards for new and significantly modified sources and National Emission Standards for Hazardous Air Pollutants (NESHAP).[11] Application of emission standards as a strategy is based on a desire to control contaminants when it is most cost-effective to do so and to limit emissions of toxic/hazardous pollutants to community air. Emission standards are also used as a tactic in achieving ambient air quality standards. In such instances emission standards will vary, depending on existing air quality.

The emission standards concept can also be applied to the control of indoor air pollution. Limits can be placed on emissions of toxic substances from products. Such standards are product standards. Product standards have been applied to hardwood plywood paneling and particleboard to be used in the manufacture of mobile homes in the United States.[10] They are also used in West Germany and other Western European countries in order to achieve compliance with indoor air quality standards and guidelines.[12]

Beyond formaldehyde, product standards could conceivably be applied to control emissions of VOCs from a variety of products. As an example, limits could be placed on total VOC emissions from paints, varnishes, and lacquers designed for indoor use and from adhesives and coating materials used in arts and crafts.

Product standards have the advantage of relative simplicity. The burden of compliance is placed on manufacturers, who must provide proof of compliance before their product can be sold. However, product standards can only reduce future potential exposures and would not be applicable to the reduction of existing problems unless new products are used to replace older, higher-emission materials.

Application Standards

Significant indoor air contamination can occur when products are misapplied. Standards of performance and certification can be required of individuals and corporations who install products with the potential for causing significant indoor contamination by misapplication. The most notable example of this is pest exterminator services. The state of New Jersey, for example, adopted regulations to reduce misapplication of termiticides.[13] They require the presence of a certified applicator onsite and specify conditions under which organochlorine compounds cannot be used. Application standards vary considerably from state to state; many states have no standards at all.

In the United Kingdom, urea-formaldehyde foam has been used to insulate the walls of over 1.5 million households. Unlike the United States and Canada, where the product is seen as inherently hazardous, the U.K. view is that problems associated with the product are due to misapplications. To minimize misapplication, a relevant British standard specifies the formulation of the foam and mandates a code of practice for its installation. Companies installing the foam are required to have proper expertise, suitably trained personnel, and properly formulated foam.[14]

Application standards can be set by a regulatory agency with authority to promulgate standards and enforce compliance. They can also be voluntary, established by the affected industry or trade association.[15] Voluntary application standards are self-enforced and depend on the integrity of individual installers and corporate management.

Prohibitive Bans

Prohibitive bans are used as tactics in regulatory efforts to achieve compliance with air quality standards. For example, bans have been used to prohibit the use of high-sulfur coal in the city of Chicago, the use of apartment house incinerators in New York and other U.S. cities, and open burning of trash and leaves.

Bans can also be applied to products that may cause significant indoor air contamination and health risks. Several notable bans have been already applied to products in order to eliminate future exposures. These include the NESHAP ban[16] on friable asbestos materials in building construction, bans on urea-formaldehyde foam insulation (UFFI) in the United States and Canada, a ban on kerosene heaters in California,[17] and use restrictions on methylene chloride in paint strippers and on chlordane for termite control.

The original NESHAP ban on friable asbestos materials in building construction was initiated to minimize future emissions of asbestos fibers to community air during building demolition. The application of the ban before indoor air quality concerns about asbestos arose was fortunate, since it significantly reduced the risk of asbestos exposure to occupants of schools and other public access buildings and even residences constructed after 1978.

Urea-formaldehyde foam insulation was banned for use as a thermal insulating material in walls and ceilings of residences in Canada in 1980. A similar ban was promulgated in the United States by the CPSC.[18] It was subsequently voided after it was appealed to the Fifth Circuit Court of Appeals by the formaldehyde industry. A ban remains in effect on the product in Massachusetts.[19]

Partial or complete bans can be applied to products whose use in buildings is discretionary and a matter of lifestyle. Prohibitive bans are widely used to regulate tobacco smoking. They may completely prohibit smoking in a building or space such as an airplane cabin; more commonly, they limit it to certain areas.

Prohibitive bans, like product standards, have the advantage of simplicity. Their apparent simplicity is even greater than that of product standards since compliance with numerical limits does not have to be assessed.

Prohibitive bans are an attractive tool for preventing future indoor exposures from contaminants associated with specific products. The application of a ban can, however, have significant economic repercussions. As a consequence, an industry faced with a proposed ban on one of its products (or, as is the case with urea-formaldehyde foam, its only product) can be expected to use all legal means to prevent the ban from taking effect. In their regulatory activities, federal agencies must conform to the Administrative Procedures Act,[20] which is designed to ensure that parties with an interest in the proposed regulatory activity are accorded full due process of the law. Such parties have the right to appeal a regulatory action. As a consequence, imposition of a ban and its final affirmation by the courts takes years. As was the case with urea-formaldehyde foam, the promulgation of a ban (or any other regulatory requirement) does not guarantee that it will continue to remain in effect.

Warnings

If a product is known to be hazardous or potentially hazardous, the manufacturer has a legal responsibility to warn consumers. Such warnings are usually in the form of a label on the product describing the conditions under which the product can be safely used and/or some of the health risks associated with the use of the product. Typically, products such as paint strippers, oil-based varnishes, lacquers, and paints have warning labels advising consumers to apply them in ventilated spaces. Labels on kerosene heaters also warn consumers to use adequate ventilation. Warnings are also applied to wood products that release formaldehyde and to structures such as mobile homes that have elevated formaldehyde levels associated with them.[10]

In theory, a consumer reading the warning will heed it by using the product in the recommended manner or avoiding the product entirely. In practice, few consumers are likely to read the warning, and fewer yet will respond to it in any positive way. For example, sales of new mobile homes have apparently not diminished since HUD required mobile home manufacturers to prominently display warning labels on their new products and to include information on formaldehyde hazards in the owner's manual. Warnings are applied to all pesticide formulations, yet misapplications and personal injuries associated with home pesticide use are not uncommon.[21] The use of warnings to prevent or reduce indoor contamination or exposures of sensitive individuals to toxic contaminants such as formaldehyde appears to have been ineffective. At best, they reduce manufacturers' legal liability in personal injury suits.

NONREGULATORY STRATEGIES

Health Guidelines

An alternative approach to the use of air quality standards to achieve and maintain acceptable air quality in public access buildings and residences is the specification of health guidelines by governmental agencies or professional groups. Health guidelines do not have the force of law; compliance with them is entirely voluntary. Their principle value is that their development and publication are not as cumbersome as the standard setting process and are, in theory, less subject to the latter's inherent political and economic compromises. Therefore, health guidelines are more likely to reflect true health risks and public health needs.

Though not enforceable, health guidelines do have considerable weight. They have the power of scientific consensus, and, in the case of governmentally promulgated guidelines, they carry the sense that the government believes that levels above the guidelines are unsafe. This sense is strongly felt in the context of the EPA recommended guideline for indoor radon. Many individuals will take remedial steps, or at least seriously consider them, based on results of their personal monitoring activities as compared to the EPA guideline. Health guidelines are also likely to affect manufacturing decisions. Corporations, though they may disagree with a guideline, are under both moral and marketing constraints to produce safe products.

The most common indoor air quality health guidelines are those that have been promulgated for radon by the United States, Canada, and countries in Western Europe. The promulgation of radon guidelines has been relatively simple, since radon is a natural contaminant and its regulation affects no one adversely. In the United States, the guideline was set in response to radon contamination of residences associated with uranium mill tailings and phosphate spoils.[22] It reflected both the need to provide health protection and the practicalities of mitigation. The 4-pCi/L guideline, below which most lay individuals believe they are safe, ironically still carries a very high risk of developing lung cancer, on the order of 2–5%.

Guidelines recently developed and issued by the Canadian Department of National Health and Welfare[23] provide an example of governmentally sanctioned health guidelines. Numerical exposure limits were set for formaldehyde, radon, and relative humidity. For other contaminants or contaminant sources, such as biological agents, consumer products, fibrous materials, lead, polycyclic aromatic hydrocarbons, and tobacco smoke, guidelines for limiting or reducing exposure were prescribed. The health or exposure guidelines for formaldehyde were formulated in terms of what was considered to be the practicable action level of 0.10 ppm (120 $\mu g/m^3$) and a longer-term target level of 0.05 ppm (60 $\mu g/m^3$).

Health guidelines for indoor air have been developed by the American Society of Heating, Refrigerating, and Air-Conditioning Engineers (ASHRAE) and published in ASHRAE Standard 62–1981.[24] Exposure guidelines for contaminants of indoor origin include $CO_2 - 4.5$ mg/m^3 (2500 ppm); chlordane $- 5$ $\mu g/m^3$; formaldehyde $- 120$ $\mu g/m^3$ (0.10 ppm); ozone $- 100$ $\mu g/$ m^3 (0.05 ppm); and radon $- 0.01$ Working Level. Guidelines were also recommended for contaminants that might be drawn into a building as part of ventilation air. These are the seven contaminants for which national ambient air quality standards have been promulgated and an additional 27 contaminants regulated under occupational health and safety programs.

Health guidelines have been recommended for exposures to pesticides by a committee of the National Research Council.[25] Exposure guidelines include 5 $\mu g/m^3$ for chlordane, 2 $\mu g/m^3$ for heptachlor, and 10 $\mu g/m^3$ for chlorpyrifos. These were developed in response to health concerns associated with the application of pesticides for termite control. Their values are 10% of the limits applicable to occupationally exposed persons.

Ventilation Guidelines

Acceptable indoor air quality can be achieved for certain air quality problems in mechanically ventilated buildings by application of ventilation guidelines or standards. In the United States, ventilation guidelines have been recommended by ASHRAE. These guidelines or standards do not have the force of law, unless they are included in state or local building codes. The ASHRAE guidelines for ventilation are widely employed by architects whether they are specified in a code or not. Ventilation guidelines usually specify some minimum ventilation rate that provides occupants with a reasonably comfortable environment with minimal sensory perception of human odor by those who enter the building.

Both ventilation guidelines and the day-to-day operation of ventilation systems in buildings have been compromised to some degree in the past 15 years by energy conservation concerns and practices. The provision of ventilation air from the ambient environment imposes significant heating and cooling costs. Reducing the amount of outside air used for ventilation minimizes such costs. In 1973, in response to energy needs, ASHRAE reduced a previously recommended ventilation guideline for office environments from 10 CFM (5 L/sec)/person to 5 CFM (2.5 L/sec)/person. This ventilation rate was contained in ASHRAE Standards 62-73 and 62-1981. In the revision process (currently under way), ASHRAE is expected to recommend a ventilation rate of 15 CFM (7.5 L/sec)/person. These recent changes reflect the fact that reductions in ventilation rates to satisfy energy conservation concerns had the unintended effect of significantly compromising indoor air quality and comfort.

In the United States, ventilation guidelines address only mechanically ventilated buildings. They specify a volumetric flow rate of outdoor air per person. Outdoor air needs are determined by occupancy. According to ASHRAE Standard 62-1981, sufficient outdoor air must be provided per person to insure that metabolically augmented CO_2 levels do not exceed 2500 ppm. In the anticipated revisions, the volumetric flow rate (15 CFM/person) will ensure that CO_2 levels do not exceed 1000 ppm. In the five Scandinavian countries where considerable emphasis has been placed on constructing very energy-efficient houses, building codes require provision of 0.5 ACH.[27] These Nordic guidelines were adopted as minimum ventilation rates by Canadian authorities in their R-2000 Energy Efficient Home Program.[28]

The biggest limitation to ventilation guidelines is the lack of awareness by building owners of the need for providing adequate outdoor air when the building is being operated. As was discussed in Chapter 1, NIOSH staff conducting over 300 health hazard evaluations of public access buildings concluded that inadequate ventilation was the cause of indoor air quality complaints in 50% of the buildings investigated.[29] In many instances, the outdoor air supply had been cut off completely. Even though building systems are designed to provide sufficient outdoor air, the problem is often one of operation. There are temptations to limit outdoor air to reduce energy costs. In many instances, building operating personnel do not have the technical understanding to operate the ventilation system properly.

Public Information/Education

A low-cost, nonregulatory approach to dealing with indoor air quality is the development and operation of public information and education programs. One such program is already in effect in the United States for radon. It assumes that individuals who wish to determine whether they have excessive radon levels in their homes (and what mitigation techniques are available) can make informed decisions from information contained in public service bulletins prepared by EPA and by contacting EPA staff.

The public information approach has been used since 1978 in EPA's Asbestos-in-Schools Program. In addition to the documents for school officials and service personnel, EPA staff are available to respond to citizen questions about asbestos by means of an asbestos hotline in each of the regional offices and in Washington. Each regional office maintains an asbestos coordinator, who among other things is responsible for responding to information requests.

Because of the reluctance of governmental regulatory agencies to intrude on individual privacy and property rights, the burden for identifying and mitigating indoor air quality problems in residences is likely to fall on homeowners and tenants. A need exists to inform such persons as to the nature of the problems, how they can be identified, and cost-effective solutions. They also need to know what governmental and private resources (e.g., testing services, consultants, etc.) are available. A further need exists to train individuals who would provide such services. Training would also be desirable for personnel of local health agencies where consumer requests are first directed.

GOVERNMENTAL RESPONSE

Federal Regulatory Authority and Programs

A variety of federal agencies have limited regulatory authority and involvement in indoor air quality. For the most part, programs are piecemeal, evolving in response to problems as they arise. No single agency has comprehensive regulatory authority. Because of its long-term experience and expertise in air quality, EPA has been recommended to be the lead agency in federal responses to the problem.[30]

The EPA, under the 1970 Clean Air Act Amendments,[11] has authority to conduct research on any aspect of air quality and in theory can regulate any substance with the potential for contaminating the ambient (outdoor) air. It is under the latter authority that it banned the use of friable asbestos materials in buildings. Under the Federal Environmental Pesticide Act, it can regulate the registration and use of pesticides, both indoors and outdoors.[31] Under the Toxic Substance Control Act (TSCA), it can regulate the use of any toxic substances, including those that may contaminate indoor air.[32] Under TSCA, EPA has been reviewing the scientific data base to determine whether it should regulate the use of formaldehyde. Under the Uranium Mill Tailings Radiation Control Act, it has authority to regulate the use of radiation-bearing wastes as fill for residential areas and in the construction of buildings;[33] EPA health guidelines for radon were established under authority provided in this legislation.

The Occupational Safety and Health Administration (OSHA) sets and enforces health and safety standards for workers. Its authority covers most occupational environments, including offices. Presumably, such building concerns would come under OSHA's purview, but OSHA involvement has been negligible. Occupational health and safety concerns have primarily been addressed by the National Institute of Occupational Safety and Health. Its involvement has been focused on conducting health hazard evaluations of problem buildings and advising OSHA.

The DOE is involved in conducting research on indoor air quality problems, particularly as they relate to energy conservation. Its involvement has developed out of concerns that energy conservation measures recommended and even financed under DOE programs may exacerbate indoor air quality problems and increase health risks to building occupants. A major concern of BPA and TVA is the potential adverse effects that their subsidized home weatherization programs may have on indoor air quality and human health.[34]

The CPSC has statutory authority to regulate the use of consumer products that may pose unreasonable health and safety risks to consumers. Under this authority, the CPSC has banned the use of asbestos-containing spackling compounds, attempted to ban urea-formaldehyde foam insulation, and initiated investigations of formaldehyde emissions from pressed wood products, unvented combustion appliances, and contamination of humidifiers by microorganisms.

HUD is empowered to establish building standards for HUD-financed projects, as well as manufactured houses. HUD has the responsibility for ensuring that mobile homes are habitable. Under this authority, HUD has promulgated product standards for particleboard and hardwood plywood for use in mobile home construction.[10] HUD also has limited the availability of FHA-financed loans for home construction in high-radium areas of Montana and South Dakota and phosphate lands in Florida.1 In 1983, Congress mandated establishment of the Interagency Committee on Indoor Air Quality. Under the chairmanship of a representative from EPA, this committee was charged with coordinating federal activities. Other agencies represented are the Departments of Health and Human Services, Housing and Urban Development, Education, Defense, Justice, and Transportation; GSA; the Federal Trade Commission; the National Aeronautics and Space Administration; OSHA; the Tennessee Valley Authority; and the National Bureau of Standards.[1]

State and Local Programs

Homeowners and occupants of other buildings who suspect that health complaints are associated with their building environment usually seek assistance from local and state health departments. Such agencies are in the forefront of response to citizen indoor air quality concerns and needs. These agencies have conducted investigations of complaints, provided testing services, and directed consumers to sources of information.

The first contact for an indoor air quality complaint or request for information is the local health department, which is probably ill-equipped to respond. As a consequence, the local health department usually refers such individuals to a state agency. At the state level, indoor air quality responses are typically the domain of industrial hygiene, consumer protection, and air quality programs. Because of their background and experience in monitoring air quality in industrial buildings, industrial hygienists are often chosen to direct state indoor air quality response programs. This has the benefit of having experienced individuals conduct investigations of complaints. On the negative side, many industrial hygienists have been conditioned by training and experience to embrace the premise that contaminant levels below occupational standards or Threshold Limit Values (TLVs) pose few, if any, health risks. With few exceptions, nonindustrial contaminant levels are substantially below occupational standards and TLVs.

Notable state indoor air quality response programs have been developed in Minnesota, Washington, California, Texas, West Virginia, Wisconsin, Ohio, New Jersey, and Massachusetts. Minnesota has led the nation in developing indoor air quality standards for formaldehyde, requiring warning labels on formaldehyde-releasing products, and developing a comprehensive program to regulate smoking in public buildings.[7,35] Minnesota also has one of the most active consumer response programs related to formaldehyde complaints.

California has one of the largest indoor air quality programs. Though described as comprehensive, it has no regulatory authority or responsibilities. Its primary mission is to conduct indoor air quality research, to respond to building occupant complaints, and to serve as the primary resource for indoor air quality information in the state of California.[36]

Indoor air quality programs at the state level vary widely. In the states described above, full-time staff is assigned to the problem, but in most states, indoor air quality is a part-time activity for one or more individuals who have other major responsibilities. Activity levels are usually determined by the "crisis" of the moment, whether it be formaldehyde, sick buildings, pesticides, or radon. Radon programs may be separated from other indoor air quality activities, as radon responsibilities fall to offices or programs of radiation protection. As a consequence, few state programs are comprehensive in scope.

A significant need exists for the development of indoor air quality programs at the state and local levels. This is where individuals are more likely to seek assistance and where such assistance, in theory, would be sufficient to solve specific problems.

CIVIL LITIGATION

In the absence of governmental regulatory programs, individuals who have been personally injured or suffered property damage and diminished property value may seek legal redress through the civil court system. They can, under a number of legal theories, file suit seeking monetary damages for the injuries claimed. These theories include expressed warranties, implied warranties, negligence, and strict liability.[37]

Expressed warranties are made by sellers of products and appear in sales contracts, labels, advertising, or samples. They are positive representations of the product. Liability depends only on the falsity of the representation and not on any particular knowledge of fault by the seller. Filing liability claims for breach of expressed warranty is an attractive way to pursue a personal injury claim because of its relative simplicity.

Implied warranties are interpreted by courts to exist even when a product seller makes no expressed warranties. Products are implicitly warranted to be fit for the ordinary purposes for which they are to be used.

Negligence is one of the most widely used product liability principles. Negligence is defined as a failure to exercise due care, the degree of care that would be exercised by a "reasonable person." Individuals or corporate entities may be found negligent in the performance of a service or in the manufacture of a product. Claimants or plaintiffs must prove that the defendant's actions or conduct was unreasonable.

Strict liability applies to claims of defective products. Such products may be defective by either manufacture or design. It does not depend on "fault," as does negligence. It does not question the conduct of the manufacturer but the safety of the product itself. It typically applies only to products, but in some jurisdictions it applies to buildings as well. It is one of the easiest theories under which damages can be recovered.

A variety of toxic torts claiming personal injury and/or property damage have been or may be filed.[37-39] These may include claims against:

1. architects who may be deemed negligent in designing a building without adequate ventilation or specifying the use of an unsafe product
2. manufacturers and vendors of urea-formaldehyde–based products that have high formaldehyde emission rates and cause health effects
3. manufacturers of unvented space heaters who fail to warn that the heater should be used only in well-ventilated spaces
4. owners and operators of public access buildings who operate ventilation systems without adequate outdoor air, use pesticides unsafely, etc.
5. sellers of residences who fail to warn buyers of a known high radon level

The use of toxic torts to recover damages for personal injury or property damage associated with indoor air pollution is fraught with problems. Litigation is very expensive and the outcome is very unpredictable, particularly when the case goes to trial. In most instances, injured parties cannot afford the costs. Access to the courts, however, is often facilitated by the use of contingency fee arrangements by the trial attorney. In the contingency fee system, the plaintiff's counsel may agree to try the case with his/her fees contingent on the outcome of the claim. Such arrangements vary, but typically the trial attorney receives 25–50% of settlements or jury awards after expenses are deducted. Expenses of experts and costs of legal discovery may be borne "up front" by the plaintiff's counsel or by the plaintiff.

On the positive side, most claims (90 + %) are settled out of court. Settlement reduces the burden on courts and costs to defendants and trial attorneys. In most cases it ensures an equitable recompense of the plaintiff's claim. It also reduces the lengthy time period (up to six years or so) often required to conclude a case and relieves the plaintiff of the emotional pain and vagaries of the trial process.

Based on the author's experience as an expert witness in personal injury cases involving formaldehyde, a successful outcome in court requires more than the truth and justice inherent in one's claim. Other factors come into play, including the relative skills and preparation of the plaintiff's and opposing counsel; the plaintiff's ability to testify, respond to cross-examination, and evoke jurors' empathy and sympathy; the abilities of the plaintiff's experts; the judge's rulings on the admissibility of evidence; and prevailing attitudes of the citizenry toward toxic torts in a given judicial district.[40]

Claims brought by smokers are especially difficult to try. Cigarette smoking is a significant, known health hazard. Its effects may be difficult to distinguish from those claimed to be associated with a toxic indoor contaminant. Moreover, jurors may hesitate to award damages to an individual who knowingly subjects him- or herself to a health hazard that was designated as such by the Surgeon General over 30 years ago.

A major risk to the plaintiff in trial is that burdens of proof are different for plaintiffs and defendants. The plaintiff must prove causation. The defendant, on the other hand, only has to create uncertainty in the minds of jurors. In addition to this, defense tactics usually focus on limiting the admissibility of evidence and the scope of an expert's testimony. How well a plaintiff does in this area depends on the trial judge and the skills of the plaintiff's counsel.

The most significant areas of toxic tort litigation associated with indoor air quality have been claims of homeowners against manufacturers, distributors, and installers of urea-formaldehyde foam insulation; manufacturers and retailers of mobile homes; and wood product manufacturers and users. Approximately a thousand or so claims alleging personal injury or property damage associated with formaldehyde in residences have been filed in the past decade. Attorneys have used all forms of the previously described legal theories in pursing plaintiffs' claims.

Taken collectively, personal injury claims have had a significant effect on products and on manufacturers. The litigation against formulators and installers of urea-formaldehyde foam was one of the main factors (the other being the attempted regulation by CPSC) in the demise of the U-F foam industry in the United States. Litigation against manufacturers of mobile homes and makers of U-F resin–based wood products has given impetus to the significant reductions in emissions of formaldehyde from wood products that occurred from 1977 to 1984. It has pushed manufacturers to improve their products and apply warning labels to them. Ironically, wood product manufacturers advocated the relatively weak product standards subsequently promulgated by HUD for new mobile homes and warning labels as a defense against future litigation. The intent and effect of the regulations, therefore, were not necessarily to protect consumers, but to provide a mode of legal protection to formaldehyde-emitting product makers and other wood product fabricators.

SELECTED POLICY AND REGULATORY HISTORIES

The development and implementation of indoor air quality policy and regulatory programs by governmental agencies has occurred in response to specific contaminant problems. Most notable of these are asbestos in schools, formaldehyde in residences, and cigarette smoking in public access buildings. Each has had a unique history. They reflect a variety of policy and regulatory approaches.

Asbestos

Asbestos has a regulatory history that precedes that of indoor air quality. In 1973, EPA officially designated asbestos as a hazardous substance under Section 112 of the Clean Air Act[16] and promulgated regulations to reduce community exposures. One of the areas of concern was asbestos release from asbestos-containing building materials during demolition. Under the NESHAP asbestos regulations, friable asbestos-containing materials (ACM) have to be removed prior to building demolition or renovation. Such removal must be conducted with wet techniques.[16] In addition, the use of friable ACM was banned under NESHAP.

Occupational exposures to asbestos are regulated by OSHA, which has established three individual sets of regulatory requirements.[41] The OSHA construction industry standard covers workers engaged in demolition and construction projects and related activities such as removal, encapsulation, alteration, repair, maintenance, insulation, transportation, disposal, and storage of ACM. General industry standards cover all work conditions under which asbestos exposure may occur, including simple occupation of buildings containing ACM. OSHA regulations cover all private sector employers and employees. The present exposure limit is 0.2 fibers per cubic centimeter (f/cc), averaged over an eight-hour period. The OSHA respirator rule specifies the type of respirator that must be worn to protect employees from excessive asbestos fiber exposure.

In 1978, concern over asbestos exposures from damaged and deteriorating ACM in schools prompted the development and implementation of an EPA program that provided guidance and technical assistance to state and local governments in identifying and mitigating asbestos hazards. This program was conducted in cooperation with the Department of Health, Education and Welfare and with OSHA. Guidance documents explaining how to identify and control deteriorating friable ACM in schools were prepared and distributed. A reporting system was established to collect information on the results of inspections and corrective actions taken.[42]

After EPA had the technical assistance program in place, it began to initiate a rulemaking action in response to citizen petitions and a lawsuit filed by the Environmental Defense Fund. In an advance notice of proposed rulemaking, EPA suggested (1) surveys of all schools to determine whether they contained friable ACM, (2) exposure assessments for all friable ACM identified, (3) the marking of all friable ACM, (4) corrective actions for friable ACM for which exposure assessments revealed an unreasonable risk, and (5) periodic reevaluation of the friable ACM to determine whether additional corrective action was required.[42]

The advance notice of proposed rulemaking was published in 1979, a few months before a new presidential administration with a strong anti-regulatory bias was to take office. The rules were, as a consequence, substantially modified before promulgation in 1982.

In 1980 Congress passed the Asbestos School Hazard Detection and Control Act.[43] It empowered the Secretary of Education to establish procedures making federal grants available to (1) assist state and local educational agencies in identifying ACM in school buildings and (2) provide interest-free loans to local educational agencies to correct asbestos hazards. The Act's provisions were never implemented because Congress did not appropriate the necessary funds to support the programs it had mandated.[44]

In 1982,[4] EPA completed the rulemaking process on asbestos hazards in schools begun in 1979. The Friable ACM in Schools Rule required that all public and private elementary and secondary schools identify friable ACM, maintain records, and notify employees of the location of friable ACM. It also required that instruction be provided to employees on reducing exposures to asbestos and that the school's parent-teacher association be notified of the results. EPA dropped its initial proposal to require corrective action if exposure assessments indicated that friable ACM posed an unreasonable health risk. It was apparently the view of EPA administrators at the time that pressure from teachers and parents would compel local education associations (LEAs) to initiate abatement actions when they were necessary.

The response of LEAs to the 1982 school asbestos rule was mixed. Serious questions arose concerning the adequacy of inspection procedures and a need for managing asbestos problems. As a consequence, Congress amended the Toxic Substances Control Act in 1986. The new amendments, entitled the Asbestos Hazard Emergency Response Act (AHERA), required EPA to promulgate rules regarding:

1. inspection of all public and private school buildings for ACM
2. identification of circumstances requiring response actions
3. description of appropriate response actions
4. implementation of response actions
5. establishment of a reinspection and periodic surveillance program for ACM
6. establishment of an operations and maintenance program for friable ACM

7. notification of state governors of asbestos management plans (governors having the authority to approve or disapprove them)
8. transportation and disposal of waste ACM[45]

EPA promulgated final rules to implement AHERA on October 17,
1987.[5] Notable requirements beyond those specifically addressed in the Act
included (1) visual inspection and assessment of the physical condition of
friable ACM by accredited or certified inspectors, (2) bulk sampling of ACM
according to a specified protocol, (3) development of asbestos management
plans by accredited management planners, (4) identification of a designated
person in a school system responsible for implementation of an asbestos management plan, and (5) minimum training requirements for the LEA-designated
person and custodial staff.

The AHERA rules required that all LEAs be inspected by accredited
personnel and submit an asbestos management plan to a responsible state
agency by October 2, 1988. Because of the dearth of certified asbestos inspectors and managers, and slow response by LEAs, Congress amended AHERA
to extend the inspection and management plan deadline to May 1989 with an
implementation date for management plans of October 1989.[46]

EPA also evaluated the much larger problem (in terms of building numbers) of asbestos in all public buildings and the need for inspections and
management plans similar to school AHERA requirements. The review indicated that ACM was present in such buildings as well, but that it was less
prevalent than in schools.[47] EPA deferred action on this part of the asbestos
problem. Unofficially, EPA administrators reasoned that the extension of
AHERA requirements to buildings other than schools would result in competition for inspection and management planning services, which were badly
needed to complete the higher-priority school program.

Formaldehyde

Regulatory agency involvement in the formaldehyde contamination problem in the United States began in the late 1970s, when a rash of consumer
complaints of odor and ill health associated with urea-formaldehyde foam
insulation, mobile homes, and wood products bonded with urea-formaldehyde
resins occurred. Complaints were directed to state and local health agencies
and, at the federal level, to the CPSC. Health authorities in many states
responded by conducting investigations of complaints and providing formaldehyde testing services. Significant complaint investigation programs were
conducted in Minnesota, Wisconsin, Washington, New York, Texas, and
Ohio. Minnesota conducted formaldehyde investigations of almost 1000
homes between 1979 and 1981. As a result of such investigations, and activities
on the part of citizen groups, agencies in several states initiated attempts to
regulate the problem.

In 1980 the Commissioner of Health in Massachusetts issued regulations banning the sale of urea-formaldehyde foam insulation and requiring install-ers, distributors, or manufacturers to repurchase the product, based on the request of the owner of a UFFI building.[19] The Massachusetts ban was appealed and overturned at the appellate court level, but eventually was sus-tained on appeal to the Massachusetts Supreme Court.

A formaldehyde statute passed by the Minnesota state legislature in 1980 gave the Commissioner of Health authority to promulgate rules governing the sales of building materials and housing units made with urea-formaldehyde–containing building materials. It also required written disclo-sure prior to the sale of new housing units and building materials containing urea-formaldehyde resins. The legislation specified the language of the prepur-chase notification.[48] An attempt in 1981 to enact legislation establishing a minimum allowable level of 0.10 ppm in new homes proved unsuccessful. The Minnesota Health Commissioner then adopted an indoor air quality standard of 0.50 ppm for new housing units and UFFI installations. The standard was appealed by the manufactured housing industry. The health department's authority to issue an indoor air quality standard was upheld by the Minnesota Supreme Court, but the Court remanded the 0.50-ppm level back to the health department for reconsideration. The Health Commissioner, after reviewing the record, established a standard of 0.4 ppm that took effect in 1985. Subse-quently, the Minnesota formaldehyde statute was amended to establish prod-uct standards and repeal health department rules establishing the 0.4-ppm indoor air quality standard. Though an indoor air quality standard existed in Minnesota, the uncertainty associated with legal challenges made it unenforce-able. In 1980 Wisconsin also promulgated an indoor air quality standard which was applicable to new mobile homes,[8] but the standard was apparently never enforced.

The CPSC was receiving complaints from homeowners as early as 1976. In 1977 a consumer petitioned it to ban formaldehyde-releasing particleboard products. CPSC denied the petition on the basis that it had insufficient infor-mation on which to act. Subsequently, CPSC received numerous complaints associated with UFFI installations. In response, CPSC began its own investi-gations, which led to proposed rules in 1980 that would have required potential purchasers of UFFI to be given notice of some of the possible adverse effects associated with the product. However, the Commission subsequently con-cluded that labeling or information disclosure would not adequately protect the public and banned the product in 1982.

The regulatory ban of UFFI was opposed by the foam insulation industry as well as the larger formaldehyde industry, united under the organizational title of the Formaldehyde Institute. The Formaldehyde Institute was primarily organized to protect the economic interests of formaldehyde producers and users of urea-formaldehyde resins. The broader industry concern was that a regulatory ban on one formaldehyde product would precipitate a ban or severe regulatory actions on other formaldehyde and urea-formaldehyde containing products.

In promulgating the ban, CPSC denied a petition by the Formaldehyde Institute to establish a mandatory standard for UFFI resin formulation and application, reasoning that the product was inherently dangerous and application standards would be insufficient to protect public health.[18] The Formaldehyde Institute appealed the ban to the historically industry-friendly Fifth Circuit Court of Appeals in New Orleans, where the ban was voided on procedural and technical grounds. The Commission elected not to appeal to the Supreme Court. The ban, adverse publicity, and litigation caused the UFFI industry to collapse. It has never recovered.

As a consequence of its regulatory work on UFFI, the CPSC began an investigation of pressed wood products, which have even higher formaldehyde contamination potentials than UFFI. It has not, however, followed up these activities with regulatory action and there are apparently no plans to do so.

The EPA was also petitioned to regulate formaldehyde under authority granted to it in TSCA. EPA administrators in the early '80s were adverse to regulating economically important toxic substances. As a consequence, EPA decided in 1980 that formaldehyde did not merit consideration under Section 4(f) of TSCA. In response, the Natural Resources Defense Council and the American Public Health Association filed a lawsuit against the EPA, seeking to force it to reconsider its decision. After a significant upheaval in EPA administrative ranks and a court order, the agency reversed itself in 1984 and designated formaldehyde for priority consideration under Section 4(f) of TSCA. It further issued an advance notice of proposed rulemaking, which initiated an investigation of regulatory options on formaldehyde.[49] Section 4(f) applies only to health effects, such as cancer, gene mutations, and birth defects. Chemicals given priority status for possible regulation under Section 4(f) are considered to pose potentially high risks to people. To date, EPA has taken no regulatory action against formaldehyde under Section 4(f) of TSCA nor under any other authority.

At the behest of the wood products and manufactured housing industries and public interest groups, HUD began to develop product standards for particleboard and hardwood plywood paneling to be used in the manufacture of mobile homes. The product standards specified that under standardized large chamber conditions of product load (loading factors similar to those of mobile homes), environmental temperature (25°C), and relative humidity (50% RH), emissions from particleboard and hardwood plywood could not exceed a concentration of 0.3 ppm for particleboard and 0.2 ppm for hardwood plywood.[10] Under combined loading conditions, it was projected that the formaldehyde concentration in new mobile homes would not exceed a target level of 0.4 ppm, a level that HUD administrators concluded would provide a reasonable degree of health protection. The HUD product standard reflected what the wood products industry was technically capable of achieving. (Minnesota product standards are similar to the HUD standards with the exception that product standards in Minnesota also apply to medium-density fiberboard [MDF].[48] HUD chose to defer action on MDF.)

In addition to product standards, the HUD rule required that mobile home manufacturers display a prominent health notice in the kitchen and include a notice in the owner's manual as well. The health notice was to read, "Some of the building materials used in this home emit formaldehyde. Eye, nose, and throat irritation, headache, nausea, and a variety of asthma-like symptoms have been reported as a result of formaldehyde exposure. Elderly persons and young children, as well as anyone with a history of asthma, allergies, or lung problems, may be at greater risk. Research is continuing on the possible long-term effects of exposure to formaldehyde."[10]

Recognition of the problem of formaldehyde contamination in residences occurred in several Western European countries before North America. By 1980, four countries[50] had indoor air quality standards for formaldehyde: Denmark (0.12 ppm), Holland (0.10 ppm), West Germany (0.10 ppm), and Sweden (0.4 ppm). These countries, as well as Finland, Norway, and Great Britain, limited emissions from particleboard as determined by the Perforator Method. Denmark and West Germany had emission limits that described product classes. In Denmark the maximum permissible emission value for particleboard and hardwood plywood corresponded to 0.12 ppm, the indoor air quality standard. West German product standards described three classes: E1-Perforator value: < 10 mg/100 g; E2-Perforator value: between 10 and 30 mg/100 g; and E3-Perforator value: between 30 and 40 mg/100 g. The different product emission classes correspond to the following use limitations:

Class E1: Boards can be used without limitations.
Class E2: Allowed for indoor use if the boards have approved surface treatments on both surfaces.
Class E3: Allowed for indoor use if the boards have approved surface treatments on both surfaces and all edges are sealed.

By 1985, indoor air quality standards were issued in Finland (0.25 ppm) and Italy (0.10 ppm), and Sweden was anticipated to reduce its indoor air exposure limit to 0.20 ppm.[12] West Germany planned to limit marketing of emission classes E2 and E3 and reduce emissions on all products to 0.1 ppm. With the possible exception of Denmark, indoor air quality standards in most European countries using them have served, more or less, as guidelines.

Smoking in Public Places

Over the past decade or so, there has been a rising tide of regulatory activities to limit smoking in public places, including a variety of both public and private sector initiatives. These initiatives have reflected changing attitudes about smoking since 1964 and the acceptability of smoking in public places. More recently, antismoking efforts have been fueled by growing evidence of adverse health effects associated with passive or involuntary smoking and the publication and authority of the 1986 Surgeon General's Report.

Private polls show that a majority of American citizens now support the rights of nonsmokers and favor policies that ensure those rights.[3]

With few exceptions, restrictions on smoking in public places have come as a result of state and local activities. Until the recent restrictions on smoking in commercial aircraft, there were no federal restrictions on smoking in public places.

Regulation of smoking in public places by state and local governments has evolved from permitting a no-smoking section to requiring one to making nonsmoking the principal or assumed condition. Legislative language increasingly has made it clear that the specific intent of regulation was the safety and comfort of nonsmokers.

One of the most notable pieces of smoking legislation was Minnesota's Clean Indoor Air Act of 1975. It prohibited smoking in public places and meetings except in designated smoking areas. It covered restaurants, private worksites, and a large number of public places. It has served as a model for other state legislation and local ordinances.[51]

State laws regulating smoking vary significantly in their language and comprehensiveness. As of 1986, at least 41 states and the District of Columbia had some type of statewide restrictions on smoking in public places. A public place is generally defined as any enclosed area in which the public is permitted or invited. It includes public facilities such as governmental offices, schools, libraries, public transportation, and health care, cultural, and sport facilities. It also includes public facilities under private ownership.[3]

Most state laws restrict smoking to designated areas. In a few states (Florida, Georgia, Massachusetts, Washington) smoking in regulated buildings or places may be banned entirely. As of 1986, smoking at one's workplace was restricted for public sector employees in 22 states and for private sector employees in nine states. Workplace smoking laws usually contain provisions that (1) require a written policy, (2) limit smoking to designated areas, (3) require posting of signs, and (4) give preference to nonsmokers in resolving conflicts over the designation of a smoking area.

In most states the state health department (or in the case of a local smoking ordinance, the local health department) is responsible for enforcement of smoking policies.[3] Many laws include penalties for smokers who violate restrictions and for employers who fail to designate smoking areas. Typically, most laws are self-enforcing, depending on the individual's sense of duty to obey the law and on peer disapproval. Compliance with restrictions also assures the smoker the right to smoke somewhere in that enclosed environment. When a health agency takes an active enforcement role, it is usually in response to complaints.

Opposition to smoking restrictions varies. The tobacco industry has lobbied hard against antismoking legislation, in great measure because such legislation may result in reduced daily smoking, hence reduced individual tobacco consumption. In addition, smoking restrictions and official and implicit peer disapproval may cause smokers to quit. Opposition has also come from restaurateurs, some of whom believe that smoking restrictions may hurt business because of an apparent association between eating and smoking. Opposition has also come from some employers who are themselves smokers or who believe that smoking restrictions would cause labor-management problems.

Laws restricting smoking in public places have generally been implemented with few, if any, problems. They are apparently met with good compliance and have been well accepted by nonsmokers and smokers.[3]

Even prior to publication of the Surgeon General's Report on Involuntary Smoking in 1986, there was considerable legislative movement at the state and local levels to restrict smoking in public places. The Surgeon General's report, by the power of its authority, gave such activities an enormous new thrust. The increase in smoking restrictions and the comprehensiveness of such restrictions has been almost phenomenal. It certainly set the stage for the successful Congressional effort to ban smoking from commercial aircraft service on domestic flights under two hours in flight time. Remarkably, the commercial aircraft ban was passed by Congress despite opposition from powerful tobacco state senators.

Restricting smoking in public places has for the most part been a North American phenomenon. Regulation of smoking in public places in Europe and Japan is, at best, in its early infancy.

REFERENCES

1. Sexton, K. 1986. "Indoor Air Quality: An Overview of Policy and Regulatory Issues." *Sci. Tech. Human Values* 11:53–67.
2. National Research Council. 1981. "Control of Indoor Pollution." 450–504. In: *Indoor Pollutants.* National Academy Press, Washington, DC.
3. U.S. Surgeon General. 1986. "The Health Consequences of Involuntary Smoking." Department of Health and Human Services (Centers for Disease Control). 87–8398.
4. EPA. 1982. "Friable ACM in Schools: Identification and Notification Rule." *Fed. Reg.* 47:23360–23376.
5. EPA. 1987. "Asbestos-Containing Materials in Schools: Final Rule and Notice." *Fed. Reg.* 52:41826–41905.
6. EPA. 1987. "EPA Indoor Air Quality Implementation Plan. Appendix C: EPA Radon Program." EPA 600/8-81-033.
7. Oatman, L. 1985. "Minnesota's Response to Indoor Air Problems." Paper No. 85-46.3. In: Proceedings of the 78th Annual Meeting of the Air Pollution Control Association. Detroit.
8. Wisconsin Department of Industry, Labor and Human Relations. 1980. "Proposed Final Indoor Ambient Air Quality Standards for Formaldehyde as Amended."
9. Sexton, K. 1987. "Public Policy Implications of Exposure to Indoor Pollution." 105–116. In: *Proceedings of the Fourth International Conference on Indoor Air Quality and Climate.* Institute for Water, Soil and Air Hygiene. West Berlin. Vol. 4.

10. Department of Housing and Urban Development. 1984. "Manufactured Home Construction and Safety Standards." *Fed. Reg.* 49:31986–32013.
11. "Clean Air Act Amendments." 1970. USC Title 42, Sections 7403, 7640.
12. Sundin, B. 1985. "The Formaldehyde Situation in Europe." 255–275. In: T.M. Maloney (Ed.). *Proceedings of the 19th International WSU Particleboard/ Composite Materials Symposium.* Washington State University. Pullman, WA.
13. New Jersey Department of Environmental Protection. 1985. "Pest Control Code." New Jersey Administrative Code. Title 7, Chapter 30, Subchapter 10.
14. Llewellyn, J.W. and P.R. Warren. 1985. "Regulatory Aspects of Indoor Air Quality—A U.K. View." 478–487. In: D.S. Walkinshaw (Ed.). *Transactions: Indoor Air Quality in Cold Climates.* Air Pollution Control Association. Pittsburgh.
15. Hirsch, L.S. 1982. "Behind Closed Doors: Indoor Air Pollution and Governmental Policy." *Harvard Environ. Law Rev.* 6:339–394.
16. EPA. 1984. "National Emission Standards for Hazardous Air Pollutants: Asbestos Regulations." *Fed. Reg.* 49:13661–13670. Subpart M.
17. California Administrative Code. 1982. "Housing and Community Development." Title 25: Part 1, Chapter 1, Article 5, Section 32.
18. CPSC. 1982. "Ban of Urea-Formaldehyde Foam Insulation." *Fed. Reg.* 47:14366–14421.
19. State of Massachusetts. 1984. "Regulations Concerning Repurchase of Banned Hazardous Substances." 105 CMR, 650. 221–222.
20. "Administrative Procedure Act." 1966, PL 89-554. USC Title 5, Sections 551, 552.
21. Maddy, K.T. and C.R. Smith. 1984. "Pesticide-Related Human Illness as Occurring in California Between January 1 and December 31, 1983." California Department of Food and Agriculture. HS-1227.
22. EPA. 1979. "Indoor Radiation Exposure Due to Radium-226 in Florida Phosphate Lands: Radiation Protection Recommendations and Request for Comment." *Fed. Reg.* 44:38644–38670.
23. Armstrong, V.C. and D.S. Walkinshaw. 1987. "Development of Indoor Air Quality Guidelines in Canada." 553–558. In: *Proceedings of the Fourth International Conference on Indoor Air Quality and Climate.* Institute for Water, Soil and Air Hygiene. West Berlin. Vol. 3.
24. American Society of Heating, Refrigerating and Air-Conditioning Engineers. 1981. "Ventilation for Acceptable Indoor Air Quality." ASHRAE Standard 62-1981. Atlanta.
25. National Research Council. 1982. "An Assessment of the Health Risk of Seven Pesticides Used for Termite Control." National Academy Press, Washington, DC.
26. American Society of Heating, Refrigerating and Air-Conditioning Engineers. 1973. "Standards for Natural and Mechanical Ventilation." ASHRAE Standard 62-73. Atlanta.
27. Haberda, F. and L. Trepte. 1981. "Annex IX. Minimum Ventilation Rates— Literature Review, Standards and Current Research Summary 1." International Energy Agency.
28. Ficner, C., et al. 1985. "Improving Indoor Air Quality Through Energy Conservation Programs." 534–548. In: D.S. Walkinshaw (Ed.). *Transactions: Indoor Air Quality in Cold Climates.* Air Pollution Control Association. Pittsburgh.
29. Wallingford, K.M. and J. Carpenter. 1985. "Field Experience Overview: Investi-

gating Sources of Indoor Air Quality Problems in Office Buildings." 448–453. In: *Proceedings of IAQ '86: Managing Indoor Air for Health and Energy Conservation.* American Society of Heating, Refrigerating and Air-Conditioning Engineers. Atlanta.

30. General Accounting Office. 1980. "Indoor Air Pollution: An Energy and Health Problem." Report to Congress. CED-PO-111.

31. "Federal Environmental Pesticide Control Act." 1972. PL 92–516. USC Title 7: Sections 136, 136y; Title 15: Sections 1261, 1471; Title 21: Sections 321, 346a.

32. "Toxic Substance Control Act." 1981. USC Title 15, Sections 2601, 2629.

33. "Uranium Mill Tailings Radiation Control Act." 1978. PL 95–604. USC Title 42: Sections 2014, 2021–2022, 2111–2114, 2201, 7901, 7911–7925, 7941–7942.

34. Love, R. 1985. "Bonneville Power Administration's Risk Management Strategy for Mitigating Indoor Air Quality Effects in Its Home Weatherization Program." 549–560. In: D.S. Walkinshaw (Ed.). *Transactions: Indoor Air Quality in Cold Climates.* Air Pollution Control Association. Pittsburgh.

35. Oatman, L. and C.A. Lane. 1987. "Development of a Comprehensive Indoor Air Quality Program: The Minnesota Experience." 521–525. In: *Proceedings of the Fourth International Conference on Indoor Air Quality and Climate.* Institute for Water, Soil and Air Hygiene. West Berlin. Vol. 3.

36. Sexton, K. and J.J. Wesolowski. 1985. "Safeguarding Indoor Air Quality." *Environ. Sci. Technol.* 19:305–309.

37. Kirsch, L.S. 1986. "Indoor Air and the Law." *Architect Tech.* July/August. 37–43.

38. Davis, E.S. 1986. "Potential Liability for Indoor Pollution." In: Proceedings of the 79th Annual Meeting of the Air Pollution Control Association. Minneapolis.

39. Turiel, I. et al. 1984. "Legal and Regulatory Aspects of Indoor Air Pollution." 167–172. In: *Proceedings of the Third International Conference on Indoor Air Quality and Climate.* Swedish Council for Building Research. Stockholm. Vol. 1.

40. Godish, T. 1988. "Why Is Formaldehyde Litigation Declining?" *Indoor Poll. Law Rep.* February.

41. Secretary of Labor. 1986. "Occupational Exposure to Asbestos, Tremolite, Anthophyllite, and Actinolite: Final Rules." *Fed. Reg.* 51:22612–22790.

42. EPA. 1979. "Asbestos-Containing Materials in School Buildings; Advance Notice of Proposed Rulemaking." *Fed. Reg.* 44:54676–54680.

43. "Asbestos School Hazard Detection and Control Act." 1980. PL 96–270. Title 20: Sections 1411, 3601–3611.

44. Department of Education. 1981. "Asbestos Detection and Control: Local Education Agencies; Asbestos Detection and State Plan; State Educational Agencies." *Fed. Reg.* 46:4538–4558.

45. "Asbestos Hazard Emergency Response Act." 1986. PL 98–469. USC Title 15: Subchapter II, Sections 2641–2654.

46. EPA. 1988. "Asbestos-Containing Materials in Schools; Deferral of Deadline for Submission of Asbestos Management Plans." *Fed. Reg.* 53(148):29210–29213.

47. EPA. 1988. "EPA Study of Asbestos-Containing Materials in Public Buildings." Report to Congress.

48. Oatman, L. 1987. "Regulating Formaldehyde in Minnesota." 579–583. In: *Proceedings of the Fourth International Conference on Indoor Air Quality and Climate.* Institute for Water, Soil and Air Hygiene. West Berlin. Vol. 3.

49. EPA. 1984. "Formaldehyde: Determination of Significant Risk." *Fed. Reg.* 49:21869–21898.
50. Sundin, B. 1980. "Formaldehyde Testing Methods and Standards — Experience from Sweden." Consumer Product Safety Commission Technical Workshop on Formaldehyde. Washington, DC.
51. Thron, R. W., et al. 1987. "Regulating Smoking in Public Buildings and Workplaces." 589–593. In: *Proceedings of the Fourth International Conference on Indoor Air Quality and Climate.* Institute for Water, Soil and Air Hygiene. West Berlin. Vol. 3.

8   AIR QUALITY DIAGNOSTICS

Before an indoor air quality problem can be solved, its nature and cause must first be identified or diagnosed. Diagnosis requires the initiation of a process that will provide information requisite to the problem's resolution. One cannot, however, begin to identify a problem's cause without occupants of residences and other buildings first suspecting that they may have a problem or the potential for a problem.

Problem diagnosis must therefore begin with building occupants. They must become aware that a particular health problem or symptoms may be related to exposure to contaminated indoor air. Such awareness may come from printed articles, news stories communicated by the electronic media, treating physicians, or word of mouth.

Awareness of a potential indoor air quality problem may also come from an intuitive recognition of special circumstances or correlations. These may include, for example, observations that (1) complaints of irritation and other symptoms are common among other occupants, (2) an unpleasant odor is noticed on building entry, (3) symptoms are only associated with, or are more severe in, a particular building environment, and (4) symptom onset occurs with some building-associated change. Building-associated changes may include remodeling or renovation, pesticide application, acquisition of new furnishings, application of energy-conserving measures, and switching from air-conditioning to heating mode.

Once a building occupant or homeowner recognizes that an indoor air quality problem may exist, he/she must seek to have an investigation conducted. Investigations of indoor air quality complaints may be conducted by public health personnel, by private consultants, and in limited circumstances by individuals themselves. A good investigation is vital. In practice, the investigatory or diagnostic phase is the weak link in the whole process of solving an indoor air quality problem.

RESIDENCES

Relative to other indoor air quality problems, an indoor air quality investigation in a residence is a simple undertaking. This is due in part to scale. Residences are small, contain few individuals, and have relatively simple heating and cooling systems. Additionally, relatively few contaminants are responsible for most residential complaints or concerns. The most common residential indoor air contamination problems are formaldehyde, biogenic particles, and radon. Considerably less prevalent problems include carbon monoxide (CO), emissions from new carpeting, emissions from fiberglass heating duct materials, pesticides, sewer gas, and gasoline or solvent vapors.

Prior to a field investigation, it is desirable to conduct a limited telephone interview of the individual(s) requesting assistance to obtain a preliminary assessment of the problem. Most importantly, the problem should be characterized by symptom types; symptom initiation relative to time and events, seasonal patterns, and responses to weather changes; the presence of odors; the type and age of the dwelling; and other factors. Severe headaches, sleepiness, and nausea, for example, suggest CO poisoning. Upper respiratory irritation sensitive to changes in weather conditions suggests formaldehyde. A musty odor suggests a mold problem. This preliminary interview is important because it suggests to the investigator what the problem's cause may be and what air testing (if any) he/she should be prepared to do.

Formaldehyde

Formaldehyde contamination is the most common cause of residential health complaints. Its prevalence as a problem is related to the widespread use of wood products bonded with urea-formaldehyde resins used in the construction of single-family dwellings, apartments, condominiums, and cabinetry and furniture. Its complaint status is also related to the fact that the illness syndrome associated with it, unlike allergies and asthma, is not generally diagnosed and treated in traditional medical practice.

Formaldehyde as a cause of building-related health complaints is recognizable from the constellation of symptoms associated with it. Its symptom pattern is very sensitive to changes in indoor temperature and relative humidity; the presence of potent sources of formaldehyde, such as particleboard subflooring, hardwood plywood, and/or medium-density fiberboard; and the presence of elevated levels of formaldehyde.

In a residence where formaldehyde is a problem, symptoms typically include several or many of the following: eye, nose, and/or throat irritation; runny nose; sinus congestion; headaches; abnormal tiredness; difficulty in breathing; chest tightness; difficulty sleeping; diarrhea; and nausea. The illness syndrome usually affects one or more family members; it is not unusual for all to be affected. The greatest number of symptoms occurs in those who

Table 8.1 Recommended Formaldehyde Sampling Protocol for Residential Investigations of Building-Related Health Complaints

Sampling Consideration	Recommended Practice
Closure	Before sampling (minimum of 12 hrs) and during sampling
Indoor temperature	25°C 12 hr before and during sampling
Climatic factors	Sampling during moderate weather
Sample number	Three desirable, two minimum
Sampling duration	60–90 min, depending on expected concentration range
Environmental measurements	Indoor temperature, outdoor temperature, indoor relative humidity
Source identification	Identify major sources only
Sample storage	Room temperature unless temperature is unusually high (>27°C)

spend the most time at home (i.e., homemakers and very young children). The illness syndrome is easily distinguished from acute and chronic carbon monoxide exposures by its diversity of symptom response.

The higher the formaldehyde level is during the period of building-related illness, the greater the likelihood is that formaldehyde is the cause of the observed symptoms. Symptoms are not only more likely, they are also greater in number and severity. For example, individuals who live or have lived in a new mobile home complain of more symptoms and claim symptoms to be more severe than do dwellers in any other residential structure. This corresponds to the higher formaldehyde levels characteristic of mobile homes. Residents of urea-formaldehyde foam–insulated houses have, compared to residents of mobile homes, relatively few symptoms.

An indoor air quality investigation with formaldehyde as its focus consists of three distinct parts: (1) conducting an in-depth interview with a standard questionnaire, (2) air testing, and (3) an inspection of the dwelling for formaldehyde sources and sources of other potential contaminants. An example of a questionnaire applicable to a formaldehyde investigation can be found in the Appendix to this chapter.

Air testing for formaldehyde can be conducted using either passive or active sampling techniques. The former are relatively inexpensive and are useful when the homeowner wants, in the interests of saving money, to do air testing on his/her own. However, the value of passive samplers is limited; beyond a screening function, interpretation of results is often difficult for experts and almost impossible for lay individuals.

The quality and utility of test results depend on a standardization of test conditions and recording indoor and outdoor temperatures and indoor relative humidity. Because it is easier to standardize test conditions occurring at the time of the test, it is most desirable to conduct air testing by using an active sampling method, such as the NIOSH chromotropic acid method.[1] Residential formaldehyde testing protocols have been reviewed and critiqued by Godish.[2] A recommended protocol for conducting residential formaldehyde sampling is presented in Table 8.1. The most notable recommendations involve pretest closure for at least 12 hr, maintenance of a standardized temperature of approximately 25°C during the presampling and sampling closure period, and measurement of environmental conditions at the time of test.

It is very undesirable to test under conditions that produce results reflecting minimum levels of formaldehyde contamination. Test results from air sampling conducted when windows and/or doors are open are meaningless. To assess the acute health-affecting potential of formaldehyde in a residence most accurately, "near worst case" conditions of building closure and indoor temperature should be approximated. "Near worst case" refers here to the upper end of the range of normal living conditions of temperature and relative humidity. During much of the heating season and in air-conditioned houses, windows and doors are typically closed, except for normal in-and-out traffic. During testing, closure is needed to reflect such exposures and reduce variability associated with occupant practices.

Even with closure, considerable variability in formaldehyde emissions and indoor levels can be expected to occur as a consequence of indoor temperature, indoor/outdoor temperature differences, and indoor relative humidity. Of the three, indoor temperature has the greatest effect. It is, in most instances, the most easily controlled of the environmental conditions of importance. A standardized temperature at the high end of the range of normal living conditions for a particular household prior to and during sampling would represent a "near worst case" condition. Typically, this will be approximately 23–25°C, depending on regional climatic conditions and home heating/cooling practices.

In some cases, maintenance of the desired standard temperature may not be easily achieved. An example would be in a non–air conditioned house in the summertime. One solution to this problem would be to close the house up overnight and collect air samples in the very early morning hours. A second approach would be to defer testing until weather conditions are moderate enough to attain the desired test conditions. The third and least desirable approach would be to test the house under whatever temperature conditions happen to occur. Formaldehyde concentrations obtained from such testing would then be standardized by the Berge equation[3] or a similar standardizing equation. However, in a house environment where the temperature is increasing with time, formaldehyde levels would be expected to lag behind temperature changes; thus, the standardizing equation would have a tendency to overcompensate.

Even with the limitation just described, use of a standardizing equation in conjunction with formaldehyde air testing is attractive because it can be used for both temperature and relative humidity. Relative humidity can have a significant effect on formaldehyde concentrations and usually cannot be controlled by individuals conducting air testing. Within the temperature/humidity range of 20–30°C and 30–70% RH, the Berge equation has been reported to have an error rate of $\pm 12\%$ at two standard errors.[4] The Berge equation has the following form:

$$C_x = \dfrac{C}{[1 + A (H - H_o) \, e^{-R(1/T - 1/T_o)}]} \qquad (8.1)$$

where C_x = standardized concentration (ppm)
  C = test concentration (ppm)
  e = natural log base (2.718)
  R = coefficient of temperature (9799)
  T = indoor test temperature (°K)
  T_o = standardized temperature (°K)
  A = coefficient of humidity (0.0175)
  H = indoor test relative humidity (%)
  H_o = standardized relative humidity (%)

Recommended standard conditions for the above equation are an indoor temperature of 25°C and relative humidity of 60%.

In northern climates formaldehyde levels are at their lowest on cold winter days. Therefore, air testing during midwinter may not be particularly meaningful. The primary limiting factors during the winter period are low relative humidities and cold outdoor temperatures, resulting in large inflows of diluting outdoor air. The problem of low relative humidity can be accommodated using the Berge equation. The problem of low outdoor temperature can be accommodated using a standardizing equation for indoor/outdoor temperature differences.[5] It has the following form:

$$C = C_x + B\Delta T \qquad (8.2)$$

where C_x = standardized concentration (ppm)
  C = measured concentration (ppm)
  B = coefficient of the temperature difference (0.0015)
  ΔT = difference between indoor and outdoor temperatures (°F)

Air testing results only reflect concentrations present at the time of testing. In most instances, air testing is conducted after formaldehyde concentrations have declined. Concentrations may decline by as much as 50–90%, depending on the building type, source materials present and the extent of their use, prevailing climatic conditions, occupant use factors, and the time elapsed since construction or product manufacture. In general, test results do not indicate prior exposures that may have initiated the health problems or caused sensitization. Therefore, interpretation of formaldehyde test results must include, at a minimum, some qualitative or quantitative assessment of prior exposures. An expanded discussion of formaldehyde decay rates can be found in Chapter 3, pp. 129-132.

Identifying a formaldehyde contamination problem and acquiring the specific technical information necessary to recommend effective control measures requires that all major sources be identified. Major sources are limited to wood products such as particleboard, hardwood plywood, and medium density fiberboard (MDF); urea-formaldehyde foam; and urea-formaldehyde finish coatings on wood cabinetry, furniture, and flooring. Wood products bonded with urea-formaldehyde resins are by far the most important sources of indoor contamination. Source identification can be conducted at any time but is most convenient during the period of air sampling. Potent sources of formaldehyde are not always apparent; therefore, it is important that the inspection/identification process be conducted in a systematic way, with some foreknowledge of what to look for and where to look for it.

In most urea-formaldehyde foam–insulated (UFFI) houses, occupants know of urea-formaldehyde foam's presence because they contracted to have it installed. However, as a UFFI house changes ownership, new owners may not be apprised of its presence. The presence of UFFI may be suspected if the home (1) has a somewhat earthy odor to it, (2) is a bungalow-style or Victorian-style house in an older neighborhood, and (3) has port caps on the exterior cladding.

UFFI houses have a distinctive odor, even 10 or more years after installation. It is not a formaldehyde odor. UFFI was typically used in older, existing houses. Its use in new construction was relatively uncommon. In installing UFFI, holes were drilled in the exterior cladding. These entry ports were closed with plastic port caps, or with mortar in the case of brick houses. The presence of port holes and caps, however, does not by itself confirm the presence of UFFI, since cellulose insulation may have been installed in the same way, nor does the absence of entry holes and port caps preclude UFFI's presence in the building's sidewalls, because many homeowners elected to have their houses re-sided after UFFI installation.

There are several ways to confirm or rule out the presence of UFFI in a house. If port caps are present, they can be removed with a screwdriver to open up the port hole for inspection. UFFI's presence is revealed by a relatively soft, whitish to brownish material that, depending on the manufacturer and quality of installation, may vary from a spongy to granular texture. A rigid, foamy material would most likely be polyurethane. It has seen limited use in retrofit insulation applications. If ports are not present on the exterior cladding, evidence of UFFI's presence can be sought in and around electrical outlets located on exterior walls. In houses with basements, UFFI excretions are commonly found around the sole plate and other joints. In most UFFI houses, UFFI is the dominant formaldehyde source, with other sources being relatively minor. For purposes of mitigation, however, all potentially major sources need to be identified and assessed. After UFFI has been identified, the inspection process should proceed as it would for any conventional house.

Potentially major formaldehyde sources in conventional homes, apartments, and condominiums vary. The single most important source, particularly in modern construction (the past 20 years), is particleboard underlayment used as subflooring. It is a potent emitter and will result in formaldehyde levels exceeding 0.10 ppm when used in a high surface:volume ratio. Particleboard subflooring is present in millions of houses in the United States. In an individual house it may be used in a room or two, or throughout. The former is characteristic of home additions, the latter of new home construction. Particleboard underlayment is normally covered by carpeting and, in some cases, vinyl flooring, so its presence is not obvious. In homes with heating outlets present in the floor, its presence can be confirmed by removing the register and inspecting the wood materials around the outlet perimeter. Particleboard underlayment is usually placed over softwood plywood and so will not be evident when the bottom floor surface is viewed from a basement. In all-electric homes with wall-to-wall carpeting, inspection for particleboard will require partial detachment of the carpeting at inconspicuous points. Because the surface:volume ratio of particleboard subflooring will affect indoor formaldehyde concentration, it is essential that the quantity present be known so that mitigation advice appropriate to the circumstance can be provided.

Significant levels of formaldehyde in a conventional residence may also be associated with decorative wall paneling made from hardwood plywood or particleboard; Luan panels used for carpet underlayment; kitchen and bathroom cabinets made from particleboard, MDF, or hardwood plywood; and furniture made from particleboard, MDF, hardwood plywood, or solid wood.

The formaldehyde contamination level associated with decorative wall paneling depends on the extent of its use. If used throughout the house, it may result in significantly elevated formaldehyde levels (0.10–0.40 ppm). In common use, it is only applied in one or two rooms and will not result in significant contamination (< 0.10 ppm).

Occasionally, Luan panels are applied to existing flooring to provide a smooth surface for the application of new carpeting. Because of the load factor or extent of their use, the application of Luan panels for carpet underlayment in this way may result in significant indoor contamination. Inspection for such panels is similar to that for particleboard underlayment.

Identification of subflooring and decorative wall covering may prove confusing, since a variety of materials is used for such purposes. Particleboard is one of three materials described as chipped wood or board. Particleboard is distinguished from waferboard and oriented-strand board by its resemblance to ground-up sawdust, compared to the large wood flakes that characterize the other two products. Both waferboard and oriented-strand board are bonded with phenol-formaldehyde and have negligible formaldehyde emissions. These materials are currently finding use as subflooring materials and as roofing. Decorative wall panels may include hardwood plywood and paper overlays over particleboard, hardboard, and gypsumboard. Only hardwood plywood and particleboard paneling have the potential for being significant sources of formaldehyde.

Cabinetry alone has the potential for raising formaldehyde levels in a residence above 0.10 ppm. Wood cabinet construction varies widely in the types of materials used. Cabinet members are rarely made of solid wood. The most expensive models typically contain hardwood plywood or a combination of hardwood plywood and hardwood plys overlying a core of MDF. When MDF core materials are cut, the core is exposed, making the MDF recognizable. MDF materials are usually used in side members and shelving. The surface coating of such cabinets may be a urea-formaldehyde resin. With the exception of the exterior doors, less expensive cabinets commonly consist of members with a particleboard core. The core may be covered with hardwood plys, a paper overlay, or vinyl. Exterior doors are normally constructed of solid wood or hardwood plywood, but the least expensive cabinets may consist of a paper overlay over particleboard or MDF. In such low-grade cabinetry, MDF with a paper overlay is used for front and side panels as well.

As with cabinetry, little modern wood furniture is made from solid wood. For example, most hardwood furniture consists of very thin plys of a high-quality hardwood bonded to a core material of a lesser quality wood, particleboard, or MDF. Drawers may be constructed of a combination of solid wood, hardwood plywood, particleboard, or MDF. Inexpensive furniture may be constructed of particleboard or MDF with a paper overlay. Table tops are commonly constructed from particleboard or MDF with a paper or plastic overlay. The bottom surface is unsealed and may be a significant source of formaldehyde emissions. Furniture probably serves as the source of background levels in most residences where no single major source of formaldehyde can be identified. In such circumstances, formaldehyde levels are usually below 0.05 ppm. Under some circumstances furniture can raise interior formaldehyde levels above 0.10 ppm, causing irritation and symptoms. New furniture in a closed bedroom or small apartment has such a potential. In one investigation conducted by the author, five-year-old bedroom furniture constructed entirely from MDF was observed to result in formaldehyde levels in excess of 0.7 ppm. Waterbeds constructed of particleboard materials placed in a closed bedroom may also cause elevated levels of formaldehyde, particularly when the heated water mattress is supported by a particleboard panel.

Cabinetry or furniture constructed of solid wood or a combination of solid wood and hardwood plywood may be a significant source of indoor formaldehyde contamination. In such circumstances, most formaldehyde emissions are associated with the clear finishes applied to the cabinetry or furniture. Because of the hardness and durability of alkyd urea-formaldehyde finishing materials, they are widely used in the manufacture of hardwood cabinetry and furniture. They may also be used to refinish hardwood floors. Such finishing materials apparently represent the most potent sources of formaldehyde present on the market today. They may be particularly significant as a source of indoor contamination when they are applied to high–surface area materials such as cabinetry and furniture.

The inspection of mobile and modular homes for major formaldehyde sources is relatively simple, since wood products bonded with urea-formaldehyde resins are used extensively in their manufacture. With the exception of many mobile homes constructed before 1974 and relatively few (but increasing numbers) built after 1985, most mobile and modular homes are constructed with particleboard decking. Such decking is a potent source of formaldehyde (despite the fact that it presently complies with HUD product standards). Hardwood plywood decorative wall covering is of equal importance in terms of formaldehyde emissions. This material differs considerably as to the nature of the decorative face, which may be a hardwood plywood or some type of paper or vinyl overlay applied to two plys of Luan bonded with urea-formaldehyde. Increasingly, gypsumboard panels with a printed wood-grain paper overlay are being used in mobile home construction in order to reduce formaldehyde levels. Formaldehyde levels in mobile and modular homes can also be affected by emission from cabinetry constructed from materials such as particleboard, hardwood plywood, and MDF. Cabinet tops are constructed from particleboard and covered with a plastic laminate except for the bottom surface. Closet shelving is often made from particleboard. More recently, however, the trend has been to use waferboard, which is a physically stronger material and a relatively minor source of formaldehyde emissions. Furniture that comes with mobile homes is typically constructed from particleboard and/or MDF.

Biogenic Particles

Health problems (e.g., allergies and asthma) caused by inhalation of biogenic particles in indoor environments are probably much more prevalent than health problems associated with formaldehyde. Nevertheless, such problems are rarely the subject of indoor air quality investigations. This is because these problems are largely diagnosable through medical testing of the patient and amenable to various levels of treatment by the administration of drugs and other therapies. Such medical practice, however, focuses on symptom alleviation rather than identification and elimination of causes. Due to expanding research in the area by medical practitioners and indoor air quality scientists, allergies and asthma are increasingly amenable to environmental management, which can be greatly enhanced by the application of information obtained from indoor air quality investigations.

As described in Chapter 1, a variety of biogenic particles can cause sensitization reactions producing allergies and asthma. However, dust mite allergens and mold spores appear to be the predominant aeroallergens in indoor environments. Air testing techniques can quantify the concentration of both of these aeroallergens in indoor air.

Dust Mites

Air sampling for dust mite allergens requires an extended sampling period with a sampling apparatus that is noisy and intrusive. Analytical techniques are relatively complex, requiring a measure of sophistication. Additionally, dust mite allergen is associated with large particle sizes (> 10 μm) and so may be subject to relatively rapid removal and reentrainment. An alternative approach to quantifying dust mite aeroallergen content or potential is to collect dust samples from vacuum sweepings and have them analyzed for mite allergen by the RAST (radioimmunoassay) method or by the use of monoclonal antibodies.[6] In several Western European countries, dust mite allergen levels in residences are determined indirectly by semiquantitative analysis of guanine levels in dust from bed, bedroom carpet, and upholstery sweepings.[7] Guanine is the major excretory product of dust mites and other members of the arachnid family (spiders, mites, and ticks). Dust mites appear to be the dominant source of guanine in indoor dust samples. The technique was developed and patented in West Germany and is not, to the author's knowledge, available in North America at the present time.

The availability of a dust mite allergen quantification technique would be a useful tool in allergy and pulmonary medicine practice to reduce some of the guesswork involved in environmental management of allergies and asthma. Dust samples could be collected during an indoor air quality investigation or by patients themselves. The advantage of dust sample collection by an indoor air quality specialist is that it provides an opportunity for a physical inspection to identify factors affecting dust mite populations. Such factors may include shade levels, moisture conditions, the presence and quality of rugs, housekeeping practices, use of humidification equipment, and other conditions. Effective environmental management is strongly dependent on a systematic evaluation of factors contributing to dust mite population growth.

Mold

Conducting an indoor air quality investigation to identify potential or existing mold problems is relatively simple compared to a dust mite investigation. Mold growth is often visible to the unaided eye, it is often associated with a characteristic musty smell, and relatively simple quantitative and qualitative techniques are available to quantify mold spores suspended in air and identify mold genera collected with volumetric viable mold samplers.

Mold growth on interior surfaces and wall cavities of residences is facilitated by factors that cause repeated wetting of surfaces and high indoor relative humidities. Visible mold growth can be expected (1) in damp, poorly drained basements; (2) on sills and ledges where condensation occurs from uninsulated windows; (3) on bathroom and kitchen floors and walls from leaking fixtures; (4) in bathrooms where water vapor is poorly exhausted; and (5) in primary and secondary residences with high relative humidities caused by

Figure 8.1. Viable particle sampler.

low setback temperatures during unoccupied periods. Mold growth, whether it is visible or not, usually has a musty odor, which may be the first clue that a particular residence has a mold problem. Mold spore levels in indoor air must be present in sufficient concentrations to (1) induce sensitization and (2) subsequently initiate symptoms of allergic rhinitis and/or asthma. They must also include, of course, those specific species or genera to which the individual is sensitive.

There are three primary techniques used to quantify indoor mold levels. The simplest and most convenient is the gravity settling plate. Its primary advantages are its low cost and the fact that it can be exposed by the patient. However, its quantification power is very limited because of the differential settling of spores from the various common mold genera. The gravity plate is relatively insensitive to the presence of mold spores less that 10 μm; this range is characteristic of *Penicillium* and *Aspergillus,* two of the most common mold genera found in indoor environments.[8] For this reason, the gravity plate sampling method cannot be recommended.

Increasingly, mold sampling in indoor air quality investigations is being conducted using volumetric air samplers. The most popular devices are viable particle samplers similar to the one illustrated in Figure 8.1. Air is sampled at a relatively high rate (1 ft^3/min [0.028 m^3/min]) through a multi-pored orifice plate. Air jets from these pores impinge on a culture medium (e.g., malt extract agar) below the plate. After a sampling period of 1–3 min, the plates are closed and incubated for seven days. Colonies are counted and identified. Total spore concentrations are expressed as colony-forming units per cubic meter (CFU/m^3). Mold counts above 1000 CFU/m^3 in residences are considered by the author to be high, with levels below 100 CFU/m^3 representing a relatively mold-free environment. Levels in between are subject to the investigator's own interpretation.

The volumetric sampling technique just described only quantifies viable mold particles. Nonviable particles are equally allergenic and may comprise a significant portion of the mold spore population in the air at a given time.[9] A high viable mold particle level, as measured by volumetric sampling, indicates a potentially significant mold contamination problem. However, the observation of low mold levels from viable sampling cannot confirm that airborne mold spore levels are low. At best, it can only allow one to conclude that the viable mold particle count is low.

An alternative approach would be to use a volumetric sampling technique that collects and quantifies both viable and nonviable mold particles. A small portable device called the Burkhard sampler has been developed for this purpose. Molds and other particles are collected on a glass plate by impaction and quantified by counting, using visible light microscopy.

During the summer months, mold levels and genera present often reflect those of the outdoor environment. Quantification of mold spores actually produced indoors requires sampling when the indoor environment is closed up, typically during the heating season or in air-conditioned houses during the summer. Because indoor mold levels are affected by both indoor and outdoor sources, sampling should be conducted in both environments during an indoor air quality investigation.

It is essential that, in addition to volumetric sampling, factors causing liquid water to remain on surfaces be assessed, as well as conditions resulting in high indoor humidity levels. An assessment should also be made of potential pathways of movement from mold-infested environments such as basements and crawl spaces. For example, leaky cold air returns have a significant potential for drawing air from such spaces and distributing it throughout the living environment.

Radon

Investigations for radon focus on indoor air quality monitoring conducted for the most part by building occupants themselves. Once unacceptably high radon levels have been confirmed in a building environment, additional investigation by professionally competent individuals is required to identify pathways of radon entry and to assess what specific radon mitigation measures would be appropriate.

In theory, building occupants can easily determine if their living or working environments have radon levels above health guidelines. Several inexpensive passive sampling techniques are available. These include charcoal canister, electret, and track-etch monitors.

Charcoal Canister Monitors

In the charcoal canister method, radon is adsorbed on charcoal contained in a metal or plastic container.[10] Sampling is typically conducted for a 2- to 7-day period with the concentration expressed as an average for the sampling period. For analysis, the exposed charcoal canister is placed in a gamma ray detector, by which gamma ray energies associated with the decay of radon progeny are detected and quantified. The charcoal canister method is the most widely used radon monitoring technique because of its relatively low cost ($15-$30/sample) and its quick turnaround time (results are usually available in a few weeks). Charcoal canisters are useful for screening purposes, particularly for very high levels (> 20 pCi/L). For levels in the range of guidelines, charcoal canister monitoring has significant limitations. Guidelines are based on annual average exposures. Since radon levels can vary significantly over the course of the year, a single measurement over a 2- to 5-day period is insufficient to assess whether radon exposures exceed 4 pCi/L, the most commonly used radon guideline.

Though most homeowners and other building occupants are fully capable of conducting radon monitoring with charcoal canisters, they usually fail to understand what their test results mean, particularly when they are slightly above or below the 4-pCi/L EPA radon guideline. The average lay person, for example, will interpret a test result of 3.95 pCi/L as being safe (breathing a sigh of relief!) and view with alarm a reading of 4.05 pCi/L. Given the range of variability in radon levels in a building, these numerical differences are meaningless.

In interpreting the results of charcoal canister testing, it is important to recognize that radon levels will vary over a range of concentrations. In research conducted by the author, in residences where average radon concentrations are less than 10 pCi/L, the range of variation in radon levels in closed houses is seen to be about 2.5–6 times. This means that the highest measured concentration is approximately 2.5–6 times greater than the lowest. If one assumes that a given radon test value is the lowest value for a given residence, then the highest value would be 2.5–6 times greater, with the average exposure concentration halfway in between. For example, let us assume that a charcoal canister test result were 3.0 pCi/L. The typical homeowner would conclude that this was a safe level, put the radon concern to rest and go on with his or her life. Within the laws of statistical probability, we could, however, expect that approximately 5% of the time 3.0 pCi/L was actually the lowest value and not the average. If it were the lowest value, then the highest value would be expected to be 7.5–18 pCi/L, with an average concentration of approximately 5–9 pCi/L, a level above the EPA guideline.

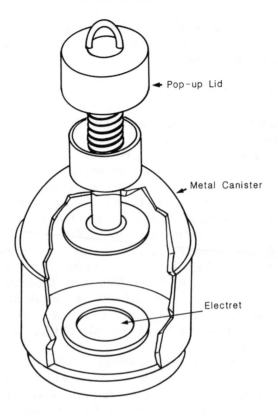

Figure 8.2.   Electret radon sampler.

Electret Monitors

The electret technique has only recently been introduced, so it has a limited history of use. It employs an electret suspended in a plastic housing. The electret is a disk of Teflon® that has been given a permanent electrostatic charge. Air enters through a particulate trap that removes radon progeny but allows radon to pass through. The subsequent decay of radon produces negatively charged progeny, which are attracted to the positively charged electret.[11] After a given period of exposure, the average radon concentration in the tested environment can be determined from the change in electrical charge on the electret medium, using a specifically designed voltmeter. An electret sampling device is illustrated in Figure 8.2. The electret testing method can be used for both short-term (2–7 days) and long-term (1–12 months) radon testing.

Track-Etch Detectors

The charcoal canister method has the disadvantage of being only suitable for making short-term radon measurements, whereas radon risks are associated with long-term exposures. The method of choice for determining long-term radon exposures has been the track-etch detector. It is based on the principle that alpha particles (associated with radon progeny) striking a plastic film leave tracks that can be counted microscopically after film development.[12] From these track counts, radon concentration can be calculated. The recommended sampling period of a track-etch detector is three months. It can be exposed for as little as two weeks or up to a year. Its principal disadvantage is that it has a relatively long turnaround time. It usually takes many months before the individual homeowner knows what radon levels are in his/her particular residential environment. So it is generally not suitable for use in real estate transactions unless a three-week waiting period can be accommodated. The principal advantage of the track-etch method is that results are readily interpretable, and, if sampling is conducted for approximately three months, retesting is less likely to be necessary.

Test results obtained from the various methods employed do not as a rule reflect actual exposures. In most instances, they represent exposures occurring under heating or cooling season conditions. For many homes without year-round climate control, actual annual exposures referred to in guidelines may be significantly less than measured or calculated long-term (3–6 months) averages. In many homes in or near northern climates, normal living conditions include the use of natural ventilation by opening windows or doors for 3- to 6-month periods. Under such conditions, radon levels could be expected to be in the range of ambient levels (approximately 0.25 pCi/L). Calculation of time-weighted average exposures for a given household should be based on both background levels and those existing under closure conditions. Such time-weighted averaging will not, however, be required in those households with year-round climate control or with minimal natural ventilation (opening of windows and doors).

Increasingly, radon testing is being sought by prospective home buyers prerequisite to real estate closings. This is particularly the case in Pennsylvania and New Jersey, and for those moving from such areas to other parts of the country. Because of the imperative of a short turnaround time, such transaction-contingent testing is being conducted by the use of charcoal canisters and by testing companies who employ quasi–real-time radon detecting devices. The use of such radon testing for real estate transactions may not be reliable because of large diurnal and seasonal variability in radon concentrations. Quasi–real-time testing is also particularly vulnerable to misinterpretation of test results. Of comparable concern is the fact that, by simply opening the windows, individuals with a significant financial interest in the outcome can intentionally lower radon levels by an order of magnitude. Based on the author's experience in testing UFFI houses for real estate sales, an intentional manipulation of contaminant concentrations by 30% of sellers can be expected.

Once a residence or other building has been tested and unacceptably high radon levels confirmed, building owners should seek professional assistance from reputable radon mitigation consultants or contractors. An inspection or investigation is prerequisite to the implementation of any mitigation measures. The appropriateness of a mitigation measure will depend on such factors as the type of substructure (basement, slab, crawl space, or combination of these), the presence of a subfloor drainage system, location and accessibility of cracks and other openings in slabs and basement walls, building use characteristics, and other factors. Obviously, it will also depend on measured radon levels and calculated exposures.

Carbon Monoxide

Conducting an investigation of a residence for excessive carbon monoxide contamination is usually a relatively simple matter. Typically, such investigations are conducted by gas companies, heating contractors, local and state health agencies, and private consultants. Investigations are usually initiated in response to homeowner complaints of "gas" odors or are in response to diagnosed CO poisoning.

Though it is colorless, odorless, and tasteless, CO is easily measured with simple and inexpensive gas syringes and detector tubes and with light, portable, real-time instruments. The former are widely used by gas companies, local health departments, and some heating contractors. For purposes of residential investigation, measurements for CO are typically made in the general living area, at heating outlets, and in the near vicinity of combustion appliances, such as gas or oil furnaces and water heaters.

Occasionally in residences, significant CO exposures occur on an episodic basis and therefore can be easily missed by air testing on a one-time or even a multiple basis. In such circumstances, the best approach is to advise building occupants upon suspected exposure to immediately have a blood sample taken by a medical laboratory for carboxyhemoglobin analysis.

Sublethal CO exposures can be detected by observing the pattern of symptoms experienced by building occupants. Such exposures are characterized by symptoms of headache, drowsiness, fatigue, nausea, and vomiting. In many cases, carbon monoxide poisoning is progressive, corresponding to a continuing deterioration of combustion appliance parts such as heat exchangers and flue vents, and to the clogging of chimneys with debris. The initial symptom may only be headaches. In its more severe stages, CO poisoning may cause vomiting and extreme (if not life-threatening) drowsiness. Exposure to CO typically occurs during the heating season, but it may occur at any time of the year, particularly if flue gas spills from a poorly vented or malfunctioning gas hot water heater.

Carbon monoxide poisonings also occasionally occur with unvented, old, or poorly designed space heaters. Carbon monoxide contamination is most commonly associated with residences whose occupants are described as low-income. Such individuals are more likely to use unsafe combustion appliances, flue vents, and chimneys, due to their financial circumstances and lack of awareness of the risks associated with using combustion appliances that are old or are poorly designed, installed, and/or maintained. Combustion appliances and flue venting systems require maintenance, particularly as they age. Heat exchangers may crack or corrode so badly that holes develop, allowing flue gases to cross into supply air. Old houses with chimneys once used for coal or oil heating may have a buildup of soot, which upon falling may cause the flow of flue gases to be obstructed.

Though CO cannot be detected by sight, taste, or scent, flue gas contains other by-products that may have a burned gas or oil odor. It also contains a substantial quantity of water vapor that may condense in a cool environment such as a basement. An unexplained condensation of water near combustion appliances in a basement may suggest that flue gases are not venting properly and that CO contamination and exposures are probable.

Significant CO contamination can result from episodic occurrences, even when combustion appliances are properly designed, installed, and maintained. Episodes may result from downdrafts during very unstable weather conditions. They may also result when a house is depressurized from the operation of exhaust systems and fireplaces. Flue gases that are already exhausted may also reenter through open flues that are in close proximity, particularly under certain conditions of ambient air flow. For example, flue gases from a furnace can be drawn back in through an adjacent open fireplace chimney. The potential for such contamination should be assessed in an investigation, particularly when occupant complaints suggest CO exposures, but CO monitoring results are negative.

In investigating a suspected CO complaint, it is important to inspect all combustion appliances and flue systems for signs of corrosion, improper installation, lack of maintenance, and other problems. Repairs conducted by heating contractors in response to a significant case of CO exposure should be followed by retesting for CO.

Miscellaneous

A variety of miscellaneous indoor air quality problems are occasionally experienced and observed in residential environments.

Pesticides

Pesticide exposures that cause acute symptoms are usually apparent immediately after application. Investigations should include an evaluation of the application, the type and persistence of pesticides used, the nature of occupant complaints, and an evaluation of whether control measures are needed and applicable to the specific circumstance. Investigations may also include the collection of air and/or surface samples. Because of the low concentrations often encountered, use of such sampling may not be particularly revealing.

Sewer Gas

Sewer gas is not an uncommon indoor air quality problem in residences and other building environments. For the most part, offensive odors associated with it are merely a nuisance, but in some circumstances ammonia levels associated with sewer gas may be sufficiently high to cause severe upper respiratory irritation. Its presence is usually obvious because of the odor. The entry of sewer gas into a building is usually associated with a dry trap in a sink or other water line connected to the wastewater pipe. Dry traps are often associated with some fixture that is rarely or no longer used but remains attached to the building's wastewater lines. Investigations of sewer gas problems should include a search for traps (whether obvious or not) with a high probability of drying out. Sewer gas entries occur in episodes. They may be precipitated by events causing pressure changes in sewer lines or even in the building itself. House depressurization associated with the use of exhaust fans, a fireplace, and even open ducts running from a basement to an upper story can cause an inflow of sewer gas. Sewer gas episodes may also be associated with rainfall events.

Even though the presence of sewer gas can be confirmed by the odor, sampling for sewer gases may be desirable in some cases. Ammonia and reduced sulfur gases (like hydrogen sulfide) are good markers for sewer gas.

Volatile Chemicals

Indoor air quality problems have occasionally been observed with fiberglass materials used to line the inside of heat pumps, the heat-transfer box of combustion appliances located external to residences, and heating duct trunk lines. In such instances, the associated high temperatures (~ 43–$65°C$) cause volatilization of surface coatings applied to reduce friction and fiber erosion. Such surface coatings may include ethylene vinyl acetate, methacrylate, or neoprene. Volatilization of the resin binder may also occur to a limited extent. Though fiberglass is commonly used to insulate flexible supply air ducts, such ducts are normally lined with a plastic barrier and therefore do not constitute a problem.

Volatile chemicals associated with unlined fiberglass insulating materials used in heating applications are often both odoriferous and irritating. Some odors are sweet, others are rubber-like (neoprene), and some appear to be metallic in both odor and taste. They are characteristically associated with activation of the heating system. The more the heating system runs, the more intense the odor and irritation. The problem is most intense just before sunrise and on the coldest winter days. Odor and irritation increase with intensity in direct proximity to heat outlets in forced-air systems.

Since the toxic and odor-producing substances associated with the above problem have not been identified, air sampling is of limited value. On investigation, the problem is recognizable by the presence of strong odors described as sweet, rubbery, or metallic when hot air is flowing through supply registers in a forced-air–heated residence. Odors and irritation become more intense with increased running time of the forced-air system.

Confirmation of the true nature of the problem can be achieved by inspecting heating appliances and trunk lines. On opening a heat pump lined with offending fiberglass insulation, an odor similar to that experienced in other parts of the building should be noticeable at a more intense level. Samples of suspect materials can also be collected, placed in glass jars, and then heated to temperatures characteristic of the heating appliance or supply air. If a sample material treated in this way is responsible for the problem, the investigator will detect an odor similar to that of the contaminated building environment on opening the heated glass sample jar. To assure better discrimination, the samples should be allowed to cool in the direction of room temperature for 15–30 min. Because VOCs associated with such materials are often very irritating, direct sniffing of sample jar air is not advised.[13]

Attic Insulation

Indoor contamination problems in residences have also been associated with cellulose, fiberglass, and rock wool insulation used in attic areas. Such contamination usually occurs when materials are placed over or near a leaky cold air return, permitting an inflow of insulating fibers and attic air to occur due to negative pressures. The problem can be confirmed by inspection of attic cold air returns and application of air quality sampling. Fiberglass and rock wool fibers can be identified and quantified by microscopy. The presence of cellulose insulation can be confirmed from air or surface samples by analysis for carbon black and/or boron. Cellulose insulation is made from shredded newsprint and treated with boric acid as a fire retardant. Occasionally, home owners complain of the onset of illness symptoms after the installation of cellulose insulation. How real this association is has yet to be determined.

New Carpeting

Complaints of irritation associated with the installation of new carpeting are common. This is particularly the case when carpeting is installed immediately after delivery from the factory. Because of the multitude of VOCs associated with new carpeting and the expense involved, air testing is not recommended. On investigation, the problem is usually identifiable by the presence of carpeting odor and/or upper respiratory irritation. It is recognizable by the fact that the onset of symptoms occurs within days of installation.

Heating/Cooling Systems

Complaints of odors have also been associated with heat pumps and air-conditioning systems.[13] These odors have been likened to "sweaty gym socks" or described as musty. Such odors suggest some form of microbial contamination of heating/cooling systems. Investigation of this type of complaint should include an inspection of cooling coil drip pans, cooling coils, and ducts for visible mold growth and the presence of odors more intense that those experienced in the general indoor building environment.

One of the best tools available to an indoor air quality investigator is his or her sense of smell. Many problems are readily recognizable by the odor present. Characteristic odors are associated with urea-formaldehyde foam–insulated houses, particleboard subflooring, mold growth, new carpeting, and fiberglass insulating materials used in heating appliances and supply duct trunk lines. The effectiveness of the olfactory approach in conducting an indoor air quality investigation is compromised by the masking effect of other strong odors, especially tobacco smoke and cooking odors. In households with smoking occupants, odors other than tobacco are virtually imperceptible. Heavy cooking odors such as bacon or apple pie also render other odors imperceptible. In some cases it may be desirable to schedule an investigation when cooking odors are not likely to be pervasive.

PUBLIC ACCESS BUILDINGS

Conducting indoor air quality investigations in large buildings is much more complex and difficult than it is in residences. This is, in part, due to scale and to the nature of heating, ventilation, and air conditioning (HVAC) systems and parameters of operation. It is also because contamination problems common to large mechanically ventilated buildings are not as well understood. In many cases, those conducting investigations are at a loss to identify with any certainty the specific causal factor or factors.

Investigations of building-related health complaints in large public access buildings are usually conducted by state health departments, private consulting firms, in-house corporate industrial hygienists, and/or health hazard investigation teams of the National Institute of Occupational Safety and Health (NIOSH). Approaches to conducting investigations are varied. They may involve extensive and sophisticated air testing, inspection and testing of mechanical systems, epidemiological surveys, or a simple walk-through. Their nature and extent of comprehensiveness depend on resources available and the technical background and experience of those engaged to conduct the investigation.

Because sick building situations have often involved complaints by office workers, investigations, at least initially, have been conducted according to standard industrial hygiene practice. In such cases, air testing is conducted to identify and quantify contaminants in order to determine whether they exceed permissible occupational exposure limits. This approach has proven fruitless, since contaminant levels involved in sick building complaints are usually an order of magnitude less than those permitted under occupational safety and health standards.

In general, measurement of contaminant levels should not be the primary focus of an investigation. Rather, the first action should be the assessment of health complaints, including the nature of specific symptoms, their prevalence among the building population, relationship of symptoms to specific building locations, onset patterns, etc. In an initial walk-through investigation, this assessment can be conducted by limited interviewing of building occupants. Information derived from such limited interviews may often be sufficient to identify the nature of the problem. If the problem is not amenable to easy identification, the investigation may require systematic interviews of occupants using a well-designed questionnaire. Both questionnaires and limited interviews should seek information on specific symptoms, onset and resolution patterns, etc.

In addition to an assessment of occupant complaints, the investigation should include an evaluation of the building's history relative to construction, renovation, maintenance practices, the operation and maintenance of HVAC systems, and unusual occurrences. The importance of asking appropriate questions of building supervisors and maintenance personnel cannot be overstated. The physical plant supervisor, maintenance men/women, and the head of custodial services can reveal important clues that may lead to the problem's identification.

Ventilation

Because it is apparently the major cause of health complaints in large nonresidential buildings, the adequacy of ventilation should be one of the first factors to be evaluated, particularly if the results of occupant interviews suggest that ventilation may be a problem. Clues may include a pattern of symptoms characterized by headaches and lethargy, stuffiness, a progressive pattern of perceived air quality deterioration, information from physical plant personnel indicating an inadequate provision of outdoor air, etc.

Previously, it was indicated that air testing should not be the primary focus of an investigation. There is one exception, and that is measurement of carbon dioxide (CO_2). Because building CO_2 levels above background concentrations of approximately 325 ppm are primarily the result of human exhalations, CO_2 is a good indicator of human bioeffluents and the adequacy of ventilation to dilute them. They are, as previously described in this book (see Chapter 5, pp. 208–212), the basis for establishing ventilation standards. Unless the problem, as discerned from occupant interviews, has been determined to be due to a specific contaminant or contamination problem, it is desirable to test for CO_2. Such testing can be conducted using gas syringes and sampling tubes or portable real-time CO_2 analyzers. Since CO_2 levels are a function of building occupancy and the amount of outdoor air provided, it is important to remember that peak CO_2 levels will occur sometime after initial daily building occupancy. If the building is operated on 100% recirculated air, CO_2 levels may not peak until late afternoon. Measurements in the early morning hours may not reveal CO_2 levels exceeding the proposed ASHRAE standard of 1000 ppm, whereas they may later in the day.

In addition to measurements of CO_2 levels, other aspects of ventilation should be assessed. These include capacity, system design, return air provision, operating cycles, percentage of outdoor air, and building odors. In most cases, the design capacity of the HVAC system is sufficient to provide adequate ventilation. It can be determined by a variety of means, including, in theory, reviewing blueprint plans, analyzing fan curves, and measuring air flow velocities.

System design is an important consideration. Variable air volume systems, for example, are notorious for frequency of ventilation complaints. In schools, ventilation is often provided through unit ventilators. Problems associated with their use may be different from other types of air handling systems.

In many building HVAC systems, conditioned air is supplied through a diffuser grill located in the ceiling. Since a large percentage of this air is recirculated, provision must be made for its return through fans and conditioning units. Air return occurs passively through grills located in the ceiling, through grills in doors, or through open doorways. Ventilation problems may be experienced in spaces where no provision has been made to facilitate air return. This may occur in offices constructed with movable walls and in those spaces where perforated ceiling panels are apparently used for return air passage. Factors inhibiting the free flow of return air can affect the performance of ventilation systems. Therefore, their assessment is an important factor in conducting a building investigation.

As described in Chapter 5, performance of ventilation systems is severely compromised by short-circuiting air flows that result when supply and return air grills are located short distances from each other in the ceiling. Although the investigator should note the proximity of supply and return air grills and be aware of the potential for short-circuiting, there are no measuring methods to easily quantify the magnitude of short-circuiting and the efficiency of ventilation in a space. However, the adequacy of air flow into the space can be easily assessed by the use of velometers and velocity measuring hoods.

The percentage of outdoor air used for ventilation is probably the most significant factor in determining whether a problem of system-wide ventilation inadequacy will occur. Under various conditions of operation, it may vary from 0 to 100%. Problems generally occur only when the percentage of outdoor air is less than 10%. On investigation, it is common to find buildings being operated on 0% outdoor air by either design or negligence. Such operation may be based on energy conservation considerations. Many buildings are operated on diurnal energy-conserving cycles in which outside air dampers are closed during the unoccupied period. Problems associated with this practice occur when the conservation cycle begins before the end of the workday and ends after the workday has already begun. Seasonal energy-conserving ventilation cycles are also commonly employed. In such applications, outside air dampers are automatically closed when outdoor air temperatures fall below a minimum value (e.g., 6°C) or rise above some maximum value (e.g., 26°C). These seasonal cycles are a common cause of ventilation problems and occupant complaints. Fortunately, a zero–outside air condition can readily be determined from visual inspection of air intake louvers and from information provided by physical plant personnel. When dampers are partially opened, actual air flows must be determined from velocity measurements.

Ventilation problems are often suggested when the building has a very perceptible malodor that is characteristic of humans and/or a sense of stuffiness is perceived.

Contaminants Associated with Building Fabric, Furnishings, and Equipment

In addition to ventilation, a major source of building-related health complaints is contaminants generated inside the building by portions of the building fabric, furnishings, and equipment. As described in Chapter 1, this represents a wide array of contaminants and contaminant problems. Toxic effects associated with formaldehyde, total VOCs, spirit duplicators, carbonless copy paper, and ammonia associated with blueprint machines have been widely recognized and accepted as causing health problems. Other contamination problems are more speculative. These include organic dusts around video display terminals, plasticizer emissions from vinyl floor tile, organic dust associated with fleecy surfaces, air ions, and others.

In conducting an investigation of a sick building complaint, it makes sense to look for the obvious. The obvious may include potential sources of formaldehyde, such as furnishings made from urea-formaldehyde–based wood products (e.g., cabinetry, room dividers, bookcases, table tops, desks, etc.). Typically, such furnishings have core materials consisting of particleboard or MDF. While formaldehyde is a causal factor in sick building complaints, it is a relatively uncommon problem in public access buildings as compared to residences.

VOCs associated with paints, varnishes, sealing caulks, adhesives, carpeting, and other materials are likely to be a major cause of health complaints associated with new buildings. Their effects may be additive and even synergistic. This potential problem can be assessed by collecting air samples and determining the total VOC concentration (expressed as hexane). Concentrations of approximately 5 mg/m^3 are considered sufficient to cause sick building symptoms, particularly mucous membrane irritation.[14] Health complaints may also be due to specific VOCs associated with, for example, carpeting and vinyl floor covering adhesives.[15]

Office equipment and materials may also cause building-related health complaints. Such problems are distinguishable by the fact that they are almost always localized and therefore are unlikely to cause building-wide problems. Potential causes of such indoor air contamination include spirit duplicators (methyl alcohol), blueprint machines (ammonia), photographic equipment (acetic acid), photocopying machines (ozone), signature machines (methacrylate), carbonless copy paper (formaldehyde), and printed materials (toluene, xylene, total VOCs).

Contaminants sufficient in concentration to cause complaints or discomfort may occasionally be associated with building maintenance materials such as disinfectants, cleaning solutions, floor waxes, and carpet shampoos.

Cross Contamination

One of the most easily identifiable building contamination problems occurs when a contaminant generated in one part of a building where it is normally tolerated is transported to another part of the building where it is not. This phenomenon is called cross contamination. It may occur in a building with multiple purposes, such as a hospital, research laboratory, or industrial building with attached office complexes. Such contamination can occur when air from one air handling unit physically leaks into another, when one part of a building is under a greater degree of negative pressure as compared to another, or when strong upflows occur through elevator and other building shafts from parking garages and other contaminant-generating spaces. The problem is very often distinguished by odors that are "out of place." It can be evaluated by using tracer techniques, including both olfactory techniques and those determined from air sampling. Odoriferous substances such as wintergreen or banana oil are useful aids in identifying and mitigating a cross contamination problem.

Carbon monoxide is one of the most common cross contaminants found in office spaces of small industrial and warehouse operations. Its primary sources are propane-fueled forklifts operated within the building or air drawn in from idling vehicles at the loading dock. In these circumstances, it is not uncommon to have interior CO levels on the order of 20–50 ppm or higher. An excessive CO level is suggested by employee complaints of headache, fatigue, nausea, and vomiting. It is detectable by air sampling. When management is unwilling to have an investigation conducted, employees can have their CO exposures assessed through blood tests for carboxyhemoglobin.

Entrainment/Re-Entry

Contaminants generated external to the building, or those which have been exhausted from it, can become entrained in air entering the building's ventilation system or air entering by infiltration. When contaminants previously in the building return, the problem is described as re-entry. Re-entry can occur when contaminants are directly vented to the ambient environment and when vapors from volatile materials disposed of in sinks re-enter through plumbing connections. Re-entry in the latter case is usually the result of an unused (and therefore dry) trap. Building owners/managers usually become aware of such traps only after a re-entry problem is recognized. Sewer gas and chemical vapors produced external to the building would be more accurately described as entrainment problems.

Entrainment or re-entry of contaminants can occur when (1) local exhaust ducts are vented upwind of outside air intakes, (2) the building is under negative pressure, (3) idling motor vehicles are parked near outside air intakes, and (4) short boiler stacks are located upwind of outside air intakes. Combustion gases represent one of the most common entrainment or re-entry problems. It is not unusual, for example, to find on investigation that outside air intakes are located close to loading docks. Vehicles are not uncommonly parked in an idling mode during the entire period of loading and unloading.

In assessing the potential for re-entry, it is essential that the investigator have access to the roof of the building since this is where most exhaust stacks and vents are located. Air handling units are often located on building roofs as well. It is not unusual to have exhaust vents and stacks in close proximity to air handling units and intake air louvers. They may even be in direct line with the prevailing wind direction. Because of architectural considerations, boiler stacks may only protrude a few feet above the roofline. Emissions from such stacks, depending on wind and other meteorological conditions, can be drawn into outside air intakes of air handling units located upwind. When re-entry occurs from a boiler stack, the odor of flue gases may be noticed by some building occupants, particularly when oil heat is being used or when, as commonly occurs during the heating season, there is a change from gas to oil heat or from oil to gas heat.

Entrainment and re-entry are episodic phenomena, so it is only a matter of luck if an episode occurs during the time that the building is being investigated. Therefore, the entrainment/re-entry potential on roof tops must be assessed from an inspection of exhaust stacks and vents, the proximity of outside air intakes, and orientation relative to prevailing or local wind flow patterns. At ground level, this potential must be assessed in the context of the proximity of outside air intakes to motor vehicle exhaust from loading docks and heavy street traffic.

Because entrainment and re-entry are episodic phenomena, gases associated with them are difficult to monitor. Monitoring will only prove viable if it is conducted while an episode is in progress. This requires immediate notification by the building manager and an immediate response by the investigator. Carbon monoxide is the best indicator of combustion gases. It can be sampled using gas syringes and detector tubes and monitored using portable real-time analyzers. If a quick response to an episode is not possible, as is often the case, it may be desirable for those affected to have blood samples taken by medical technologists and have them analyzed for carboxyhemoglobin.

Microbial Contamination

Illness syndromes caused by microbial contaminants associated with air handling systems and building materials are distinctive. In general, they can be easily identified vis-a-vis other building-related health complaints. Hypersensitivity pneumonitis, for example, is characterized by symptoms of chest tightening, coughing, chills, fever, muscular aches, and headache, with the onset of symptoms occurring within 5–6 hr after exposure. It can be confirmed by serological testing.

Investigation of a hypersensitivity pneumonitis complaint commonly involves an inspection of components of HVAC systems and interior building materials and surfaces for evidence of microbial contamination. It may also include air sampling for mold, actinomycetes, bacteria, or amoebae using a viable particle sampler.

Common sources of microbial contamination of buildings leading to outbreaks of hypersensitivity pneumonitis include (1) microbial slime developing on poorly drained cooling-coil condensate drip pans, (2) spray air washers associated with some HVAC systems, and (3) water incursions. The presence of water incursions is evident in the tell-tale stains observed on ceiling tiles, walls, and floor covering.

Mold-induced allergic rhinitis and asthma can also be associated with large nonresidential buildings. Such illness syndromes may be caused by water-damaged building materials and furnishings, contaminated HVAC system components, indoor conditions of high humidity, and other conditions. Mold sensitivities can be confirmed by allergic testing. Mold levels can be determined using volumetric sampling techniques.

Outbreaks of Legionnaires' disease are recognizable from symptom patterns and serological testing of patients' blood. Investigations are directed toward possible sources, including cooling towers and evaporative condensers from which water samples are collected to determine the presence of *Legionella pneumophilae* in sufficient concentration to cause a disease outbreak. Since the organism is difficult to grow and contamination of outside air drawn in for ventilation purposes is likely to be episodic, air testing would not prove practical in most instances.

REFERENCES

1. National Institute of Occupational Safety and Health. 1984. *Manual of Analytical Methods,* 3rd ed. Method 3500. Department of Health and Human Services (NIOSH) Pub. No. 84–100. Cincinnati, OH.

2. Godish, T. 1985. "Residential Formaldehyde Sampling—Current and Recommended Practices." *Am. Ind. Hyg. Assoc. J.* 46:105–110.
3. Berge, A. et al. 1980. "Formaldehyde Release from Particleboard—Evaluation of a Mathematical Model." *Hals als Roh-und Werkstaff* 38:251–255.
4. Godish, T. and J. Rouch. 1985. "An Assessment of the Berge Equation Applied to Formaldehyde Measurements under Controlled Conditions of Temperature and Relative Humidity in a Mobile Home." *JAPCA* 35:1186–1187.
5. Godish, T. 1988. "A Standardizing Equation for Formaldehyde and Indoor/ Outdoor Temperature Differences." Unpublished studies.
6. Reed, C. 1987. Personal Communication. Mayo Clinic. Rochester, MN.
7. Bischoff, E. 1987. "Sources of Pollution of Indoor Air by Mite Allergen–Containing House Dust." 742–746. In: *Proceedings of the Fourth International Conference on Indoor Air Quality and Climate.* Institute for Water, Soil and Air Hygiene. West Berlin. Vol. 2.
8. Solomon, W.R. 1975. "Assessing Fungus Prevalence in Domestic Interiors." *J. Allergy Clin. Immunol.* 56:235–342
9. Burge, H. 1987. Personal Communication. University of Michigan, Ann Arbor, MI.
10. Cohen, B.L. and E.S. Cohen. 1983. "Theory and Practice of Radon Monitoring with Charcoal Adsorption." *Health Physics* 45:501–508.
11. Rad-Elec, Inc. 1988. Product Literature. Gaithersburg, MD.
12. Alter, H.W. and R.A. Oswald. 1983. "Results of Indoor Radon Measurements Using the Track Etch Method." *Health Physics* 45:425–428.
13. Godish, T. 1987. "Scientists Find Some Answers to Heat Pump Odor Problems." *Air Cond. Heating Refrig. News.* May 18.
14. Molhave, L. et al. 1986. "Human Reactions to Low Concentrations of Volatile Organic Compounds." *Environ. Int.* 12:167–175.
15. Lauderdale, J.F. et al. 1984. "A Comprehensive Study of Adverse Health Effects." American Industrial Hygiene Association Conference Abstract.

APPENDIX

Occupant Symptom/Ailment Checklist

Occupants' Names

Symptoms

Symptoms					
Eye irritation	( )	( )	( )	( )	( )
Eye infection	( )	( )	( )	( )	( )
Dry/sore throat	( )	( )	( )	( )	( )
Cough	( )	( )	( )	( )	( )
Excessive phlegm production	( )	( )	( )	( )	( )
Runny nose	( )	( )	( )	( )	( )
Sinus congestion	( )	( )	( )	( )	( )
Sinus infection	( )	( )	( )	( )	( )
Bronchial pneumonia	( )	( )	( )	( )	( )
Shortness of breath	( )	( )	( )	( )	( )
Wheezing	( )	( )	( )	( )	( )
Asthmatic attacks	( )	( )	( )	( )	( )
Bronchitis	( )	( )	( )	( )	( )
Headaches	( )	( )	( )	( )	( )
Disturbed concentration	( )	( )	( )	( )	( )
Dizziness	( )	( )	( )	( )	( )
Unusual fatigue	( )	( )	( )	( )	( )
Drowsiness	( )	( )	( )	( )	( )
Difficulty in sleeping	( )	( )	( )	( )	( )
Rashes	( )	( )	( )	( )	( )
Nasal irritation	( )	( )	( )	( )	( )
Nosebleed	( )	( )	( )	( )	( )
Nasal sores	( )	( )	( )	( )	( )
Nausea	( )	( )	( )	( )	( )
Vomiting	( )	( )	( )	( )	( )
Diarrhea/loose stool	( )	( )	( )	( )	( )
Chest pain	( )	( )	( )	( )	( )
Abdominal pain	( )	( )	( )	( )	( )
Menstrual problems	( )	( )	( )	( )	( )
Unusual thirst	( )	( )	( )	( )	( )
Whole body ache	( )	( )	( )	( )	( )
Fever	( )	( )	( )	( )	( )

Building-Related Illness Checklist

Yes	No	Don't Know	N/A	
___	___	___	___	1. Symptoms reported by more than one occupant.
___	___	___	___	2. Irritating symptoms most severe in occupants who spend most time in building.
___	___	___	___	3. Symptoms severe in infants or very young children.
___	___	___	___	4. Symptoms become less severe with absence from building, with longer absences resulting in marked improvement.
___	___	___	___	5. Symptoms diminish in severity when building is provided with significant continuous ventilation.
___	___	___	___	6. Symptoms exhibit a seasonal pattern.
				7. In residences, onset of symptoms can be associated in time with:
___	___	___	___	a. moving into a new conventional or mobile home.
___	___	___	___	b. recent house remodeling.
___	___	___	___	c. acquisition of new furnishings.
___	___	___	___	d. insulating home with urea-formaldehyde foam.
				8. In nonresidential buildings, onset of symptoms can be associated in time with:
___	___	___	___	a. moving into a new or recently renovated building.
___	___	___	___	b. acquisition of new furnishings.
___	___	___	___	c. implementation of ventilation system changes.
___	___	___	___	9. Symptoms reported by building visitors.

Formaldehyde Building-Related Illness Checklist

Yes	No	Don't Know	N/A	
___	___	___	___	1. Symptoms most severe during warm, humid, and/or rainy weather.
___	___	___	___	2. Symptoms least severe on cold, dry winter days (in colder climates).
___	___	___	___	3. Symptoms in eyes and upper respiratory and central nervous systems.
___	___	___	___	4. Peak formaldehyde levels of ~0.06 ppm or higher.
___	___	___	___	5. Symptom onset in environments where elevated formaldehyde levels are common (e.g., clothing stores, clothing sections of department stores, furniture stores, etc.)
___	___	___	___	6. Symptoms very severe in occupants of mobile homes or new homes with particleboard subflooring.
___	___	___	___	7. Symptoms associated with one area of building where potent sources of formaldehyde are located (e.g., closed bedroom)
				8. Presence of major formaldehyde sources:
___	___	___	___	a. particleboard subflooring
___	___	___	___	b. paneling: hardwood plywood, particleboard
___	___	___	___	c. cabinets: particleboard, MDF, hardwood plywood (with U-F finish)
___	___	___	___	d. wood furniture: particleboard, MDF, hardwood and hardwood plywood (with U-F finish)
___	___	___	___	e. urea-formaldehyde foam insulation

9   MITIGATION PRACTICE

With few exceptions, the mitigation of an indoor air quality problem depends on the identification of its specific nature and an assessment of the factors that contribute to or affect it in some way. The investigatory processes and knowledge of specific contaminant problems described in Chapter 8 provide the basis for determining the appropriate mitigation measure or series of mitigation measures for resolving a particular problem.

Successful mitigation depends on the availability of techniques to significantly reduce contaminant levels and health complaints. Ideally, mitigation techniques should have a known and predictable effectiveness and a low cost of implementation. If the cost of mitigation is considered to be excessive (despite being effective), building owners will generally decline to take action, accepting the consequences of the contamination problem. What an "acceptable cost" is depends on the building owner and the unique circumstances of the problem.

The paucity of information relative to contaminant control measures evaluated under systematic laboratory and field trials is one of the major limitations involved in the prescription of mitigation measures. Significant studies of control measures for formaldehyde and radon in residences have been conducted, so control recommendations can be made from a strong scientific base. In many instances, however, systematic studies evaluating the effectiveness of mitigation measures are not available. As a consequence, the choice of a mitigation measure must necessarily depend on the experience and intuitive skills of the investigator.

Mitigation recommendations described in this chapter represent the author's assessment of the available options for controlling specific indoor air contamination problems and the advantages and limitations of each. They are based on a combination of solid scientific research, extensive consulting practice, and strong intuitive skills.

RESIDENCES

Control recommendations for contaminants in residential environments will depend on the findings of the indoor air quality investigation. They may be generic or case-specific.

Formaldehyde

For a variety of technical reasons discussed in Chapter 3, formaldehyde is a difficult contaminant to control. It is particularly difficult to reduce levels below 0.05 ppm.

Case 1

A homeowner complained of illness symptoms in family members that began after the installation of UFFI. On inspection, the presence of UFFI was confirmed in all four sidewalls of a 1400-ft^2 bungalow. Additional sources included a single wood-paneled room and several furniture pieces manufactured from particleboard and MDF. At test, the formaldehyde concentration was 0.06 ppm. On standardizing for temperature, relative humidity, and indoor/outdoor temperature differences, the concentration was 0.12 ppm.

Recommendations. Options available to reduce formaldehyde levels in UFFI houses are limited. One of the most obvious approaches is to have the UFFI removed and replaced with a safer insulating material. Typically, removal costs would be on the order of $7,000-$12,000 for a conventionally clad house and upwards of $15,000-$25,000 for a brick- or stone-clad one.

The effectiveness of removal on reducing formaldehyde levels is determined by the quality of the removal procedure and the presence of other sources. Because UFFI is applied as a semisolid, it impregnates materials in surface contact with it. As a consequence, a residue will remain in wall components after UFFI's physical removal. This residue can significantly affect post-removal formaldehyde levels. It is therefore necessary to treat it with a sodium bisulfite solution, as recommended by the National Research Council of Canada.[1]

Even after the application of the best UFFI removal practices, formaldehyde levels in the range of 0.03-0.05 ppm can be expected, primarily because other less potent sources previously suppressed when UFFI was present are now able to emit formaldehyde freely. Levels of 0.05 ppm may still be sufficient in magnitude to cause symptoms in some sensitive individuals.

A second mitigation approach would be to install an air-air heat exchange ventilation system. If the system is properly sized and installed, formaldehyde levels can be reduced to approximately 0.03 ppm. On a capital and installation cost basis, the use of an air-air heat exchange ventilation system is attractive. A typical cost for a 1400-ft^2 house would be about \$1200-\$1500. This is considerably less expensive than foam removal.

A variety of practical problems, however, are associated with the use of heat exchange ventilation systems for UFFI and other types of formaldehyde-contaminated houses. One of these is sizing. Based on efficacy studies,[2] a delivered air exchange rate of 1-1.5 ACH is needed to reduce formaldehyde levels to 0.03 ppm. In actual use, heat exchange ventilation systems produce only about 50% of the air exchange calculated from design airflows and building volume. As a consequence, the volume flow rate of the system has to be oversized by a factor of two. In northern climates, the exchanger is likely to freeze up under very cold conditions; therefore, it must be equipped with a defrosting system. If the ventilation system is turned off in cold weather, it may be a source of annoying drafts, and in fact, many homeowners do turn such a unit off out of concern about electrical consumption and additional heating requirements. Moreover, if the unit is operated only intermittently, it may not provide the degree of formaldehyde reduction desired.

One of the practical disadvantages of using an air-air heat exchange ventilation system for formaldehyde control in a UFFI house is that, for purposes of a real estate transaction, it is still a UFFI house. The presence of an air-air heat exchange ventilation system is a signal that a toxic hazard exists and, as such, may deter buyers. If UFFI has been removed, it is no longer physically a UFFI house. However, a few northeastern states require notification to prospective buyers that the house once had UFFI in it.

In addition to favorable cost, the use of an air-air heat exchanger for formaldehyde control in a UFFI house is made attractive because it reduces other contaminants as well, most notably radon.

Case 2

A middle-class family recently purchased a new 2400-ft^2 single-story ranch-style home. Soon after its occupation, several family members began to complain of illness symptoms characteristic of formaldehyde exposure. Air testing was conducted under near-worst-case conditions, revealing a formaldehyde concentration of 0.18 ppm. On inspection, the house was observed to have particleboard subflooring throughout, a hardwood plywood–paneled family room, and pressed wood cabinets in the kitchen and bathrooms.

Recommendations. Because of the multitude of sources and house size, this is a difficult problem to solve. Interestingly, it describes many newly constructed single-family dwellings.

The dominant source in this house is likely to be the particleboard subflooring. It alone has the potential for producing formaldehyde levels of 0.13-0.14 ppm. On the other hand, cabinetry alone may produce concentrations of 0.10-0.12 ppm; the paneling may produce concentrations of 0.06 ppm. However, formaldehyde concentrations are not additive. As a consequence, formaldehyde levels may be 20–30% higher than those observed for the most potent source alone. This relationship must be understood before this problem can be solved, that is, before formaldehyde concentrations can be brought down below 0.05 ppm.

In theory, the best approach is to remove all three major sources. For the particleboard, this means (1) taking up the carpets, (2) removing the subflooring and replacing it with a low-emitting material (softwood plywood), and (3) reinstalling the carpeting. Such removal and replacement may well cost over $1000.

In some cases, removal of the particleboard subflooring may not be practical. This is the case when it has been glued to flooring below or applied with an excessive number of nails, or when walls have been constructed on top of it. In such cases, it may be advisable to apply a formaldehyde-sealing coating instead. Examples of such coatings include specially formulated coatings or varnishes with a nitrocellulose base. While the latter are effective, they have a petroleum distillate vehicle and consequently have high VOC emissions, which may be a problem when application is made on large surface areas. An unacceptably long drying time (several weeks) may be required before occupants can comfortably live in the house again. Specially formulated formaldehyde-sealing coatings are water-based and have a rapid drying time. These are the materials of choice in sealing particleboard subflooring. Unfortunately, because of low demand, such sealants are not easily obtained.

Removing the particleboard underlayment can be expected to reduce the formaldehyde concentration from 0.18 ppm to approximately 0.12 ppm. The application of a sealant would be expected to reduce it to about 0.13 ppm. These levels are still excessive.

The next major source of formaldehyde in this house is the pressed wood cabinetry. Most homeowners would be reluctant to remove it because of the cost (thousands of dollars) and social expectations associated with wood cabinetry. In theory, it is desirable to seal cabinetry joints, counter tops, and other such sources of formaldehyde emissions. Unfortunately, the effectiveness of such treatment measures is unknown and is probably quite small, since many unsealed joints are not accessible and formaldehyde may penetrate the coating materials used to finish wood surfaces. Additionally, in many cases, the finish coating on both interior and exterior surfaces may contain a urea-formaldehyde resin. As a consequence, the finish coating may itself be a significant source of formaldehyde.

If one chooses to remove and replace the cabinets, the only way to assure that they do not pose a formaldehyde problem is to use metal cabinets. However, metal cabinets have a low level of social acceptance in middle-class households. Solid wood cabinets, in theory, do not have significant formaldehyde emissions. With the exception of pine, solid wood is rarely used in cabinet making. Pine is, in many instances, an acceptable alternative. But even solid wood cabinets may produce excessive formaldehyde emissions if they are finished with urea-formaldehyde–based coating material.

Hardwood plywood paneling can be easily removed and replaced with hardboard or gypsumboard decorative panels. Replacement of the paneling without first controlling emissions from the particleboard subflooring and cabinetry will not result in any significant reduction in formaldehyde levels.

Because of the complexity and difficulties of using source removal and/or treatment mitigation techniques in a multisource environment, other options should be considered. A relatively inexpensive (several hundred dollars) approach is to conduct an ammonia fumigation (Weyerhauser method). A permanent reduction in formaldehyde levels of approximately 50 + % can be expected. Postfumigation formaldehyde levels, though considerably lower than initial levels, may still exceed health and comfort levels. Another alternative is to install an air-air heat exchanger (oversized as previously described). Because of the larger house size (as compared to the UFFI house in Case 1), a very large heat exchanger unit would be required. A heat exchange ventilation system with a delivered air exchange rate of 1 ACH would be expected to reduce formaldehyde levels by about 60%, or to about 0.07 ppm. The same system applied after an ammonia treatment could be expected to reduce formaldehyde levels to approximately 0.03–0.04 ppm, the minimum practicable level of control.

Case 3

A young couple with an infant child purchased a new mobile home. The infant repeatedly became ill with upper respiratory symptoms and respiratory and ear infections. The adults also reported symptoms but were not as severely affected. On investigation, the formaldehyde concentration (under near-worst-case conditions) averaged 0.30 ppm. Formaldehyde sources included particleboard decking and particleboard wall cabinets and shelving.

Recommendations. Options for controlling formaldehyde levels in mobile homes are very limited. The major source of formaldehyde in this instance is the particleboard decking. Because interior and exterior walls are built on it, it cannot, as a practical matter, be removed and replaced. In theory, it can be treated with a formaldehyde sealant similar to those applied to decking by some mobile home manufacturers during the construction process. Commonly, sealants are applied to the upper surface with the reasoning that emissions from the lower surface do not enter the living space. This is unlikely to be the case, though, since the cold air return is beneath the particleboard decking. Therefore, the efficacy and desirability of sealing the particleboard decking is in doubt. If emissions from the particleboard decking cannot be reduced, there is then no purpose in attempting to reduce emissions from the cabinetry and shelving.

The only viable option for significantly reducing formaldehyde levels (to the Canadian target level and California guideline of 0.05 ppm) is to conduct an ammonia fumigation followed by the installation of an air-air heat exchange ventilation system with a delivered air exchange of 1 ACH. A 48-hour treatment with ammonia using the Weyerhauser method could be expected to reduce formaldehyde levels to 0.10 ppm, with a further reduction to 0.05 ppm by the use of an air-air heat exchange ventilation system.

An ammonia treatment is relatively inexpensive and therefore practical. Because mobile home dwellers as a group have limited finances, the additional step of installing and operating an air-air heat exchange ventilation system may not prove practical. It also assumes that the mobile home is equipped with an air conditioning unit or system.

Though effective, the ammonia fumigation method has limitations. Because of the need for high ($\sim$1000 ppm) and very toxic ammonia levels, special care must be taken in its application. Professional application is recommended. Efforts must be taken to remove plants, pets, food, light oak furniture, and clothing from the indoor environment. To ensure safety on reoccupancy, the fumigated environment must be well ventilated for a week or more.

Case 4

A 25-year-old ranch-style house was renovated with the addition of a large family room. The family room added an additional area of 400 ft^2 to an existing 1400-ft^2 house. In the renovation, particleboard underlayment was used on the flooring and all walls were paneled with quarter-inch hardwood plywood. A week after completion, family members began to notice symptoms of eye and upper respiratory irritation. Air testing for formaldehyde indicated housewide formaldehyde levels of 0.10 ppm.

Recommendations. Because of the relatively low volume of urea-formaldehyde–bonded wood products present with respect to the total area, this problem can be easily mitigated. Removal of both the underlayment and wall paneling is the preferred mitigation strategy. Alternatively, the particleboard underlayment could be sealed with a specially formulated sealant or with nitrocellulose-based varnish. The paneling could also be removed, sealed from the back side, and reinstalled. If the paneling were to be removed entirely, it could be replaced with decorative wall panels made of hardboard or gypsumboard.

Case 5

A homeowner complained of illness symptoms occurring after the purchase of new bedroom furniture. The symptoms of upper respiratory system irritation were most noticeable on awakening. Symptoms tended to resolve somewhat during the day. Results of monitoring indicated formaldehyde levels in the closed bedroom of 0.10 ppm, and 0.04 ppm in the rest of the house. Air sampling within furniture cavities revealed formaldehyde concentrations of 0.30–0.50 ppm.

Recommendations. As most contemporary wood furniture is constructed from pressed wood materials, the case described is common. It is easily resolved by removing and replacing the furniture with older, previously used furniture or with any furniture older than five years.

Case 6

A couple in their mid-50s moved into their recently constructed 1800-ft^2 home. Within weeks, they were plagued by upper respiratory symptoms, headaches, unusual fatigue, etc. The symptoms became more severe on warm, humid days and during rainy weather. The house was tested six months after construction and found to have a formaldehyde concentration of 0.12 ppm. On investigation, hardwood plywood kitchen and bathroom cabinets were the only sources identified as potential sources of significant formaldehyde contamination. Counter tops were constructed from exterior-grade plywood. The formaldehyde concentration in the cabinet interior tested out to be 0.35 ppm. The formaldehyde concentration in this house was surprisingly high, given the fact that cabinets had been constructed from low-emission hardwood plywood and were present only in a moderate volume. Further investigation, however, revealed that they were finished with an alkyd urea-formaldehyde base and top coat with a high potential for releasing formaldehyde during and after application.

Recommendations. Options for mitigating this problem are limited. The best approach is to remove the cabinets and replace them with units not coated with an alkyd urea-formaldehyde–based finish. This may prove difficult, since they are the dominant finishes used in the kitchen cabinet industry. A second approach is to conduct an ammonia fumigation. This may not be desirable, since the high ammonia concentrations required may damage the finish. A third options is to install an air-air heat exchange ventilation system (as described in previous cases).

Other Cases

Formaldehyde problems in residences can also be demonstrated by a variety of other cases, variations of those described above. These would include a conventional house constructed in 1980 with both particleboard subflooring and urea-formaldehyde foam insulation, a conventional house with wood cabinetry as the only major contaminant source, and so on. The reader is referred to Case 3, which discusses control approaches applicable to these two cases as well.

Commentary

Based on experience, homeowners on the whole are reluctant to implement any control measure that requires a commitment of either time or money. This reflects economic factors, interpersonal relations between spouses (the wife is ill, the husband is not!), the inability of the investigator or consultant to guarantee that control recommendations will resolve the problem, and psychosocial factors favoring pharmacological/medical relief of symptoms as opposed to elimination of causes. Treatment of symptoms is often reimbursable through medical insurance. Formaldehyde-induced symptoms usually affect an individual's quality of life but are not usually life-threatening. Many individuals will opt to continue living with the problem or implement some "interim" control measures that, in essence, become permanent. The most common interim measure is the opening of windows, even during cold weather. Where levels are low to moderate (0.05–0.20 ppm), this can be used effectively to reduce formaldehyde levels and to alleviate symptoms. On a long-term basis, it is not cost-effective. Another interim approach is to control indoor climate by maintaining low temperatures (18–21°C) during the heating season and by air-conditioning to temperatures of 22–24°C during the cooling season.

Radon

In studies under EPA sponsorship, specific radon control plans have been described for high-radon houses.[3] Because these are available from EPA, they will not be repeated here. Cases described below are applicable to houses exceeding the EPA guideline of 4 pCi/L by no more than a factor of 10.

Case 1

A family had their home tested for radon with a track-etch detector. The three-month average concentration was 15 pCi/L on the first-floor level and 30 pCi/L in the basement. The basement was completely finished and underlain with a good system of drain tiles and fill. The drain emptied to a sump.

Recommendations. The presence of a drain tile system and sump suggests that the best mitigation approach is to vent the sump by installing first a sump cover and then a plastic vent pipe with an in-line fan (with sufficient flow capacity and horsepower rating). All openings (such as those around toilet fixtures, etc.) should be closed. All major cracks in the basement slab and walls should be sealed with a flexible caulk to maximize the negative pressure placed on soil gas. If the basement has a concrete block wall, the block voids at the top of basement walls must be sealed with mortar or some other effective barrier. Though a very effective technique when applied properly, the application of sump ventilation is limited by the ability of the radon mitigator to locate all major openings and cracks. This may be very difficult in a finished basement.

Assuming a finished basement and a presumed difficulty in locating and sealing openings and cracks, an alternate approach is to install an air-air heat exchanger with the objective of ventilating only the basement. Radon reductions of 90 + % can be expected for a cost of about $1500 for equipment and installation.

Case 2

A 2500-ft^2 ranch-style house built on a slab was tested with a track-etch detector. The three-month average concentration was 15 pCi/L.

Recommendations. This radon contamination problem can be mitigated by the installation of a sub-slab ventilation system whose effectiveness will depend on the adequacy of fill under the slab and the ability of the mitigator to locate and seal all major openings in the slab. The latter is made difficult by the presence of floor covering such as carpeting, which must be taken up and reinstalled.

This problem can also be mitigated by the installation of an air-air heat exchange ventilation system. However, it is difficult to install such a system in this type and size of house. The unit must be placed in an unheated attic or garage and may require a special duct system. It must be sized sufficiently to provide an effective air exchange rate of 0.5 ACH. Using the rule of thumb previously described for formaldehyde, the system should have a design air flow to provide 1 ACH.

Case 3

An 1800-ft^2 ranch-style house constructed on a crawl space was tested with a long-term monitor and found to have a radon concentration of 8 pCi/L. Testing was conducted during the winter period with the crawl space vents completely closed to conserve energy and to reduce the problem of "cold floors."

Recommendations. In this house, radon levels may be reduced significantly by simply opening crawl space vents during the heating season. The effectiveness of this simple mitigation measure should be assessed by retesting with crawl space vents open. If the desired reduction has not been achieved, installation of a large fan may be in order. The fan should be installed so that air is blown into the crawl space rather than exhausted. As ground surfaces in a crawl space are often moist, the use of a fan in an exhaust mode will result in significant fan corrosion.

Biogenic Particles

Case 1

An eight-year-old child had developed asthma. The child was atopic with a strong sensitivity to dust mites. On investigation, the house was found to be approximately 25 years old and located on a heavily shaded site. It was situated on a crawl space with evidence of water seepage. The earth in the crawl space was moist. The house was fully carpeted except for kitchen and bathroom areas. The treating physician had recommended the implementation of dust control measures directed to the patient's bed. Such measures did not bring the patient much relief. The cold air return duct was in the crawl space.

Recommendations. Based on house characteristics, it can be assumed that conditions are favorable for the establishment of significant dust mite populations in the child's room and other parts of the house. Relatively speaking, it is a moist house.

There are a variety of environmental modifications which may be applied to this house to control dust mite populations, thus reducing the frequency of asthmatic attacks. The first recommended action is to remove the carpeting in the child's bedroom and replace it with wood flooring or a vinyl floor cover. The carpeting provides an excellent microenvironment for dust mites and is not very amenable to the application of dust control measures. An alternative approach to removing and replacing the carpeting is to cover it securely with a polyethylene or polypropylene transparent plastic. Vacuuming of room surfaces should only be conducted with a HEPA vacuum cleaner or one that collects particles in a liquid medium.

The next environmental management step is to control indoor moisture levels. If an air conditioning system is in place, it would be desirable to air condition the house at all times during hot/humid weather and during rainy weather as well. The latter is notable, since rainy conditions often occur during cooler seasons when air conditioning would not normally be turned on. In such instances, air conditioning can be operated only for humidity reduction. Because it may be difficult to maintain thermal comfort under such seasonal conditions, it may be more desirable to dehumidify the air with a portable dehumidifier. In a 2000-ft^2 house, a dehumidifier unit with a 40-pint capacity will be sufficient to maintain relative humidity below 60% during rainy weather. It may be desirable to install an air conditioning system in a house that is not so equipped. If used only on an intermittent basis, it is unlikely to control humidity (and therefore dust mite populations) sufficiently.

The crawl space is likely to be one of the major sources of water vapor in the house interior. It is therefore desirable to implement measures to reduce water vapor entry from the crawl space. There are several approaches that may be effective, applied singly or in combination. The average homeowner would most likely want to try those measures that would be relatively easy and inexpensive. Along this line, it would be desirable to inspect cold air return ducts to determine whether, due to leaks, there is a significant potential for moisture-laden crawl space air to be drawn into the building. All major leaks in the cold air return should be sealed with duct tape. The crawl space should also be ventilated passively by maintaining all vents in the open position and removing obstructions to air flow such as foundation plantings. It could also be ventilated by the installation of a fan. As previously described for radon, the installation should be in the positive pressure mode in order to reduce the significant fan corrosion potential associated with the exhaust mode.

Moisture ingress into the building from the crawl space can also be reduced by site modifications to reduce water seepage and moist soil surfaces in the crawl space. Site modifications would most likely include the installation of drainage tile. This may or may not be practicable, depending on house location.

One of the more extreme actions would be to reduce shade levels by cutting trees down near the house. An even more extreme action would be for the family to relocate to a house environment that would pose a lower risk to a dust mite–sensitive asthmatic child. This could involve the purchase of another house or the construction of a new house. Relocation has the potential for being very effective. Such decisions must, however, be based on realities and knowledge consistent with environmental factors contributing to the asthmatic condition.

Case 2

A child had recently developed asthma. On allergy testing, the child showed a strong positive response to a mixture of mold allergens but was not sensitive to dust mites. Air sampling with a viable particle sampler indicated mold spore levels of 3000–4000 CFU/m^3. The ranch-style house had a partially completed basement. On inspection, visible mold growth was observed on window frames and sills, metal storm doors, and basement wall surfaces. Water seepage through basement walls was also evident. Though a sump was present, the sump was blind with no drainage tile connected to it.

Recommendations. This is a severe case of mold infestation. It will require a significant mitigation effort. The first effort should be directed to resolving the drainage problem. It will require the installation of drainage tile around the basement perimeter and under the slab as well. Depending on the nature and depth of fill, the existing pump may prove suitable or it may be desirable to install one in a perimeter area.

It is likely that almost all materials in the basement are mold-infested. It would therefore be advisable to discard all biodegradable materials such as paneling, carpeting, stored boxes, books, magazines, etc. Clothing can be disinfected by washing and bleaching.

Mold-damaged materials in the living area should be discarded. If this is not practical, then attempts should be made to clean, disinfect, and seal them. Mold growth on the storm doors can be removed by cleaning with strong detergent and by the use of bleach. Mold growth on wooden window frames and sills will require the use of an abrasive to remove surface mold growth and varnish residue. Because mold is likely to have penetrated the wood by a centimeter or more, it will be desirable to apply a strong disinfectant solution and allow it to penetrate the wood. On drying, wood members should be sealed with a product with a high volatiles content, such as sander/sealer. Wood frames and sills can then be given two coats of varnish.

Even with the installation of a drainage system, this house will have a tendency to be moister than most. This is, in part, due to the fact that it has a basement and is located in shade. It would be desirable to dehumidify the basement during the warm months on a continuing basis. Dehumidification of the living environment can be accomplished by the operation of a central air conditioning system.

If the windows are single pane and subject to condensation in the heating season, they should be replaced with double- or triple-glaze types.

Case 3

Several children and adults experienced upper respiratory system irritation. On the advice of their physician, they humidified their home during the heating season. As a consequence of the humidification, condensate formed on window panes and ran onto frames and sills which became darkly stained from mold growth. Two of the children were tested by an allergist and found to be strongly allergic to mold. On inspection, no mold odor was evident. Rather, the odor was characteristic of a relatively new house with particleboard underlayment. The house had particleboard subflooring, which was approximately six years old. It had been recently remodeled, with an addition extending to within a few feet of a septic tank drainage field. The addition was constructed on a crawl space. However, the contractor neglected to install any crawl space vents.

Recommendations. In this house, the upper respiratory symptoms were probably caused initially by formaldehyde exposures due to emissions from the particleboard subflooring. The attempt to alleviate these upper respiratory symptoms by the use of humidification caused condensate to form on the windows during cold weather conditions. The condensate damaged the wooden window frames and sills, providing an excellent medium for mold growth. Subsequently, mold levels were sufficient in magnitude to cause sensitization in two of the children.

This house has both a formaldehyde problem and a mold problem. The best approach to the formaldehyde problem is to remove the particleboard or treat it with a formaldehyde sealant or two coats of nitrocellulose-based varnish. The mold-damaged window frames and sills can be treated in the same manner as described in Case 2. To prevent future condensation, it is advisable to discontinue humidification and to take steps to minimize the inflow of moisture-laden air from beneath the crawl space. The latter is a unique problem. The proximity of the septic leach field to the addition will by capillary action result in a significant quantity of moisture under the crawl space. Due to negative pressures in the house, this moisture can be expected to be drawn into the house in the form of water vapor. The lack of crawl space ventilation can be expected to further exacerbate the problem.

Because of the lot size and the absence of a city sewer, high soil moisture levels near this house cannot be corrected. The most practical approach is to install vents in the crawl space and to install and operate a fan in a positive pressure mode to provide crawl space ventilation.

Case 4

A family lived on a heavily shaded lot in a house that was approximately 90 years old. The house was originally built on a crawl space that had been dug out sufficiently to provide space for a furnace and a root cellar. Air returned to the furnace through grills in the basement door. The cold air return was the stairwell, which was partially isolated from the rest of the basement by walls and two doors. Return air was drawn into the furnace from the stairwell through a plenum along the floor joists. The homemaker, who was 55 years old, had been plagued by a variety of upper respiratory symptoms including those characteristic of hypersensitivity pneumonitis. On visible inspection, the basement area had a noticeable musty odor, as did the upstairs living area. Viable mold sampling revealed living area mold levels in the range of 200–300 CFU/m³, with levels in the basement stairwell of 600 CFU/m³. There was evidence of water standing in the basement area and damaged joists below the bathroom. Visible mold growth was observed on bathroom walls, though viable mold levels appeared to be low. An electronic air cleaner had previously been installed in a return air duct immediately preceding the blower fan. Inspection was conducted on a cold day, and a strong flow of air from the basement area through cracks in the hardwood floor of the master bedroom was noticeable when a hand was placed near the floor.

Though viable mold sampling did not reveal high mold spore counts, there was considerable evidence that this house did in fact have a mold problem. The homemaker was sensitive to mold on allergy testing and had other complaints suggestive of hypersensitivity pneumonitis. The house had a musty odor and an earthen basement. The basement had been subject to water intrusion from an upstairs bathroom, and standing water was common during the spring months. There was evidence of visible mold growth on some of the floor joists. Additionally, the use of the stairwell as a return air plenum provided an avenue of basement mold spore entrainment in the recirculated supply air. Mold spore–laden basement air could also be readily transported into living spaces through floor cracks, as well as in upward air flow through the basement stairwell when the blower fan was not engaged.

Recommendations. There are three approaches which may be applicable to the mitigation of this mold problem. These are: (1) operating the blower fan and air cleaner 24 hours a day, (2) isolating the furnace cold air return from basement air, and (3) implementing measures to keep the basement dry.

Since a central electronic air cleaner was available, it may be desirable to operate the air cleaner continuously, as this would in theory direct both living area air and basement air (leaking into and through the stairwell plenum) through the air cleaner. Cleaning efficiencies on a per-pass basis should be on the order of 80–90%. In this case, however, continuous operation of the furnace fan and air cleaner resulted in an increase of upper respiratory system complaints. On reinspection, large gaps were observed in the duct system between the furnace blower and the air cleaner, as well as in the housing around the blower. It was evident that considerable basement air was bypassing the air cleaner, and the furnace itself significantly contributed to the transport of mold spores into living spaces.

Simply isolating the cold air return from basement air will not be sufficient. It appears that the entire heating system should be isolated from the basement. Options may include installing the furnace in a dry garage area, a closet in the living area, etc.

Even with a resolution of the furnace problem, the basement can still be expected to be a source of mold spores. The installation of a drain and sump system would be desirable, as would the covering of bare soil with polyethylene plastic.

Mold growth on the bathroom walls was also evident. Damaged wall materials would be best discarded and replaced. The bathroom should be provided with an exhaust fan, which should be operated when occupants bathe.

Case 5

A young married woman complained that the air in her recently purchased 25- to 30-year-old 1400-ft^2 brick house made her sick. She suspected some type of toxic chemical and was reluctant to live in the house. On preliminary interview, it was apparent that formaldehyde as a causal factor was unlikely to be the case, and there was no indication of a mold problem. On investigation, the house had a strong odor suggestive of mold. A window located at mid-level in the shower area allowed water to enter the wall. Carpeting in the house was old and in poor condition. The crawl space was dry and there was no evidence of standing water. The young woman was referred to an allergist, and on testing was found to be strongly allergic to mold.

Recommendations. Mitigation of this problem requires extensive renovation of the bathroom. The wall should be opened and all mold-damaged materials, including wall studs, removed and discarded. In renovation, the window should be eliminated. Damaged tile and grouting on shower walls should be replaced and an exhaust fan installed. Because the carpets are old and in poor condition, they are likely to be mold-infested; thus, removal and replacement is desirable.

Carbon Monoxide

Typically, carbon monoxide problems are identified after serious accidental poisonings. Mitigation involves the correction of the air or water heating system defect or malfunction by a heating contractor. This includes the replacement of furnace parts, flue vents, and/or the entire combustion appliance. It may also include the removal of chimney obstructions and the installation of liners.

Case 1

Carbon monoxide poisoning of an urban family occurred at approximately mid-afternoon. Two residents plus several visitors were overcome by carbon monoxide fumes from a furnace in the basement. One of the visitors died as a consequence. The poisoning resulted from a blocked flue. Occupants complained of headaches, nausea, and drowsiness for weeks prior to the episode. On inspection, the flue was observed to be blocked by soot and pieces of mortar from the chimney. No CO levels were measured in the house, but high carboxyhemoglobin levels were reported in the blood of all affected, including the decedent.

Recommendations. This poisoning could (and should) have been avoided by preventive maintenance. Since the chimney in this 80-year-old house had been designed for a coal furnace, it had a soot clean-out port below the furnace flue vent connection. The buildup of soot from coal combustion makes such clean-out ports essential. In this case, however, the heating system had been converted to oil heat more than 20 years ago, and occupants apparently did not see the need for annual inspections and soot removal.

Mitigation of future problems would require removal of the soot at the clean-out port, cleaning of the chimney by chimney sweeping techniques, and installation of a liner to prevent brick and mortar fragments from falling and accumulating.

Case 2

An elderly couple complained that they suffered episodes of CO poisoning. As a consequence, they had their gas-fired furnace replaced with a new one. They continued to complain of gas odors. Personnel from the local gas company tested the air in their home, finding no CO in detectable levels. The couple continued to complain, with each investigation showing no detectable CO levels. An air quality investigation was conducted. Measurements revealed no detectable levels of CO. However, living spaces had a strong odor of gasoline, traceable to a leaky lawnmower tank and a partially closed gasoline can. Several months later, the old couple again complained of gas odors. Measurements revealed no detectable levels of CO. The strong sulfurous odor of natural gas present in the living area was traceable to the gas furnace in the garage. After correction of this problem, complaints continued to occur. The numerous requests for investigations to gas company personnel and heating contractors resulted in the characterization of the old couple as being "crack pots." In response to an episode, the old couple had their blood sampled and analyzed. The carboxyhemoglobin content was a surprising 17%. On investigation, CO levels were, again, not detectable.

Recommendations. This case represents a situation where CO exposures were episodic. Complaints were resolved after the fireplace damper was closed. The fireplace and furnace flues were approximately six inches (15 cm) from each other, with the furnace flue upwind of the fireplace flue. Apparently, under certain weather conditions, flue gases entered living spaces through the fireplace. Because of their age, the occupants had failed to close the fireplace flue for the entire period during which they complained of so-called "gas odors."

Miscellaneous

Case 1

A young family moved into a 90-year-old two-story house that had been recently converted into apartments. Almost from their initial occupancy, they began to complain of a variety of symptoms that appeared to be, in part, characteristic of formaldehyde and carbon monoxide. Their second-floor apartment was plagued on an episodic basis by sewer gas odors. It contained several major sources of formaldehyde, including hardwood plywood paneling in a bedroom, a living room, and a hallway. Kitchen cabinets were made from medium-density fiberboard. On air testing, the formaldehyde concentration three years after occupancy was 0.06 ppm. Measured ammonia levels were 18 ppm. The second-story apartment was connected to the basement by blind ducts.

Recommendations. This case is a complex one. From the description of symptoms and known sources present, it appears that the problem was at least partially due to formaldehyde. The sewer gas odors and measured ammonia levels indicate that sewer gas was a significant factor, with ammonia levels in sufficient magnitude to cause upper respiratory symptoms. Interestingly, the symptom complex suggested that CO exposures may have also been occurring.

The blind ducts appear to be a significant factor in the problem. They had in the past been connected to a coal furnace, which was replaced by two gas furnaces and baseboard water heat when the renovation took place. The blind ducts served as a conduit between basement air and the upper-level apartment. They presented a stack effect potential that apparently placed the basement under a strong negative pressure, causing sewer gases to be drawn in from an open drain connection. It also created the potential for backdrafting of flue gases.

Resolution of the sewer gas and potential backdrafting problem requires the closing of the open drain and sealing of the blind ducts. If the complaint diminishes in intensity but still persists, consideration should be given to removal of the hardwood plywood paneling and kitchen cabinetry.

Case 2

Immediately after the installation of new carpeting, a family was plagued by a variety of upper respiratory symptoms. On investigation, there did not appear to be any other building-related circumstance which could have accounted for the problem. No new underlayment material was placed under the carpeting.

Recommendations. This is a problem that can be expected to resolve in about 3-6 months, since VOCs from carpeting diminish rapidly with time. The best approach to this short-term problem is to provide significant continuous ventilation by opening windows.

VOCs from new carpeting placed in a closed environment can occur in sufficient concentrations to cause mucous membrane irritation. The problem is most acute when carpeting comes directly from the factory and is immediately installed. Potential health complaints can be mitigated by the installation of carpeting during seasons when environmental conditions are favorable to opening windows and doors. Such natural ventilation can be expected to cause a dilution of VOC levels and to accelerate their rate of decay.

Case 3

A pesticide applicator mistakenly sprayed a mixture of chlorpyrifos and dichlorvos on the baseboard floor junctions of a two-story house. (It was actually requested by a neighbor next door.) Family members complained of chemical fumes and odors and claimed a variety of illness symptoms. Even after several months, the complaints failed to resolve. Air testing and wipe samples revealed only trace amounts of the two primary ingredients in the pesticide formulation.

Recommendations. The problem can be expected to resolve with time. Perceived chemical fumes and odor were most likely due to the volatile substances used to dissolve the pesticides and to carry them to targeted surfaces on spray application. Because of their high volatility, they can be expected to dissipate after a matter of weeks, particularly if the environment is well-ventilated by opening windows and doors.

Because of their semivolatility, the active pesticide ingredients can be expected to persist for a longer period of time. Their removal rate may be affected in great measure by scrubbing the sprayed areas with a strong alkaline detergent and by opening curtains and drapes to allow the entry of sunlight into living spaces. Pesticides such as chlorpyrifos are sensitive to ultraviolet light and the entry of sunlight into contaminated spaces can be expected to accelerate pesticide decomposition.

Case 4

A family was plagued by significant eye and upper respiratory system irritation. The house had a strong sweet odor, particularly when the heating system was operating. An investigation revealed that the odor was coming from a new heat pump installed in an unheated attic. The source of the odor, a fiberglass insulating material, was confirmed by sample testing (for odors).

Recommendations. This problem can be resolved rather simply. The fiberglass insulating material should be removed from the inside surfaces of the heat pump housing. The unit can be insulated by placing insulating materials, including fiberglass, on the exterior surfaces of the heat pump unit.

Case 5

A family complained of recurring odors, described as being characteristic of "sweaty gym socks." The source was identified as the air-conditioning unit.

Recommendations. It is likely that this problem is due to microbial growth (particularly bacteria) on the condensate drip pan. The drip pan and any other surface with visible microbial growth should be cleaned by scrubbing with a strong detergent. (Chlorine bleach may also be used. However, on operation a strong odor of chlorine bleach can be expected for a few days.)

Case 6

A family installed a high-efficiency furnace near the laundry area of their basement. Within a few weeks, supply air ducts became significantly corroded and occupants complained of headaches and a variety of irritating symptoms.

Recommendations. This problem was relatively common when high-efficiency furnaces were first developed and marketed. Because the extraction of heat from flue gases is very efficient (90 + %), vapors like water condense. If combustion air is taken from within the home, chlorine from water (the nearby laundry, in this case) or chlorinated solvents (such as trichloroethylene, etc.) can be converted to hydrochloric acid. The acid causes the corrosion of the heat exchanger, facilitating the passage of flue gases and corrosive hydrochloric acid into supply air ducts. The best solution to this problem is to modify the system so that combustion air is brought in from the outside.

Using Residential Air Cleaners

Many homeowners and apartment dwellers, in coming to realize that they have a significant indoor air contamination problem, respond by purchasing an air cleaner. Air cleaners are also recommended to allergy and asthma patients by their treating physicians. They are seen as a generic and relatively simple solution to building-related health problems.

In sound mitigation practice, air cleaners should generally not be the first choice. Source control measures are technically more desirable because they deal with the cause of the problem. Nevertheless, air cleaners can be recommended for certain limited circumstances. They should be selected and operated in a fashion that will maximize contaminant reduction.

Commercially available residential air cleaners are usually dust cleaners. They vary considerably in their ability to collect dust particles. Though most claim to remove gases and odors as well, such claims are usually without substance.

Choice of an air cleaner should be made, in most cases, on the basis that the primary purpose is to remove airborne particles. This may be for purposes of keeping residential surfaces cleaner and reducing concentrations of biogenic particles, which are responsible for allergy symptoms and/or asthma attacks.

In selecting an air cleaner, the consumer must wade through a host of models and claims. They may initially be drawn to the small desk-top panel filter cleaners because of their low cost. As described in Chapter 6, such air cleaners are worthless. Ionizers appear to be effective for tobacco particles, particularly in a small closed room. Though with some models airborne particles plate out on room surfaces (the "dirty wall effect"), newer models are designed to draw ionized particles to ionizer surfaces or through a filter. Although the "dirty wall effect" may be undesirable, ionization has the potential for providing room-wide air cleaning without the dead space and short-circuiting problems associated with fan-driven cleaners. One of the attractive features of ionizers is their low cost ($100 or less).

In addition to the small panel filters and ionizers, numerous modular fan-driven filtration units are available. These typically are powered by fans that can move 100–400 CFM of air at zero static pressure. They may employ fibrous filters or remove particles by electrostatic means. Fibrous filters may be simple, relatively densely packed panels or extended media filters, such as HEPAs. In many instances, fibrous filters employed with such devices have high particle collection efficiencies (>90%). Many commercially available devices, however, are equipped with blower fans that cannot tolerate the significant pressure drops associated with the high-efficiency filters employed. As a consequence, air flows claimed to be as high as 400 CFM (based on zero static pressure) are usually considerably less than 100 CFM. Additionally, the chassis of many such devices are made from particleboard or MDF. These devices are relatively expensive ($200-$800). Because of low air flow rates and high cost relative to their effectiveness under actual operation, most portable fibrous media air cleaners commercially available are not worth purchasing. Ironically, these devices are widely sold by medical supply dealers to allergy patients.

Of the portable fan-driven air cleaners commercially available, those of the electronic or electrostatic variety are likely to be the most effective under actual operation. This is due to their high-rated collection efficiencies (80–90 + %) and their minimal resistance to air flow. Because of low resistance, the rated air flow at zero static pressure is close to the delivered flow in cleaner operation. Air flows may be on the order of 100–400 CFM, depending on the model. Unit prices of such models are in the range of $250-$500.

Portable air cleaners intended for dust particle control are most effective when operated continuously in a small enclosed room. Their effectiveness is diminished when operated intermittently, when air moves through the room from forced-air heating and cooling systems, and when room doors are left open. To maximize dust cleaning effectiveness, particularly for bedrooms of allergy/asthma patients, portable electronic air cleaners should be operated continuously, with the room isolated from air flows from other parts of the building.

Whole house or apartment cleaning can be achieved by the use of extended-media fiber filters and electronic air cleaners installed in the cold air return of a central heating and cooling system. In such applications, both filter types can be very effective. In practice, however, the actual air cleaning likely to be realized is going to be small, because the air cleaner usually operates only when the heating/cooling blower fan is operating. Operating frequency is particularly low during moderate weather. For the effective particle control required by allergy/asthma patients, the furnace blower fan should be operated continuously. Because of the large size and horsepower rating of the furnace blower fan, continuous operation results in significant power consumption. It can, however, be used effectively in controlling biogenic particles such as mold in residential environments.

Beyond dust cleaning, few commercially available air cleaners are effective in removing gases and vapors. Room temperature catalyst types are limited in application to a few inorganic gases, such as CO. Those that employ densely packed thin-bed carbon filters with adequate fan power and flow capacity are effective in removing higher–molecular weight organic compounds. They are generally ineffective in controlling formaldehyde. Even when a special medium such as potassium permanganate on activated alumina is used, it rarely does an adequate job and its activity is expended within a matter of months.

Many patients diagnosed as being "chemically sensitive" use air cleaners designed to remove VOCs. Several well-designed, commercially available air cleaners based on charcoal can in theory provide effective control of VOCs when used continuously in a single isolated room. They are not likely to be effective in controlling formaldehyde and other low–molecular weight organics, even when filters contain a mixture of gas-adsorbing media.

Building a New House

Many potential indoor air quality problems can be avoided or minimized by thoughtful planning prior to construction of a new house. To prevent excessive levels of formaldehyde, the prospective home builder should (1) avoid the use of particleboard subflooring, (2) use no hardwood plywood paneling, or minimal quantities, and (3) avoid cabinetry made from particleboard and MDF components. Hardwood plywood cabinets may also be a problem, particularly if a urea-formaldehyde resin–based finish is applied. Counter tops should be constructed of exterior-grade plywood. If particleboard is used, it should be enclosed on all surfaces with a plastic laminate material. To be very confident about formaldehyde emissions from cabinets, the home builder should construct them from solid pine components. Though less aesthetically pleasing, metal cabinetry is a desirable alternative to that constructed from hardwood and composite material components. Avoidance of pressed wood materials in house construction should be extended to pressed wood furniture as well, since it can be a significant source of formaldehyde.

Radon exposures can also be minimized by thoughtful planning prior to construction. The potential of a site for radon transmission into a house can be roughly determined in many areas by a review of a Soil Survey Report, obtainable from the Soil Conservation Service or county agent. Soils with a heavy clay base typically will have low radon-releasing potential. On the other hand, soils with a sandy/gravelly base may have a considerably higher radon transport potential. If radon is a concern, the latter soils should be avoided.

Radon exposures can also be minimized simply by one's choice of substructure. A house on a basement can be expected to have higher radon levels than one on a slab. A house on a slab can be expected to have higher radon levels than one on a crawl space. A crawl space substructure is also very amenable to radon mitigation after construction. If a crawl space substructure is selected, it is desirable to locate the cold air return in the garage or attic.

Though sandy/gravelly soils as house sites have the disadvantage of higher radon exposure potential, because they are dry they do represent desirable sites for individuals whose family members are atopic for common allergens such as mold and dust mites. The selection of a dry site is particularly desirable for individuals with asthma or chronic allergic rhinitis. The site should be open with no or minimal shade. Individuals with asthma or allergic rhinitis would be best served by constructing their house on a slab or crawl space. Basements should be avoided, since they are prone to both moisture problems and high relative humidity. If the site is not dry, it should be sufficiently modified prior to construction to insure that moisture problems do not arise. To minimize dust mite populations in houses constructed for allergy or asthma patients sensitive to dust mite antigen, consideration should be given to the use of hardwood, vinyl, and ceramic flooring material in lieu of wall-to-wall carpeting, particularly in bedroom and major living areas. All bath and shower areas should be provided with an exhaust fan.

In moving into a new house, one can expect significant VOC levels associated with paints, varnishes, adhesives, and carpeting. VOC levels can be minimized by the use of water-based paints and varnishes. Carpeting vapors can be minimized by purchasing from stocks on hand at a retail outlet as compared to ordering from a factory. It is also desirable to time one's initial occupancy to coincide with moderate weather conditions when significant ventilation can be provided by opening windows or doors. This will dilute VOC concentrations and accelerate their rate of decay from household sources.

The prospective house builder can expect that the soil around the substructure will be treated for termites. It is more likely than not that the primary active ingredient in the pesticide formulation used will be chlorpyrifos. To minimize pesticide entry into living spaces, it would be desirable to locate heating/cooling return and supply ducts in the attic area.

The prospective home builder has a variety of choices of heating systems. Electric heat described as cable heat would have the least prospect of causing a contamination problem. This may not be the case for heat pumps, which are often insulated internally with fiberglass materials. Such insulated systems should be avoided, as should high-efficiency furnaces providing only for an indoor combustion air supply. Furnaces installed external to a building, functioning by heating a refrigerant that is transferred to supply air in a fiberglass-insulated box in the building, should also be avoided. Well-designed gas/oil forced-air or radiant heat systems are reliable and widely used. With proper maintenance, the indoor contamination potential is low.

PUBLIC ACCESS BUILDINGS

Mitigation practices applied to indoor air quality problems in public access buildings, like those in residences, depend on what the problem is and on its unique circumstances. In contrast to residences, it is not uncommon in large buildings to have more than one problem extant at the same time.

Case 1

Employees in a new office building complained of a variety of acute irritating symptoms, including those of the eyes and upper respiratory system. On investigation, the ventilation system appeared to be operating properly, with measured CO_2 levels in the range of 600–800 ppm. VOC levels were not measured. The building, though, had a strong solvent odor. Because of the alleged health problems, some employees refused to come to work and threatened legal action.

Recommendations. Complaints in this case are suggestive of exposure to high VOC levels characteristic of many new buildings. The problem is likely to be transient, resolving in a period of 3–6 months. The fact that employees expected immediate action by management makes self-resolution an unacceptable option. Two approaches may serve to mitigate the problem in a shorter period of time. The best approach is to operate the building continuously on 100% outside air for a period of three months or so. Such operation can be expected to accomplish two things. First, by increasing the amount of air available for dilution, contaminant levels can be expected to be significantly reduced immediately. Secondly, by increasing the ventilation rate, VOC emissions from source materials can be expected to decay at an accelerated rate. Another approach to this problem is to conduct a bakeout. The building could be heated to a high temperature (32 + °C) for several days when it is unoccupied. Significant ventilation should also be provided during the bakeout period.

Case 2

Employees in a large public library building complained of a variety of symptoms, including headaches, fatigue, upper respiratory system irritation, etc. An air quality investigation revealed that the building was operated on 0% outside air when outdoor temperatures were less than 6°C or greater than 26°C. Employees complained of episodic odors characteristic of flue gases associated with oil heat. The outdoor air intake located on the roof was directly downwind of boiler stacks, whose height was approximately one meter above the roofline.

Recommendations. It is likely that most of the complaints in this building are ventilation-related. During both cold and very warm weather, the building was operated on 0% outside air. At high occupant density characteristic of library operation on many days, bioeffluent levels would be expected to be high, causing discomfort and symptoms such as headaches and fatigue. To minimize these problems, the ventilation system should be operated at 15–20% outside air at all times when the building is occupied. Occupant complaints of oil-type flue gas odors and the physical nature of the boiler chimney and its orientation relative to the outdoor air intake suggest that flue gas re-entry caused significant indoor contamination. The most practical solution to this problem is to extend the height of the chimney above the roofline by about 5-10 m.

Case 3

An old school building was converted to office use. The building was heated throughout its history by radiant heat. It had no central HVAC system. After occupation, open stairwells were physically isolated from hallways, and an air-conditioning system was added. Occupants complained of headaches and lassitude, particularly at mid-day or later. On entry, the building had a strong odor of gym socks. Measured CO_2 levels were in the range of 1500-2000 ppm.

Recommendations. This is obviously a case of inadequate ventilation. Its resolution requires the installation of a ventilation system with sufficient capacity to provide 15 CFM/person at maximum occupancy. The building ventilation system should be operated with a minimum outside air intake of 15-20% during all times the building is occupied.

Case 4

Employees working in the data processing center of a government office building complained of hoarseness and excessive phlegm production, occurring within a few hours after work commenced each day. On investigation, it was noted that these individuals were operating video display terminals in the same area where larger computer units were maintained. The computers were supposed to be in a self-contained, environmentally controlled space. Though the air in the computer area was to be isolated from adjoining spaces, it apparently was not, and significant air flows from one space to another occurred. Further investigation of the building revealed that the building HVAC system was constantly operated at 0% outdoor air. Ammonia fumes in the engineering department were not vented to the outside. Rather, they were ducted into the supply air of the west wing of the building. Carpeting was old and worn and showed evidence of mold infestation. Custodial practice was to wet-shampoo carpets during unoccupied periods when HVAC fans were shut down. Damaged friable asbestos sprayed-on insulation was present on steel girders and adjacent structures. This material was in the return air plenum. Interviews with occupants revealed a wide variety of complaints suggesting numerous indoor air quality problems.

Recommendations. The complaints of hoarseness and excessive phlegm production by employees in the data processing area suggest particulate contaminants. They may be more susceptible to the problem because of electronic forces associated with the video display terminals. The potential for a particulate problem is suggested by the presence of damaged friable asbestos in the return air plenum, worn carpeting, visible mold growth on carpeting, and HVAC filter panels of low efficiency (< 30%).

The friable asbestos is a problem in itself. Its presence in the return air plenum and significantly damaged nature suggest that it should be removed by an abatement contractor. Carpeting should be removed and replaced. Shampooing should only be conducted with steam extraction and during periods when the HVAC system is being operated. Filtration units should be upgraded with higher-efficiency filter panels (50-70%).

Although there are no conclusive studies linking the operation of large computers with health complaints, such a link has occasionally been suggested. For this and the primary reason of good management of computer system environments, the air systems of the computer space and adjoining data processing area should be isolated as originally intended.

The building should be operated on a minimum of 15-20% outside air during all periods when it is occupied. Slotted doors should be installed in all office spaces that do not have return air grills. Fumes from the blueprint machine should be vented directly to the outside by an exhaust fan.

Case 5

A multipurpose single-story office building was plagued by complaints of odors that were noticeable on building entry. An indoor air quality investigator characterized the odor as one of rancid butter. The odor was not perceptible in an adjacent restaurant, which was on an air handling unit separate from the rest of the building. The odor was not noticeable in kitchen exhaust on the roof top. Restaurant employees confirmed that they used butter-flavored oil for cooking. On further investigation, oil stains were observed on structural members above the ceiling in the area of the large kitchen exhaust hood. Potential for leakage from one air handling system to another was evident from the observed presence of openings around piping and other utility lines.

Recommendations. This is a case of cross contamination, apparently due to cooking oils that escaped the large kitchen exhaust and became deposited on the ceiling where they became rancid. Utility line openings apparently allowed odor-tainted air to be drawn into the air handling system of the office complex. The latter was under negative pressure as compared to the former. Because of the heat buildup in the restaurant kitchen, its air handling system was being operated on a high percentage of outside air; though the restaurant was the source of the odor, it was not perceptible there.

The resolution of this problem requires the sealing of holes around utility line penetrations between the two air handling systems and the removal (by scrubbing) of the cooking oil residue in the ceiling area around the kitchen exhaust hood. Broken ceiling panels near the kitchen hood should be replaced to reduce the deposition of cooking oil residue in the plenum area above.

Case 6

Students and school employees complained of eye and upper respiratory symptoms, headaches, and fatigue. The complaints were limited to a single wing of an elementary school. On investigation, the ventilation system appeared to be providing an adequate supply of outside air. Carbon dioxide levels were in the range of 600–800 ppm. The wing contained a central open library surrounded by a number of classrooms. Most of the library furniture was found to be made from MDF. Formaldehyde concentrations were on the order of 0.10–0.12 ppm.

Recommendations. This building-related health complaint appears to be caused by elevated formaldehyde levels associated with the extensive quantity of MDF used to construct library furniture.

The best way to solve this problem would be to replace the library furniture with furniture made from solid wood and/or metal components. Because of the cost involved, school administrators will probably be reluctant to do so. The second alternative would be to increase the percentage of outside air used for ventilation. The latter is likely to be more expensive in the long term. It does, however, have the advantage of having no embarrassing capital costs.

Case 7

Complaints of solvent odors were voiced by central office staff at a vocational high school. On investigation, the odors were traced to a laboratory where printing techniques were taught. The two areas had individual air handling units. However, cross contamination occurred through an open classroom adjoining the printing laboratory. The classroom door was kept open because the space received inadequate cooling. The classroom had previously been a series of storage rooms; the conversion was somewhat of an afterthought. It had a return air grill but no supply air; the room was under strong negative pressure.

Recommendations. The first step in resolving this problem is to balance the air flow in the classroom space by providing it with supply air sufficient to cause a small positive pressure with respect to the adjoining printing laboratory. In the interim, the classroom door should be closed, particularly when solvent use is heavy.

Case 8

Numerous complaints of headaches, stuffiness, etc. were registered by occupants of a new classroom building. Even after a period of six months, the problem continued to persist. CO_2 levels varying from 1820 to 2500 ppm were reported for hallways and classrooms during the occupied period. A variable-air-volume (VAV) HVAC system was used.

Recommendations. This appears to be a ventilation problem not uncommonly associated with the peculiarities of VAV systems. Ventilation to building spaces was reduced to zero when the thermostat was satisfied. Resolution of the problem requires the system to be set so that a minimum amount of ventilation air is provided at all times. The minimum will reflect the occupancy of the space. It should be at least 20%.

Case 9

Smoking was banned in all areas of a small office building with the exception of a conference room. Several nonsmokers complained of heavy smoke odors in their working area. On investigation, it was revealed that the conference room and the smoke-contaminated work area were on the same air handling system.

Recommendations. There are several approaches to solving this problem. The best option is to install a local exhaust system in the conference room. It should be sized with sufficient capacity to minimize the flow of tobacco smoke into the return air of the air handling system. The system should be wired so that the local exhaust fan operates continuously during the work day. This mitigation approach is attractive because it will remove both gas- and particulate-phase tobacco-generated contaminants. An alternative is to install a ceiling-mounted electronic air cleaner with air flow sufficient to effectively remove smoke-generated particles. Though air cleaning will not remove gas-phase contaminants, it will remove particulate-phase materials, which are largely responsible for smoke and odor complaints. Cleaner ground plates must be washed periodically to remove deposited particles and maintain cleaning efficiency. Cleaning frequency will depend on smoking intensity. A need for weekly cleaning can be expected.

Case 10

An outbreak of hypersensitivity pneumonitis (HP) among employees of an office building was confirmed from symptom patterns and serological testing. The source of the HP antigens was determined to be the microbial slime accumulated on cooling coil drip pans.

Recommendations. The first step in mitigating this problem is to remove the microbial slime and thoroughly clean condensate drip pans. If they are rusted and fail to drain properly, they should be removed and replaced. New pans should be installed with sufficient incline to facilitate good drainage. Because of the heavy buildup of microbial slime, it was evident that a significant amount of organic particles, which can serve as a substrate for microbial growth, was passing through the air handling system and that the filters being utilized were not efficient enough. HVAC system filters should be replaced with filters of a rated efficiency of 50–70%. They should be replaced frequently and routinely. The building manager should then implement a vigorous continuing maintenance program for all components of the HVAC system.

Case 11

An outbreak of HP-type illness occurred in a large office complex. On investigation, it was evident that the building was subject to recurring water intrusions resulting from leakage around roofing flashings. Numerous ceiling tiles were stained, as were masonry walls. The building was carpeted, and evidence of water staining was evident on the carpet as well. A buildup of microbial slime was observed in the cooling coil condensate pans in 10 of 15 air handling units. HVAC system filters had a heavy buildup of dust and apparently were replaced infrequently. Viable mold particle counts in building air were in excess of 1500 CFU/m^3.

Recommendations. In this building, mold particles responsible for HP responses may come from a variety of sources. As a consequence, good mitigation practice would require attention to all major sources. As in the previous case, the microbial slime accumulating on cooling coil drip pans should be removed and the pans thoroughly cleaned. Rusted pans should be removed and replaced. All pans should be adjusted so that a sufficient incline is provided for good drainage.

It is evident that significant re-roofing or roof maintenance is required to resolve the problem of water incursions around roofing flashings. All water-damaged materials, such as ceiling tiles and carpeting, should be removed and replaced. Stains on masonry walls should be cleaned with a strong detergent.

The heavy dust accumulation on HVAC system filters may also serve as a mold substrate. Filters should be replaced frequently and routinely. Filters with rated efficiencies of 50–70% should be used.

REFERENCES

1. Bowen, R.P. et al. 1981. "Urea-Formaldehyde Foam Insulation: Problem Identification and Remedial Measures for Wood Frame Construction." Building Practice Note No. 23. National Research Council, Ottawa, Canada.
2. Godish, T. and J. Rouch. 1988. "Residential Formaldehyde Control by Mechanical Ventilation." *Appl. Ind. Hyg.* 3:93–96.
3. U.S. EPA. 1986. "Radon Reduction Techniques for Detached Houses. Technical Guidance." EPA 625/5-86-019.

INDEX